The Boiler Operator's Exam Preparation Guide

Theodore B. Sauselein, P.E.

McGraw-Hill

New York San Francisco Washington, D.C. Auckland Bogotá
Caracas Lisbon London Madrid Mexico City Milan
Montreal New Delhi San Juan Singapore
Sydney Tokyo Toronto

Library of Congress Cataloging in Publication Data
Sauselin, Theodore B.
   The boiler operator's exam preparation guide / Theodore B. Sauselein
     p.   cm.
   ISBN 0-07-057968-7
   1. Steam-boilers-Examinations, questions, etc.   I. Title.
TJ289.S33   1995
621.1′83′076—dc20
                                           95-34792
                                            CIP

McGraw-Hill

A Division of The McGraw·Hill Companies

22 23 24 25 26 27 QVS/QVS 23 22 21 20 19

ISBN   9781265829995
MHID 1265829993

The editor for this book was Joanna Turpin, the art director was Mark Leibold and the copy editor was Carolyn Thompson.

Contents

Contents

INTRODUCTION

Taking the Test

Here are a few tips when taking your licensing examination:

Select what appears to be the most correct answer.
While examiners swear that there is only one obvious correct answer to a question, you may find that either none or more than one of the choices is obviously correct. Instead of trying to figure out which is the correct answer, eliminate those that are obviously wrong and then narrow it down to the most correct answer.

Don't get hung up on one question.
Go through the test and answer the questions you definitely know. Then go back and work on the ones you skipped. The last thing you want to do is expend all your nervous energy on a hard question at the beginning of the test. Don't be surprised if, after going through the test several times, there are still a few questions that make no sense. If that's the case, just give them your best guess.

Learn definitions.
A lot of multiple choice questions are based on definitions. Take the following example:

Convection means

    a) the movement of liquids or gases created by a temperature difference.

    b) transfer of heat by direct molecular contact.

    c) the weight of a substance as compared to unit.

    d) a form of heat transfer by waves.

The correct answer is a, but change the question to *radiation means* or *conduction means*, and the correct answer is then d and b, respectively.

MATH PROBLEMS AND UNITS

When working a math problem, always assign units to the numbers in the problem. When the equation is set up, the units should work out to the units you expect. If not, then the equation is not set up properly. Consider the following problem:

The water surface in a fire tank is 135 feet above the suction of a fire pump. What is the psi at the suction of the fire pump?

You remember the conversion is 2.31, but the question seems to be, do you multiply or divide 2.31 by 135 feet. But the question is really 2.31 what?! Every number that is written down should have units attached to it. When these numbers, especially conversions, have units, then the math becomes much easier.

The correct measurement is actually 2.31 feet/psi. Let's say your first guess is to multiply 135 feet by 2.31 feet/psi (at this point don't worry about the numbers, just concentrate on the units):

$$\text{feet} \times \frac{\text{feet}}{\text{psi}} = \frac{\text{feet}^2}{\text{psi}}$$

Obviously feet^2 per psi is not the answer you were looking for. Now divide 135 feet by 2.31 feet/psi:

$$\frac{\text{feet}}{\text{feet/psi}} = \text{psi}$$

This is more like it. The unit you were expecting was psi. Now that the problem is set up properly, just add numbers to the units and let the calculator do the number crunching:

$$\frac{135 \text{ ft}}{231 \text{ ft/psi}} = 58.44 \text{ psi}$$

Some examiners insist that you show all steps to the solution. This method automatically makes you show all your work. Even if you don't come up with what the examiner thinks is the correct result, if your work is laid out in an easily followed fashion partial credit might be given.

Now after all this talk about units, there are some numbers that are dimensionless. It's not that these numbers started out lacking units; it's that all the units canceled themselves out. Percentages, efficiencies, and ratios like π (pi) are examples.

From elementary school you learned that the equation for the area of a circle is πr^2. This same equation can also be written as $0.7854d^2$. Use whichever one you feel most comfortable with.

With that in mind, let's find the area of a 6 foot diameter circle. The answer must be in square inches.

$$0.7854 \times d^2 = 0.7854 \times (6 \text{ ft})^2 \times 144 \text{ in}^2/\text{ft}^2 = 4,071.51 \text{ in}^2$$

The same problem can also be worked out as follows:

6 ft diameter circle = 3 ft radius
3 ft × 12 in./ft = 36 inches
$\pi r^2 = 3.14 \times (36 \text{ in.})^2 = 4,069.44 \text{ in}^2$

Notice the answers are close but not exact. The problem here is how many decimals to use with π. Usually two decimal places is plenty. The important thing is to show all your work.

Another advantage of using units is that you are forced to do all the conversions. In the previous example, diameter was given in feet, but the answer had to be in inches. You can bet your bottom dollar that if this were a multiple choice question, it would include the answer that did not have feet converted to inches. You would work the problem without converting, then be relieved that the incorrect answer you came up with was included in the choices.

Make sure your calculator can do square roots. There are a few problems where this feature comes in handy.

CONVERSIONS

Now for conversions. Since most tests are closed book, these will have to be learned or memorized:

1 in. Hg	= 0.491 psi
1 psi	= 2.31 feet$_{water}$
1 foot$_{water}$	= 0.433 psi
1 hp	= 0.746 kW
	= 33,000 ft-lb/min
	= 2,545 Btuh
1 kW	= 1.34 hp
	= 3,413 Btuh
1 gal	= 231 in^3
	= 8.33 lb$_{water}$
cubic foot of water	= 7.48 gallons
	= 1,728 in^3
	= 7.48 gallons
1 boiler horsepower	= evaporation of 34.5 lb$_{water}$/hr from and at 212°F
	= 33,475 Btuh
°F	= (°C x 9/5) + 32
°C	= 5/9(°F - 32)

One British thermal unit (Btu) is required to raise the temperature of one pound of water 1°F and is equal to 778 ft-lb of work

Atmospheric pressure of 14.7 psia will sustain a column of water 34 feet high and a column of mercury (Hg) 29.92 inches high.

EQUATIONS

The following are a few of the equations you will have to memorize:

area of a circle	= $0.7854d^2$
circumference of a circle	= πd
torque	= (5,252 x hp)/rpm
Btu/lb of oil	= 17,780 + (54 x API gravity)
heating value of coal	= 14,000 Btu/lb
heating value of fuel oil	= 19,000 Btu/lb
Btu/lb$_{oil}$	= 17,687 + (57.7 x API gravity)
longitudinal stress	= $\dfrac{pd}{4t}$
circumferential stress	= $\dfrac{pd}{2t}$

Boilers

The equation for maximum allowable working pressure (MAWP) is:

$$P = \frac{TS \times t \times E}{R \times FS}$$

where:

- P = maximum allowable working pressure in psi inside drum or shell
- TS = tensile strength of plate (psi - use 55,000 psi for steel)
- t = thickness of plate (inches)
- R = inside radius of drum or shell (inches)
- FS = factor of safety (ultimate strength divided by allowable working stress or bursting pressure divided by safe working pressure. It can vary between four and seven depending on age, type of construction, and condition. Use five for most calculations.)
- E = efficiency of the joint (for welded joints, use 100%)

For bursting pressure, set the factor of safety to one.

The equation for boiler horsepower is as follows:

$$\frac{lb_{steam}/hr \times Factor\ of\ evaporation}{34.5\ lb_{steam}/Boiler\ horsepower}$$

Where the factor of evaporation is:

$$\frac{Btu_{steam} - (Feedwater\ temperature - 32°F)}{970.3\ Btu/lb}$$

The equation for boiler efficiency is as follows:

$$\frac{lb_{steam}/lb_{fuel} \times [Btu/lb_{steam} - (Feedwater\ temperature - 32°F)]}{Btu/lb_{fuel}}$$

Pumps

To calculate horsepower, use the following equation:

$$\frac{gpm \times 8.33\ lb/gal \times Head\ (feet)}{33,000\ ft\text{-}lb/min \times Efficiency}$$

Or the following equation may be used to calculate horsepower:

$$\frac{\text{gpm} \times \text{Head (feet)}}{3{,}960 \times \text{Efficiency}}$$

To calculate the gpm of a reciprocating pump, use the following equation:

$$\text{gpm} = \frac{LANE}{231}$$

where:

L = length of stroke in inches
A = area of piston
N = number of strokes per minute
E = efficiency
231 = number of cubic inches per gallon

Note: A stroke is a piston moving once over its path. For a duplex pump, a stroke is both pistons moving once over their path.

Turbines

For condenser cooling water, use the following equation:

$$Q = \frac{H - (t_o - 32°F)}{T_2 - T_1}$$

where:

Q = weight of water to condense 1 lb of steam
H = heat content of exhaust steam
t_o = temperature of condensate
T_1 = temperature of cooling water entering condenser
T_2 = temperature of cooling water leaving condenser

The equation for torque is:

$$\frac{\text{hp} \times 5{,}252}{\text{rpm}}$$

where:
5,252 = constant that converts radians to rpm

For pipe expansion, use the following equation:

$$(T_1 - T_2) \times L \times \text{Coefficient of expansion} \times 12 \text{ in./ft}$$

where:
coefficient of expansion for steel = 0.0000065 in./in.-°F

To calculate pipe size, use the following equation:

$$A = \frac{144 \times Q}{V}$$

where:
- A = area in square inches
- Q = quantity of steam in cubic feet per minute
- V = velocity in feet per second
- 144 = number of square inches per square foot

Become familiar with different steam pressures and their corresponding temperatures and heat content.

in. Hg vacuum	in. Hg absolute	psia	°F	cu ft/lb	Btu/lb
29"	1"	0.491	79	652	1,104
28"	2"	0.952	100	340.4	1,105
27"	3"	1.473	115	231	1,111

psi	psia	°F	cu ft/lb	Btu/lb
15	30	250	13.7	1,164
100	115	338	3.9	1,189
150	165	366	2.75	1,196
200	215	388	2.13	1,199
250	265	406	1.8	1,202
285	300	417	1.5	1,203
585	600	486	0.77	1,203

Heat

While the concept of thermodynamics may seem a little daunting, it is necessary to know a little about it, as it lays the foundation for everything that follows. What is more important is that it also helps with some test questions. Understanding terms such as *sensible heat* and *latent heat* and knowing what happens when water changes to steam will help you understand what happens in boilers and steam traps.

HEAT TRANSFER

Let's begin by examining the process of changing water into steam. The heat energy from fuel is delivered to the water by three methods of transfer: radiation, conduction, and convection:

- **Radiation** does not require a transmission medium; it travels like light waves through a vacuum and through air. The most common example of radiation is the heat we feel from the sun.

- **Conduction** is the transfer of heat from a warm molecule to a cooler one. Some materials conduct heat better than others; for example, gases and vapors are poor conductors, liquids are better, and metals are best. Materials that are poor heat conductors, like asbestos and calcium silicate, are called *insulators*. Heat travels through insulators but at a slower rate.

- **Convection** heat transfer takes place by movement of the heated material itself. In a heated room, warm air rises and the cold air falls. In a boiler, the hot water rises and the cold water falls to the bottom.

Now it is possible to see how the three forms of heat transfer work in a boiler. The tubes in the furnace section of the boiler receive their heat by *radiation* from the visible flame. In fact, about half of the steam in an industrial boiler and all the steam in a utility boiler is generated by the furnace tubes. The part of the boiler that contains most of the tubes is called the *convection* section. This section receives its heat by convection from the hot flue gas. Heat is then transferred through the tube metal and into the water by *conduction*.

Water to Steam

To demonstrate how water is transformed into steam, pour one pound of 32°F water into a pot sitting on a stove burner. Because this demonstration takes place in an open pot, the pressure of the water and any steam produced remains at atmospheric pressure. (Standard atmospheric pressure is 14.7 pounds per square inch absolute (psia), which is explained later in this chapter. Experiments performed at different pressures yield different results.) Place a thermometer in the water to monitor its temperature. You must imagine placing a device (let's call it a heat-o-meter) in the water that measures the amount of heat absorbed by the water. The heat-o-meter would be calibrated in Btu (British thermal units). One British thermal unit is the heat required to raise one pound of water one degree Fahrenheit, Figure 1-1.

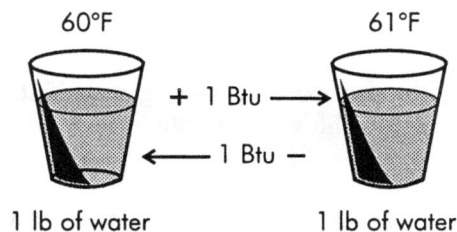

Figure 1-1. One British thermal unit raises one pound of water one degree Fahrenheit

With the dial on the heat-o-meter set to zero and the thermometer reading 32°F, turn on the burner (Point 3, Figure 1-2). The readings on both the thermometer and heat-o-meter will increase. The heat absorbed by the water that causes the temperature increase is called *sensible heat*. If you put your finger in the water, you can detect or sense the sensible heat. Sensible heat changes the temperature of a substance but not its state. This means that water absorbing sensible heat stays water and will not turn to steam.

As the water temperature reaches 212°F (Point 4, Figure 1-2), there will still be a pound of water in the pot because boiling hasn't started yet.[1] The heat-o-

---

[1] Bubbles will form on the bottom of the pot long before the water reaches 212°F. These bubbles are dissolved gases coming out of solution. Hot water holds less dissolved gas compared to cold water. This is the principle deaerators use to drive out destructive dissolved gases like oxygen from the feedwater before it gets to the boiler.

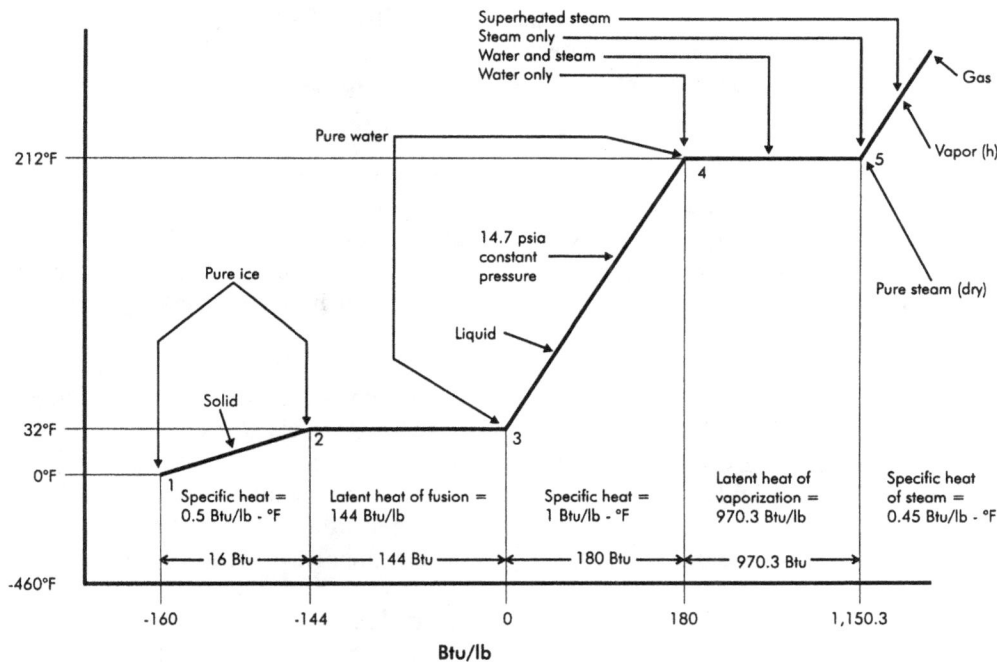

Figure 1-2. Water changing to steam

meter reads 180 Btu. This matches the definition of a British thermal unit, because we increased the temperature of one pound of water 180°F (212° - 32°F = 180°F).

Just beyond Point 4 in Figure 1-2, the water is still absorbing heat because the burner is still on, but the temperature of the water remains constant at 212°F. This is called *latent heat* (latent means hidden). Latent heat changes the state of a substance but not its temperature. It takes extra energy to change the state of a substance, and at this point, latent heat is necessary to convert water into steam.

As the water continues to boil, the heat-o-meter reading would continue to increase, but the thermometer would stay at 212°F. When the last drop of water evaporates (Point 5, Figure 1-2), the heat-o-meter would read 1,150.3 Btu. This is the total amount of heat required to evaporate one pound of water starting at 32°F. Subtract the sensible heat from this total (1,150.3 Btu - 180 Btu = 970.3 Btu) and you have the amount of latent heat required to evaporate one pound of water from 212°F at 14.7 psia.

During the evaporation process, the volume of one pound of water changes drastically. It starts as a liquid at 0.01672 ft^3 (about one pint) and changes to steam (gas) at 26.79 ft^3 (about one cubic yard), which is an increase of more than 1,600 times!

Steam may be called a gas, because that is what it is. What you see around the top of a pot of boiling water is water vapor condensing out of the steam. Condensation is the process by which steam gives up its latent heat and turns back into water.

Now that all the water has evaporated from our pot, turn off the burner. While evaporating, the water was absorbing heat fast enough to keep the metal pot at a safe level. Without the heat-absorbing water, meltdown occurs. This is an extremely important point, because the same thing can happen in a boiler; if the water gets too low, property damage or even personal injury can take place.

Steel holds its strength up to 700°F, and it weakens rapidly above that point. Since the flame temperature is over 2,500°F, how does the boiler survive under these conditions? It all has to do with how fast heat travels through the metal. If there is something on the other side of the metal that can absorb large amounts of heat quickly (water, for example), then the metal temperature will stay at a safe level. If there is nothing on the other side that can absorb a lot of heat quickly (air or steam, for example), then the metal overheats.

Events in nature may be explained by the theory just described. For example, in the geyser Old Faithful in Yellowstone National Park, water under the Earth's surface absorbs heat from rocks heated by Earth's molten core. After a time, the water absorbs enough heat so that some of it flashes off into steam. Since steam occupies so much more volume than water, pressure increases sharply. This pressure is relieved by forcing the mixture of hot water and steam out of openings and cracks in the Earth's surface. It is because of this principle that it is dangerous to *sparge* (directly inject) steam into a closed pressure vessel such as a tank or pipe. Although this might seem a good way to heat water, the water under pressure absorbs heat until some of it flashes into steam. The sudden increase in volume results in damaged equipment if the pressure has no place to go.

STEAM TABLES

Steam tables show various pressures, temperatures, heat content, and specific volumes. Notice in Table 1-1 that as water pressure increases, the corresponding boiling temperature also increases. Also notice that *water at a specific pressure always boils at a specific temperature.* For example, water at 15 psig always boils at 250°F, and water at 250 psig always boils at 406°F.

Saturation Line

The curve shown in Figure 1-3 is known as the *saturation line.* Every point below the curve is water, and every point above the curve is superheated steam.

STEAM TABLE (approx.)							
				Heat Content			Specific Volumn
Gauge	Absolute		Sensible	Latent	Total	Steam	
Pressure	Pressure	Temp.	(h_f)	(h_{fg})	(h_g)	(V_g)	
(in. Hg vac)	(psia)	(°F)	Btu/lb	Btu/lb	Btu/lb	(cu ft/lb)	
27.96	1	101.7	69.5	1032.9	1102.4	333.0	
25.91	2	126.1	93.9	1020.0	1113.9	173.5	
23.87	3	141.5	109.3	1011.3	1120.6	118.6	
19.79	5	162.3	130.1	999.7	1129.8	73.4	
15.70	7	176.9	144.6	991.5	1136.1	53.6	
9.58	10	193.2	161.1	981.9	1143.0	38.4	
7.54	11	197.8	165.7	979.2	1144.9	35.1	
5.49	12	202.0	169.9	976.7	1146.6	32.4	
3.45	13	205.9	173.9	974.3	1148.2	30.0	
1.41	14	209.6	177.6	972.2	1149.8	28.0	
psig							
0	14.7	212	180	970.3	1150.3	26.8	
1	15.7	216	183	968	1151	25.2	
5	19.7	227	195	960	1155	20.1	
10	24.7	240	207	953	1160	16.5	
15	29.7	250	218	946	1164	13.9	
20	34.7	259	227	939	1166	11.9	
25	39.7	267	236	934	1170	10.6	
30	44.7	274	243	929	1172	9.5	
35	49.7	281	250	924	1174	8.5	
40	54.7	287	256	920	1176	7.8	
45	59.7	293	262	915	1177	7.2	
50	64.7	298	267	912	1179	6.7	
60	74.7	307	277	906	1183	5.8	
70	84.7	316	286	898	1184	5.2	
75	89.7	320	290	895	1185	4.9	
85	99.7	328	298	889	1187	4.4	
95	109.7	335	305	883	1188	4.1	
100	115	338	309	880	1189	3.9	
125	140	353	325	868	1193	3.2	
150	165	366	339	857	1196	2.7	
175	190	378	351	847	1198	2.4	
200	215	388	362	837	1199	2.1	
225	240	397	372	828	1200	1.9	
250	265	406	382	820	1202	1.8	
275	290	414	391	812	1203	1.6	
300	315	421	398	805	1203	1.5	
325	340	429	407	797	1204	1.4	
350	365	435	414	790	1204	1.3	
375	390	442	421	784	1205	1.2	
400	415	448	428	777	1205	1.1	
450	465	460	439	766	1205	1.0	
500	515	470	453	751	1204	0.9	
550	565	479	464	740	1204	0.8	
600	615	489	475	728	1203	0.7	

Table 1-1. Properties of saturated steam

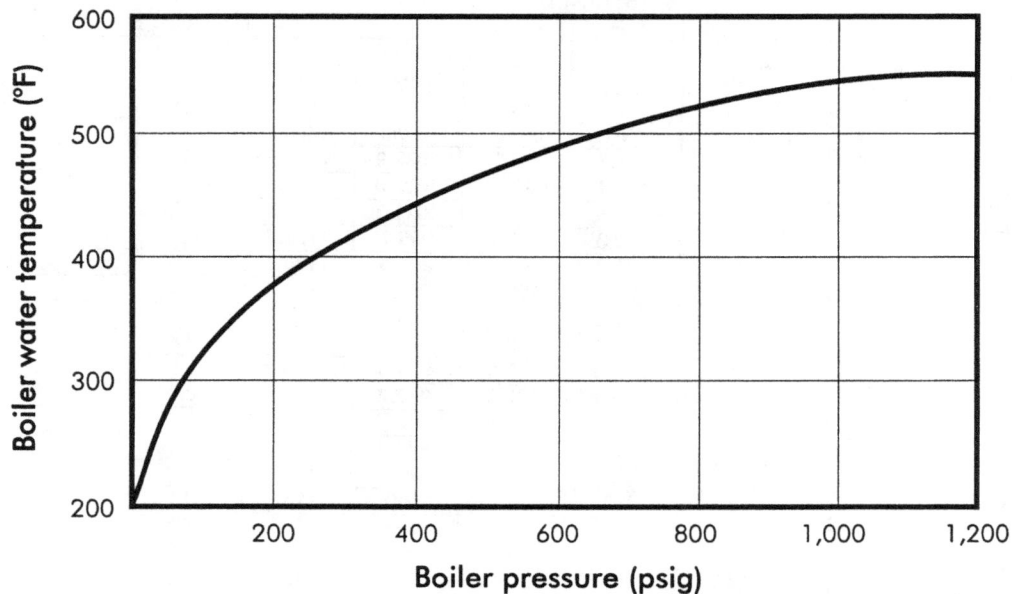

Figure 1-3. Saturation line

Every point on the line is water and/or steam at its saturated temperature and pressure (where the gas is at the same pressure/temperature as the liquid it contacts). A point on this saturation curve is actually on the line between Points 4 and 5 in Figure 1-2. Point 4 is 100% water at its saturation temperature. Halfway to Point 5 is 50% water and 50% steam. Point 5 is 100% steam. Point 5 is *dry steam* (all gas and no liquid), while *wet steam* is anywhere in between, but not including, Points 4 and 5. The closer to Point 4, the wetter the steam. The steam and water are both at the same temperature and pressure, but instead of having 100% gas, there is a mixture of liquid and gas at the same temperature.

Refrigerants also have pressure-temperature tables and saturation curves. Refrigeration gauges have dials calibrated to measure pressure and temperature. In the steam table shown in Table 1-1, both gauge pressure and absolute pressure are given. (Absolute pressure and gauge pressure are explained later.) In unabridged versions, only absolute pressures are listed. Absolute pressure must be used in most engineering calculations; gauge pressure is listed on the steam table for convenience.

Absolute and Gauge Pressures

A *vacuum* is any pressure less than atmospheric pressure. Zero pounds per square inch absolute (psia) is a perfect vacuum. Zero pounds per square inch gauge (psig or psi) is atmospheric pressure. When stating absolute pressure, psia must be used. When stating gauge pressure, psi is all that is required, but sometimes psig is used. At sea level, absolute pressure is 14.7. When converting absolute pressure to gauge pressure, subtract 14.7. When converting gauge pressure to absolute pressure, add 14.7. Remember, gauge pressure plus atmospheric pressure equals absolute pressure. Figure 1-4 shows a comparison between absolute and gauge pressures.

Superheated Steam

Unabridged steam tables are divided into two parts. The first part lists the saturated pressure-temperature relationships between liquid and vapor phases. In the first part, the first column lists temperature, while the second column lists pressure. The second part of the steam table describes the properties of superheated steam.

Although it was stated that for a given pressure there is only one temperature for steam, the temperature of superheated steam is higher than saturated steam at a given pressure. This is because after steam is produced, additional heat can

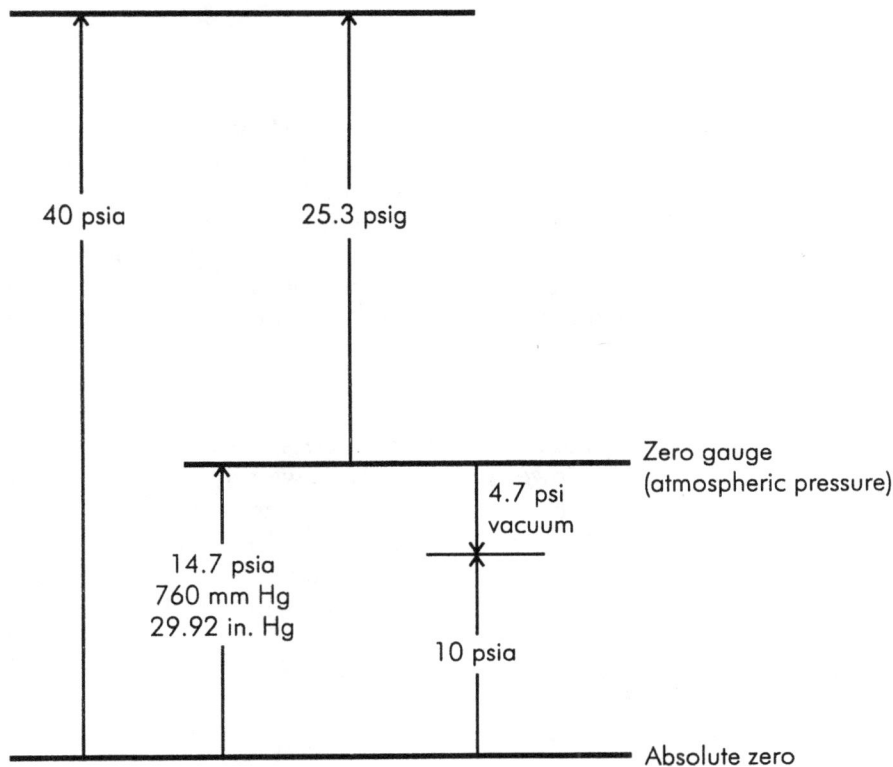

Figure 1-4. Comparison between absolute and gauge pressures

be added to increase its temperature and heat content. Any point beyond Point 5 in Figure 1-2 is superheated steam.

Superheated steam is used for two main reasons: 1) to provide extra energy that is used for driving a steam turbine; and 2) its higher temperature means that less of it condenses when transported over long distances, such as airports, military bases, and large campuses. Superheated steam is rarely used for space and process heating. To understand why, let's look at the steam cycle and how steam is used.

Consider that 15 psi steam is transported to its point of use (for example, a heating coil in an hvac system, a kettle in a kitchen, or a fuel oil heater). As heat is transferred from the steam and into the process, the steam condenses back into water. For every pound of steam that condenses, 946 Btu are delivered (see Table 1-1). This is the great advantage of steam; a lot of heat can be transferred with a little amount of effort.

If we add 50°F to a pound of superheated steam at 250°F and 15 psi, around 25 Btu will be added. This is a very small percentage compared to the 946 Btu already contained in the saturated steam. The superheat is used very quickly, and most heating is still done at the saturated temperature of 250°F.

Flash Steam

Nature will neither permit water to remain in the liquid state at temperatures higher than 212°F, nor to contain more than 180 Btu/lb at atmospheric pressure. Saturated water at 0 psi at 212°F contains 180 Btu/lb. Saturated water at 150 psi at 366°F contains 339 Btu/lb. In the latter case, the Btu/lb exceeding 180 Btu/lb must be jettisoned. Nature takes care of this surplus by converting a fraction of the water to flash steam. *Live steam* is generated in a boiler, while *flash steam* is produced when hot water at its saturated temperature is released to a lower pressure.

The percentage of flash steam can be calculated by finding the difference in heat content between the high and low pressure waters, then dividing by the latent heat of the steam at the lower pressure. To convert this decimal answer to a percentage, multiply by 100. For example, a steam trap in a 100 psi system discharges condensate to atmospheric pressure (0 psi). The steam table (Table 1-1) shows that 100 psi water contains 309 Btu/lb, 0 psi water contains 180 Btu/lb, and that the latent heat of 0 psi steam is 970.3 Btu/lb. What is the percentage of water flashed into steam?

$$\left(\frac{309 \text{ Btu/lb} - 180 \text{ Btu/lb}}{970.3 \text{ Btu/lb}} \right)(100) = 13.29\%$$

Wasting 13% of the water is bad enough, but now calculate how much heat is lost by letting the flash steam get away. What percentage of heat does the flash steam contain compared to the saturated water at 100 psi?

$$\frac{\text{Heat in flash steam}}{\text{Heat in 100 psi water}} = \left(\frac{309 \text{ Btu/lb - } 180 \text{ Btu/lb}}{309 \text{ Btu/lb}}\right)(100) = 41.7\%$$

Figure 1-5 is a graphic representation of the flash steam equation.

Unabridged Steam Tables

For those of you who must take more advanced examinations, steam table questions are fair game. It is sometimes asked how the tables are organized. The tables are divided into three sections: saturated steam listed by temperature, Table 1-2; saturated steam listed by pressure, Table 1-3; and superheated steam, Table 1-4. The complete steam tables may be found in the appendix.

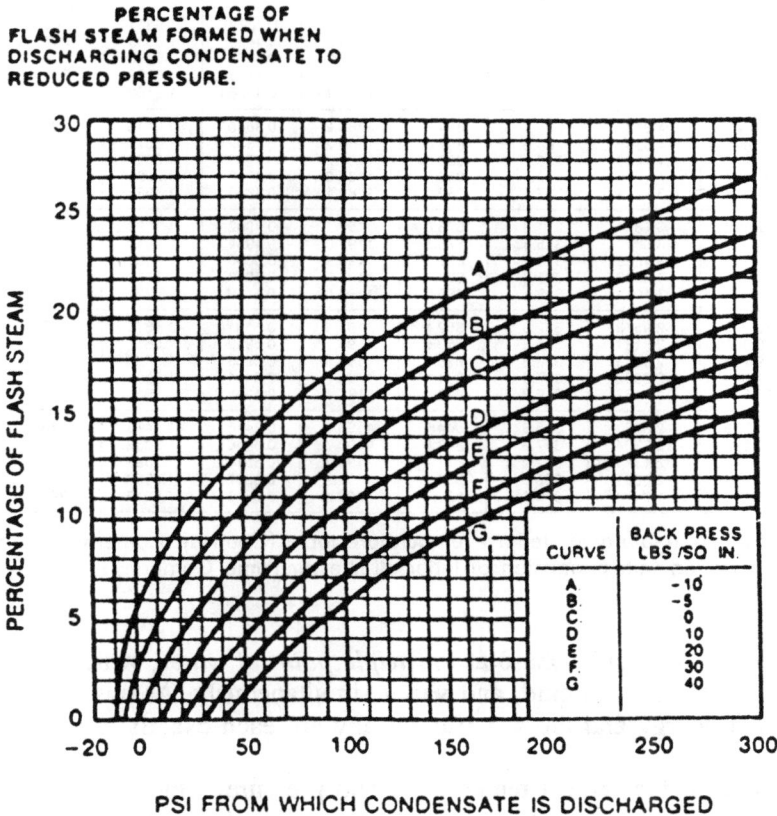

PERCENTAGE OF FLASH STEAM FORMED WHEN DISCHARGING CONDENSATE TO REDUCED PRESSURE.

CURVE	BACK PRESS LBS /SQ IN.
A	-10
B	-5
C	0
D	10
E	20
F	30
G	40

Figure 1-5. Percentage of flash steam formed when condensate is discharged to a lower pressure (Courtesy, Armstrong International, Inc.)

Temp F	Abs press lb/in²	Specific volume ft³/lbm V_g	Enthalpy, Btu/lbm			Entropy, Btu/lbm × F		Temp F
			Sat liquid h_f	Evap h_{fg}	Sat vapor h_g	Sat liquid s_f	Sat vapor s_g	
212	14.696	26.799	180.17	970.3	1150.5	0.3121	1.7568	212
213	14.990	26.307	181.17	969.7	1150.8	0.3136	1.7552	213
214	15.289	25.826	182.18	969.0	1151.2	0.3151	1.7536	214
215	15.592	25.355	183.19	968.4	1151.6	0.3166	1.7520	215
216	15.901	24.894	184.20	967.8	1152.0	0.3181	1.7505	216
220	17.186	23.148	188.23	965.2	1153.4	0.3241	1.7442	220
224	18.556	21.545	192.27	962.6	1154.9	0.3300	1.7380	224
228	20.015	20.073	196.31	960.0	1156.3	0.3359	1.7320	228
232	21.567	18.718	200.35	957.4	1157.8	0.3417	1.7260	232
236	23.216	17.471	204.40	954.8	1159.2	0.3476	1.7201	236
240	24.968	16.321	208.45	952.1	1160.6	0.3533	1.7142	240
244	26.826	15.260	212.50	949.5	1162.0	0.3591	1.7085	244
248	28.796	14.281	216.56	946.8	1163.4	0.3649	1.7028	248
252	30.883	13.375	220.62	944.1	1164.7	0.3706	1.6972	252
256	33.091	12.538	224.69	941.4	1166.1	0.3763	1.6917	256
260	35.427	11.762	228.76	938.6	1167.4	0.3819	1.6862	260
264	27.894	11.042	232.83	935.9	1168.7	0.3876	1.6808	264
268	40.500	10.375	236.91	933.1	1170.0	0.3932	1.6755	268
272	43.249	9.755	240.99	930.3	1171.3	0.3987	1.6702	272
276	46.147	9.180	245.08	927.5	1172.5	0.4043	1.6650	276
280	49.200	8.6439	249.2	924.6	1173.8	0.4098	1.6599	280
284	52.414	8.1453	253.3	921.7	1175.0	0.4154	1.6548	284
288	55.795	7.6807	257.4	918.8	1176.2	0.4208	1.6498	288
292	59.350	7.2475	261.5	915.9	1177.4	0.4263	1.6449	292
296	63.084	6.8433	265.6	913.0	1178.6	0.4317	1.6400	296
300	67.005	6.4658	269.7	910.0	1179.7	0.4372	1.6351	300
304	71.119	6.1130	273.8	907.0	1180.9	0.4426	1.6303	304
308	75.433	5.7830	278.0	904.0	1182.0	0.4479	1.6256	308
312	79.953	5.4742	282.1	901.0	1183.1	0.4533	1.6209	312
316	84.688	5.1849	286.3	897.9	1184.1	0.4586	1.6162	316
320	89.643	4.9138	290.4	894.8	1185.2	0.4640	1.6116	320
324	94.826	4.6595	294.6	891.6	1186.2	0.4692	1.6071	324
328	100.245	4.4208	298.7	888.5	1187.2	0.4745	1.6025	328
332	105.907	4.1966	302.9	885.3	1188.2	0.4798	1.5981	332
336	111.820	3.9859	307.1	882.1	1189.1	0.4850	1.5936	336
340	117.992	3.7878	311.3	878.8	1190.1	0.4902	1.5892	340
344	124.430	3.6013	315.5	875.5	1191.0	0.4954	1.5849	344
348	131.142	3.4258	319.7	872.2	1191.9	0.5006	1.5806	348
352	138.138	3.2603	323.9	868.9	1192.7	0.5058	1.5763	352
356	145.424	3.1044	328.1	865.5	1193.6	0.5110	1.5721	356
360	153.010	2.9573	332.3	862.1	1194.4	0.5161	1.5678	360
364	160.903	2.8184	336.5	858.6	1195.2	0.5212	1.5637	364
368	169.113	2.6873	340.8	855.1	1195.9	0.5263	1.5595	368
372	177.648	2.5633	345.0	851.6	1196.7	0.5314	1.5554	372
376	186.517	2.4462	349.3	848.1	1197.4	0.5365	1.5513	376
380	195.729	2.3353	353.6	844.5	1198.0	0.5416	1.5473	380
384	205.294	2.2304	357.9	840.8	1198.7	0.5466	1.5432	384
388	215.220	2.1311	362.2	837.2	1199.3	0.5516	1.5392	388
392	225.516	2.0369	366.5	833.4	1199.9	0.5567	1.5352	392
396	236.193	1.9477	370.8	829.7	1200.4	0.5617	1.5313	396

Table 1-2. Properties of saturated steam - temperature table (Courtesy, *Cameron Hydraulic Data* book. Reproduced with permission of Ingersoll-Dresser Pump Company.)

The tables start at the liquid phase at the *triple point of water*, which is where water can exist as a solid, liquid, and vapor simultaneously. At this point, the specific internal energy and the specific entropy are each exactly zero.

The tables go up to the critical temperature and pressure, which are 705.47°F and 3,208.2 psia, respectively. At these points, the figures for specific volume, enthalpy, and entropy are the same for liquid as they are for vapor.

Abs press lb/in²	Temp °F	Specific volume ft³/lbm		Enthalpy btu lbm		Entropy btu/lbm × F		Abs press lb/in²
		Water v_f	Steam v_g	Water h_f	Steam h_g	Water s_f	Steam s_g	
.08865	32.018	0.016022	3302.4	0.0003	1075.5	0.0000	2.1872	.08865
0.25	59.323	0.016032	1235.5	27.382	1067.4	0.0542	2.0967	0.25
0.50	79.586	0.016071	641.5	47.623	1096.3	0.0925	2.0370	0.50
1.0	101.74	0.016136	333.60	69.73	1105.8	0.1326	1.9781	1.0
3.0	141.47	0.016300	118.73	109.42	1122.6	0.2009	1.8864	3.0
6.0	170.05	0.016451	61.984	138.03	1134.2	0.2474	1.8294	6.0
10.0	193.21	0.016592	38.420	161.26	1143.3	0.2836	1.7879	10.0
14.696	212.00	0.016719	26.799	180.17	1150.5	0.3121	1.7568	14.696
15.0	213.03	0.016726	26.290	181.21	1150.9	0.3137	1.7552	15.0
20.0	227.96	0.016834	20.087	196.27	1156.3	0.3358	1.7320	20.0
25.0	240.07	0.016927	16.301	208.52	1160.6	0.3535	1.7141	25.0
30.0	250.34	0.017009	13.744	218.9	1164.1	0.3682	1.6995	30.0
35.0	259.29	0.017083	11.896	228.0	1167.1	0.3809	1.6872	35.0
40.0	267.25	0.017151	10.4965	236.1	1169.8	0.3921	1.6765	40.0
45.0	274.44	0.017214	9.3988	243.5	1172.0	0.4021	1.6671	45.0
50.0	281.02	0.017274	8.5140	250.2	1174.1	0.4112	1.6586	50.0
55.0	287.08	0.017329	7.7850	256.4	1175.9	0.4196	1.6510	55.0
60.0	292.71	0.017383	7.1736	262.2	1177.6	0.4273	1.6440	60.0
65.0	297.98	0.017433	6.6533	267.6	1179.1	0.4344	1.6375	65.0
70.0	302.93	0.017482	6.2050	272.7	1180.6	0.4411	1.6316	70.0
75.0	307.61	0.017529	5.8144	277.6	1181.9	0.4474	1.6260	75.0
80.0	312.04	0.017573	5.4711	282.1	1183.1	0.4534	1.6208	80.0
85.0	316.26	0.017617	5.1669	286.5	1184.2	0.4590	1.6159	85.0
90.0	320.28	0.017659	4.8953	290.7	1185.3	0.4643	1.6113	90.0
95.0	324.13	0.017700	4.6514	294.7	1186.2	0.4694	1.6069	95.0
100.0	327.82	0.017740	4.4310	298.5	1187.2	0.4743	1.6027	100
105.0	331.37	0.01778	4.2309	302.2	1188.0	0.4790	1.5988	105
110.	334.79	0.01782	4.0306	305.8	1188.9	0.4834	1.5950	110
115.	338.08	0.01785	3.8813	309.3	1189.6	0.4877	1.5913	115
120.	341.27	0.01789	3.7275	312.6	1190.4	0.4919	1.5879	120
125.	344.35	0.01792	3.5857	315.8	1191.1	0.4959	1.5845	125
130.	347.33	0.01796	3.4544	319.0	1191.7	0.4998	1.5813	130
135.	350.23	0.01799	3.3325	322.0	1192.4	0.5035	1.5782	135
140.	353.04	0.01803	3.2010	325.0	1193.0	0.5071	1.5752	140
145.	355.77	0.01806	3.1130	327.8	1193.5	0.5107	1.5723	145
150.	358.43	0.01809	3.0139	330.6	1194.1	0.5141	1.5695	150
160.	363.55	0.01815	2.8386	336.1	1195.1	0.5206	1.5641	160
170.	368.42	0.01821	2.6738	341.2	1196.0	0.5269	1.5591	170
180.	373.08	0.01827	2.5312	346.2	1196.9	0.5328	1.5543	180
190.	377.53	0.01833	2.4030	350.9	1197.6	0.5384	1.5498	190
200.	381.80	0.01839	2.2873	355.5	1198.3	0.5438	1.5454	200
210.	385.91	0.01844	2.18217	359.9	1199.0	0.5490	1.5413	210
220.	389.88	0.01850	2.08629	364.2	1199.6	0.5540	1.5374	220
230.	393.70	0.01855	1.99846	368.3	1200.1	0.5588	1.5336	230
240.	397.39	0.01860	1.91769	372.3	1200.6	0.5634	1.5299	240
250.	400.97	0.01865	1.84317	376.1	1201.1	0.5679	1.5264	250
260.	404.44	0.01870	1.77418	379.9	1201.5	0.5722	1.5230	260
270.	407.80	0.01875	1.71013	383.6	1201.9	0.5764	1.5197	270

Table 1-3. Properties of saturated steam - pressure table (Courtesy, *Cameron Hydraulic Data book*. Reproduced with permission of Ingersoll-Dresser Pump Company.)

The explanations for the subscript abbreviations are as follows:

f = saturated liquid state
g = saturated vapor state
fg = difference between the liquid and vapor state

In the superheated steam table (Table 1-4), the top row of temperatures represent the total steam temperature, while the temperatures to the left of the abbreviation "sh" show the amount the steam is superheated.

Abs press lb/in² (sat temp-F)		Sat water	Sat steam	Temperature— degrees Fahrenheit						
				300	400	500	600	700	800	900
1 (101.74)	sh			198.26	298.26	398.26	498.26	598.26	698.26	798.26
	v	0.01614	333.6	452.3	511.9	571.5	631.1	690.7	750.3	809.8
	h	69.73	1105.8	1195.7	1241.8	1288.6	1336.1	1384.5	1433.7	1483.8
	s	0.1326	1.9781	2.1152	2.1722	2.2237	2.2708	2.3144	2.3551	2.3934
5 (162.24)	sh			137.76	237.76	337.76	437.36	537.76	637.76	737.76
	v	0.01641	73.53	90.24	102.24	114.21	126.15	138.08	150.01	161.94
	h	130.20	1131.1	1194.8	1241.3	1288.2	1335.9	1384.3	1433.6	1483.7
	s	0.2349	1.8443	1.9369	1.9943	2.0460	2.0932	2.1369	2.1776	2.2159
10 (193.21)	sh			106.79	206.79	306.79	406.79	506.79	606.79	706.79
	v	0.01659	38.42	44.98	51.03	57.04	63.03	69.00	74.98	80.94
	h	161.26	1143.3	1193.7	1240.6	1287.8	1335.5	1384.0	1433.4	1483.5
	s	0.2836	1.7879	1.8593	1.9173	1.9692	2.0166	2.0603	2.1011	2.1394
14.696 (212.00)	sh			88.00	188.00	288.00	388.00	488.00	588.00	688.00
	v	0.0167	26.799	30.52	33.963	38.77	42.86	46.93	51.00	55.06
	h	180.17	1150.5	1192.6	1239.9	1287.4	1335.2	1383.8	1433.2	1483.4
	s	0.3121	1.7568	1.8158	1.8720	1.9265	1.9739	2.0177	2.0585	2.0969
20 (227.96)	sh			72.04	172.04	272.04	372.04	472.04	572.04	672.04
	v	0.01683	20.087	22.356	25.428	28.457	31.466	34.465	37.458	40.447
	h	196.27	1156.3	1191.4	1239.2	1286.9	1334.9	1383.5	1432.9	1483.2
	s	0.3358	1.7320	1.7805	1.8397	1.8921	1.9397	1.9836	2.0244	2.0628
40 (267.25)	sh			32.75	132.75	232.75	332.75	432.75	532.75	632.75
	v	0.01715	10.497	11.036	12.624	14.165	15.685	17.195	18.699	20.199
	h	236.14	1169.8	1186.6	1236.4	1285.0	1333.6	1382.5	1432.1	1482.5
	s	0.3921	1.6765	1.6992	1.7608	1.8143	1.8624	1.9065	1.9476	1.9860
60 (292.71)	sh			7.29	107.29	207.29	307.29	407.29	507.29	607.29
	v	0.1738	7.174	7.257	8.354	9.400	10.425	11.438	12.446	13.450
	h	262.21	1177.6	1181.6	1233.5	1283.2	1332.3	1381.5	1431.3	1481.8
	s	0.4273	1.6440	1.6492	1.7134	1.7681	1.8168	1.8612	1.9024	1.9410
80 (312.04)	sh				87.96	187.96	287.96	387.96	487.96	587.96
	v	0.01757	5.471		6.218	7.018	7.794	8.560	9.319	10.075
	h	282.15	1183.1		1230.5	1281.3	1330.9	1380.5	1430.5	1481.1
	s	0.4534	1.6208		1.6790	1.7349	1.7842	1.8289	1.8702	1.9089
100 (327.82)	sh				72.18	172.18	272.18	372.18	472.18	572.18
	v	0.01774	4.431		4.935	5.588	6.216	6.833	7.443	8.050
	n	298.54	1187.2		1227.4	1279.3	1329.6	1379.5	1429.7	1480.4
	s	0.4743	1.6027		1.6516	1.7088	1.7586	1.8036	1.8451	1.8839
120 (341.27)	sh				58.73	158.73	258.73	358.73	458.73	558.73
	v	0.01789	3.7275		4.0786	4.6341	5.1637	5.6813	6.1928	6.7006
	h	312.58	1190.4		1224.1	1277.4	1328.2	1378.4	1428.8	1479.8
	s	0.4919	1.5879		1.6286	1.6872	1.7376	1.7829	1.8246	1.8635
140 (353.04)	sh				46.96	146.96	246.96	346.96	446.96	546.96
	v	0.01803	3.2190		3.4661	3.9526	4.4119	4.8588	5.2995	5.7364
	h	324.96	1193.0		1220.8	1275.3	1326.8	1377.4	1428.0	1479.1
	s	0.5071	1.5752		1.6085	1.6686	1.7196	1.7652	1.8071	1.8461
160 (363.55)	sh				36.45	136.45	236.45	336.45	436.45	536.45
	v	0.01815	2.8336		3.0060	3.4413	3.8480	4.2420	4.6295	5.0132
	h	336.07	1195.1		1217.4	1273.3	1325.4	1376.4	1427.2	1478.4
	s	0.5206	1.5641		1.5906	1.6522	1.7039	1.7499	1.7919	1.8383

Table 1-4. Properties of superheated steam (Courtesy, *Cameron Hydraulic Data* book. Reproduced with permission of Ingersoll-Dresser Pump Company.)

TEMPERATURE

In addition to absolute pressure, another term you need to know is *absolute temperature*. Absolute temperature is measured on the Rankine scale (°R). Absolute zero temperature is the point at which all molecular motion ceases, and 0°R is equal to about -460°F. Figure 1-6 shows a comparison of the temperature scales.

It is also important to know how to convert from Fahrenheit to Celsius or vice versa. To convert Celsius to Fahrenheit:

$$°F = (9/5 \times °C) + 32$$

To convert Fahrenheit to Celsius:

$$°C = 5/9 \times (°F - 32)$$

A few other points to know:

- The freezing point of water is 32°F or 0°C; the boiling point of water is 212°F or 100°C.

- The ratio 100/180 (5/9) is the number of graduations from freezing to boiling on the Celsius and Fahrenheit scales, respectively. The reciprocal is 9/5, and its decimal equivalent is 1.8.

- If you think you are using an incorrect formula, try it out by plugging in the value for a conversion you know. For example, 0°C or 32°F should convert to 32°F or 0°C, respectively, and 100°C or 212°F should convert to 212°F or 100°C, respectively.

- -40°F = -40°C

Kelvin
ABS C

Celsius

Fahrenheit

Rankine
ABS F

373	100	Boiling temperature of water	212	672
353	80		176	636
333	60		140	600
313	40		104	564
293	20		68	528
273	0	Freezing temperature of water	32	492
253	-20		-4	456
233	-40		-40	420
213	-60		-76	384
33	-240		-400	60
13	-260		-436	24
0	-273	Absolute zero	-460	0

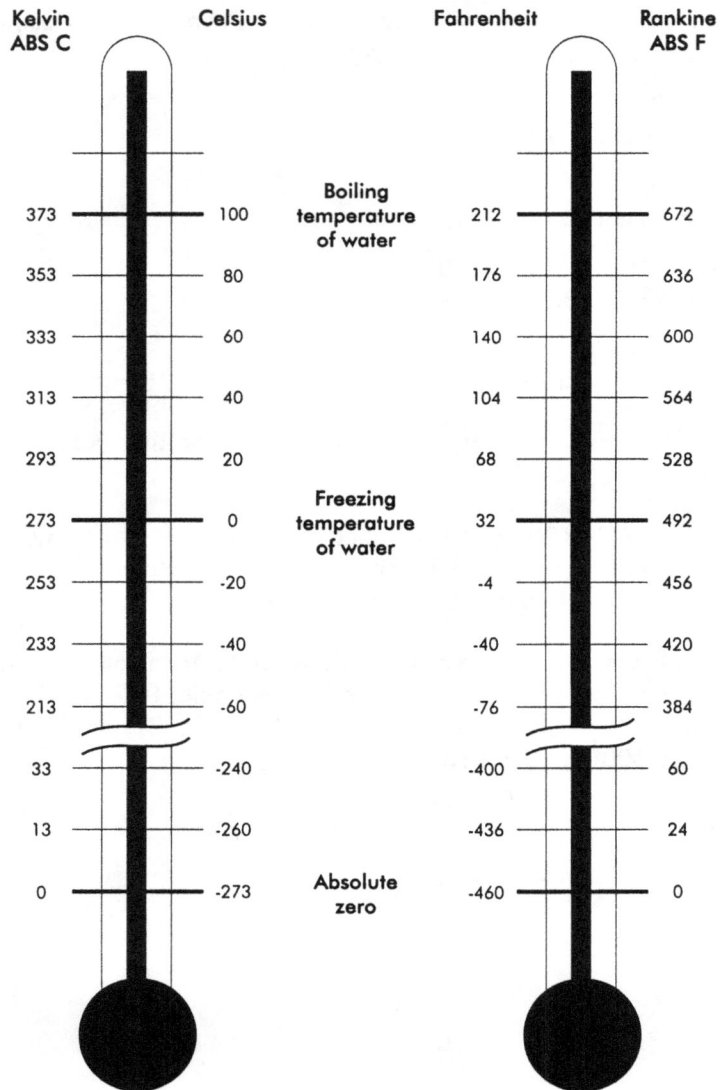

Figure 1-6. Celsius, Fahrenheit, Kelvin, and Rankine thermometer scales

Boilers

The purpose of any boiler is to transfer heat produced by burning fuel to water under pressure to generate steam. The hot water and steam in even the smallest boiler has a large amount of energy.

Early boilers were prone to explode due to defective design, fabrication, and materials. An organization concerned about the loss of life and property that resulted from boiler explosions was the American Society of Mechanical Engineers (ASME). This professional society developed standards for boilers and other pressure vessels. Known as the ASME Code, or simply the Code, it is not a legal document, however, most states have written the Code into their statutes, so it is the law of the land.

Manufacturers have developed different boiler designs to maximize efficiency and make their products competitive in the marketplace. There are three basic boiler types: *fire-tube*, *water-tube*, and *cast iron* boilers. In a fire-tube boiler, the products of combustion pass through tubes surrounded by water. In water-tube boilers, water circulates through the tubes, which are surrounded by the products of combustion. Cast iron boilers are discussed later in the chapter.

FIRE-TUBE BOILERS

The fire-tube boiler is the most common type used in industrial and heavy commercial applications. It is popular because it is less expensive to make and easier to repair than water-tube boilers. The horizontal return tubular (HRT) and the Scotch Marine are the two types of fire-tube boilers.

Horizontal Return Tubular (HRT) Boilers

The HRT was the first fire-tube boiler, and at one time, it was the most popular boiler type. HRTs consist of a cylindrical tank-like vessel with tubes secured in its head, or tube sheet, Figure 2-1. The HRT does not contain a fire box; therefore, the boiler is placed over an external furnace. Hot combustion gases rise from the furnace and move along the bottom of the shell towards the back of the combustion chamber. The gases rise, move through the horizontal fire tubes, then discharge into the breeching. The gases then dissipate through the stack. This vessel is mounted over its furnace by suspending it from rods connected to hanger brackets riveted or welded to the shell. It is a big project to install an HRT. Installing the footings, structural steel, and rigging; building the furnace or setting brick by brick; and installing the burner and various controls on site is very labor intensive, requiring many trades.

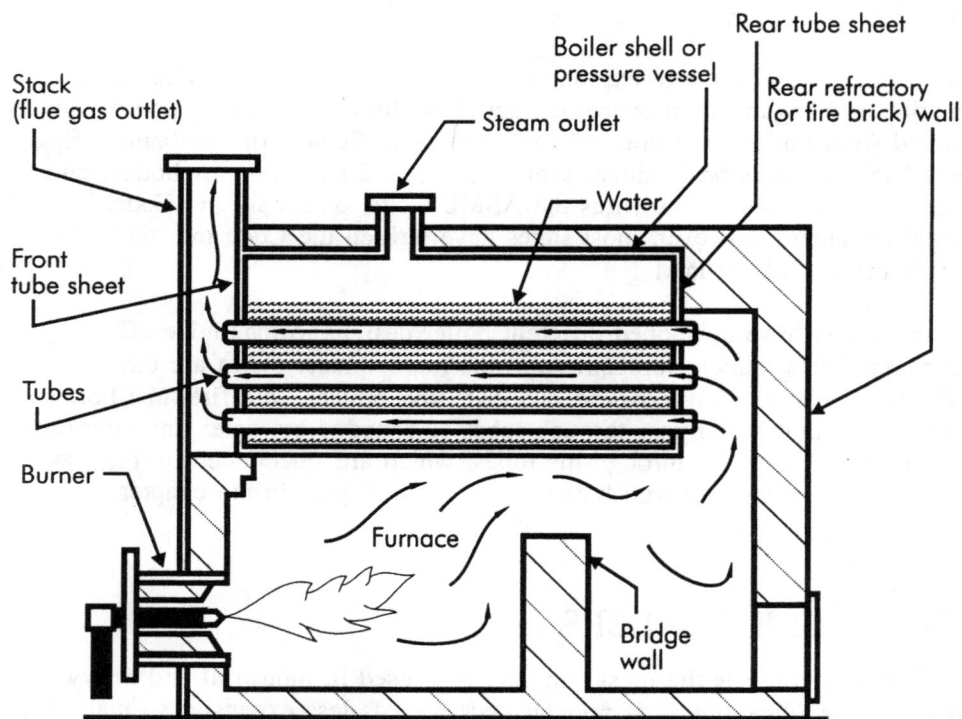

Figure 2-1. HRT boiler

The term *boiler setting* originally applied to the brick wall enclosing the furnace and heating surface of the boiler. Now the term comprises all the walls that form the boiler and furnace enclosure, including the insulation and lagging of these walls.

Scotch Marine Boilers

Today, most fire-tube boilers are shop-constructed *Scotch Marines*, Figure 2-2. The main design feature of a Scotch Marine boiler is its self-contained, built-in furnace. Because it requires no on-site structural steel or setting, the Scotch Marine boiler with all its auxiliary equipment can be completely assembled in a factory under controlled, efficient conditions. Frequently called *packaged boilers*, they are shipped complete with burners, oil and gas trains, combustion controls, feedwater controls, and the trim (water column, gauge glass, pressure gauge, low-water cutoffs, etc.). At the end of the production line, packaged units are filled with water, test fired, and brought up to operating pressure. When delivered to the customer, they are ready to run after connection of the electric power, exhaust stack, and steam, feedwater, blowdown, and fuel lines.

Figure 2-2. Scotch Marine boiler (Courtesy, Williams & Davis Boilers, Inc.)

In the Scotch Marine boiler, the hot gases from combustion move into a brick-lined combustion chamber at the back of the boiler. The gases pass into the fire tubes, and these tubes return the gases to the front of the boiler. The gases then pass through large furnace flues, into the breeching, and finally out the chimney. The furnace is surrounded by water, thus capturing more of the flames' heat for higher efficiency. With this design, the boiler's pressure can crush the furnace. With low-pressure boilers, ordinary large-diameter pipe can be used for the furnace. By definition, **the safety valve setting is limited to 15 psi on low-pressure boilers. Anything above 15 psi is considered a high-pressure or power boiler.**

With high-pressure boilers, the furnace must be reinforced to prevent collapse. Figure 2-3 shows an Adamson ring furnace, which uses flanges or reinforcing rings on the tubes, while Figure 2-4 shows the Morison furnace, in which the furnace tubes are corrugated. With modern fabrication techniques, it is easy to corrugate standard pipe to make a Morison furnace.

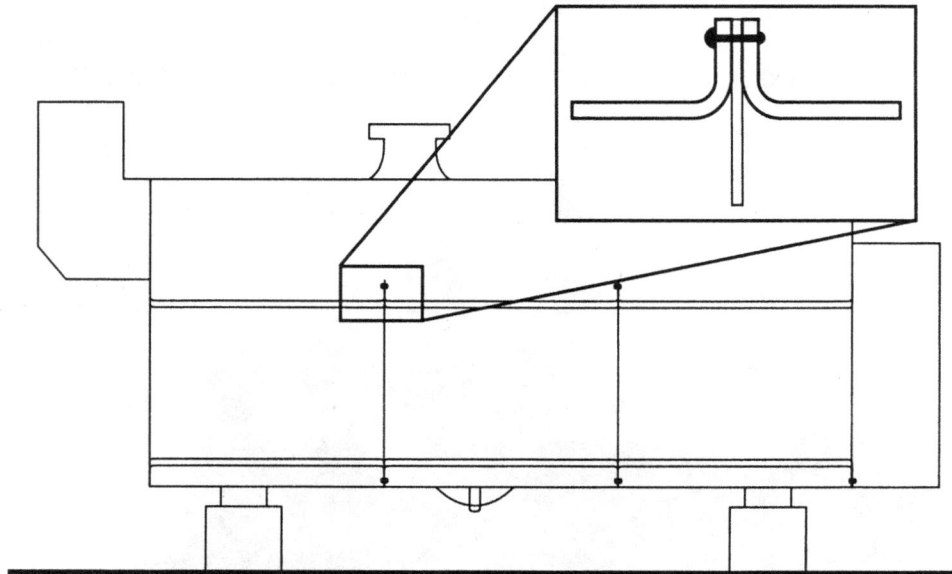

Figure 2-3. Adamson ring furnace

When pressure is applied to a flat surface, it tends to bulge out. This force is always present on the tube sheets of fire-tube boilers. The lower 7/8 of the tube sheet is supported by the tubes and furnace; however, tubes are omitted from the upper section of the boiler to provide steam space. Welding diagonal stays in place is the most common method to support the upper area, Figure 2-5. Boiler inspectors make sure they are intact for safe operation.

Figure 2-6 shows the internal construction of a Scotch Marine boiler. The large circular opening is the furnace, and the steel plate that looks like a pair of bird wings is a baffle that separates the third and fourth passes. (Passes are discussed later in this chapter.)

To secure the furnace and tubes to the tube sheet, the furnace is welded, and the tubes are expanded by rolling. Tube expanding or rolling is a simple and economical way of cold working the end of a tube into contact with the circular opening of a tube sheet, drum, or header. The expanded joint provides an easy,

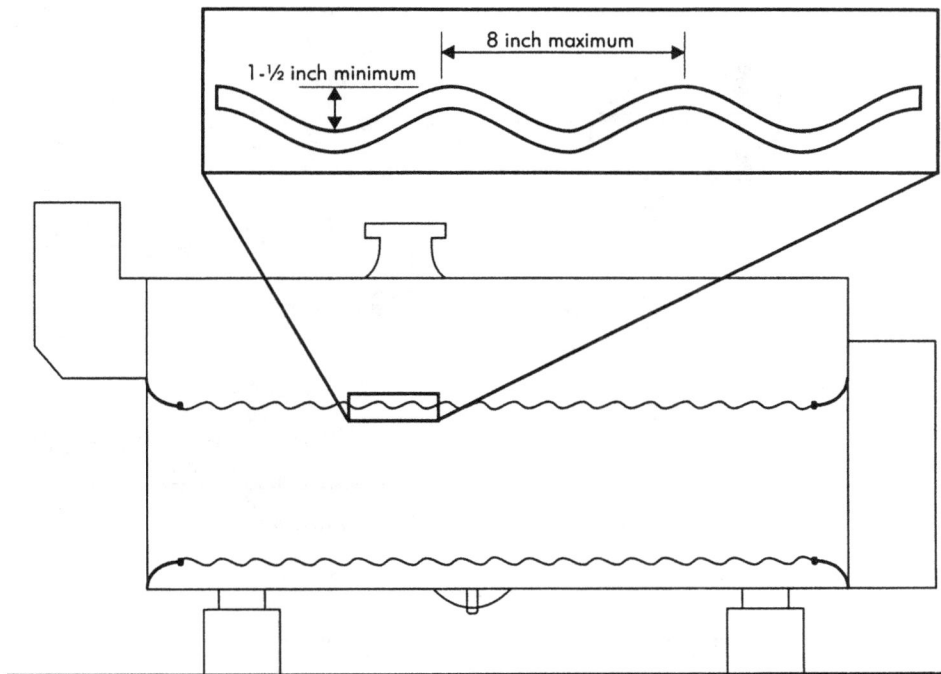

Figure 2-4. Morison furnace

secure method to fasten tubes. Under axial loading, the expanded joint is almost as strong as the tube itself (axial loading is the force trying to pull the tube out of the tube joint).

Figure 2-7 shows an expanded tube and an expander. The tapered mandrel is rotated by a high-torque, slow-speed motor and is pushed into the expander. The rotating mandrel turns against the tube with great force. This force increases the tube's diameter and forms a strong, watertight connection to the tube sheet. A bell roll flares the end of the tube. Tube sheets are sometimes grooved to give the joint more strength. On fire-tube boilers, Figure 2-8, this flare is exposed to hot flue gases and would eventually be damaged. For this reason, and not for additional strength, the ends of the tubes are beaded over back to the tube sheet.

On older boilers, such as locomotive types, stay bolts are an important part of the pressure vessel. They are used when two adjacent flat surfaces must be held in place, or stayed, to resist the bulging effect of pressure. Since they are rigidly attached, any parallel movement of the two adjacent flat surfaces places considerable stress on these bolts. If one of the two plates makes up part of the furnace and thus is exposed to considerable heat, it expands more than the adjacent plate. After a while, these bolts crack. It is like bending a metal coat hanger back and forth; eventually it breaks. The ASME Code requires a *crack detector*, which is nothing more than a small hole drilled in the middle of the

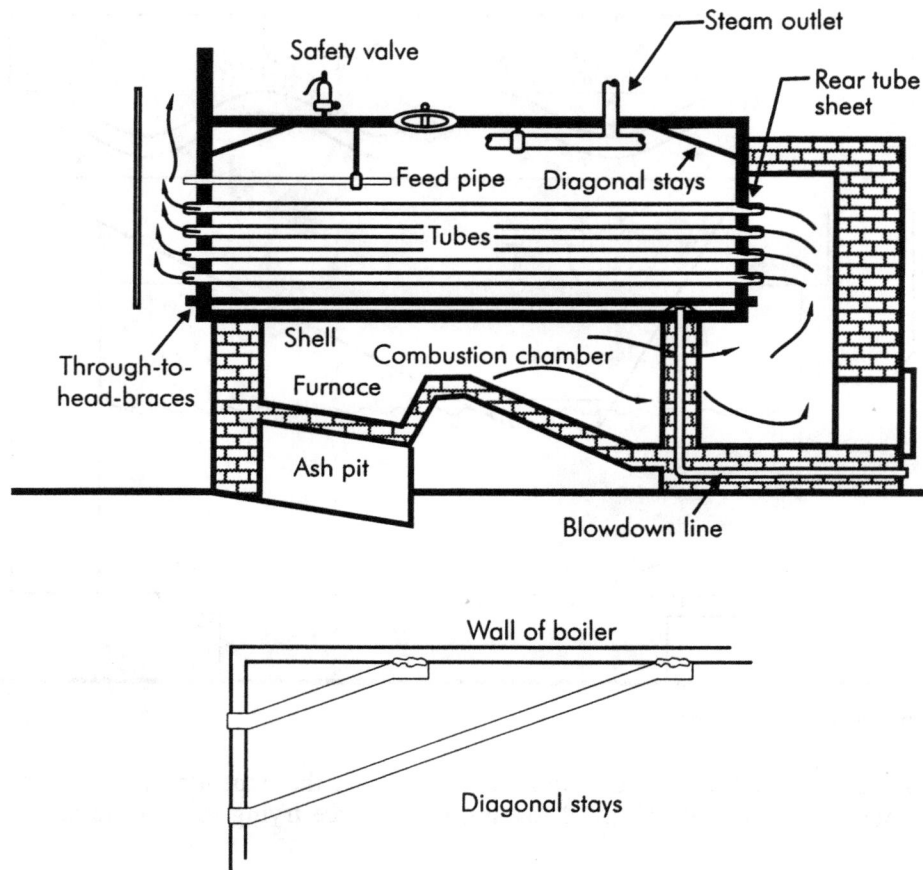

Figure 2-5. Diagonal stays support the upper portion of fire-tube boilers

bolt. When a crack reaches the drilled hole or *telltale hole*, steam or water escapes, alerting the operator to the problem, Figure 2-9.

Dry-Back and Wet-Back Boilers

Fire-tube boilers are classified either as dry-back or wet-back.

In a dry-back boiler, the rear part of the boiler consists of a refractory-lined door that is hinged or hung from a davit, allowing it to be opened for inspection or maintenance. This lining also forms the rear baffle in three- and four-pass boilers.

In a wet-back boiler, the rear part of the furnace is a water-cooled steel jacket instead of a refractory. This jacket is part of the pressure vessel. Though more expensive than a dry-back, a wet-back boiler reduces temperature stress on the rear tube sheet, adds heating surface, and forms a leakproof gas baffle.

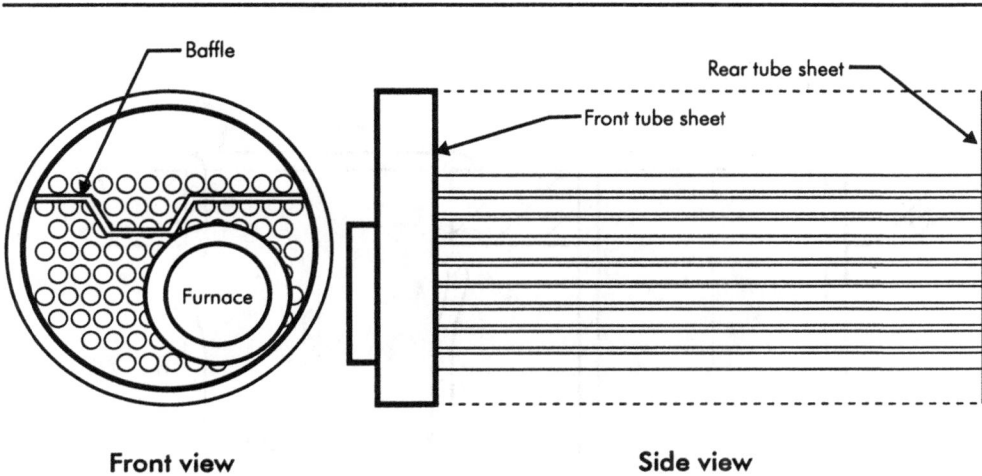

Front view   **Side view**

Figure 2-6. Internal construction of a Scotch Marine boiler

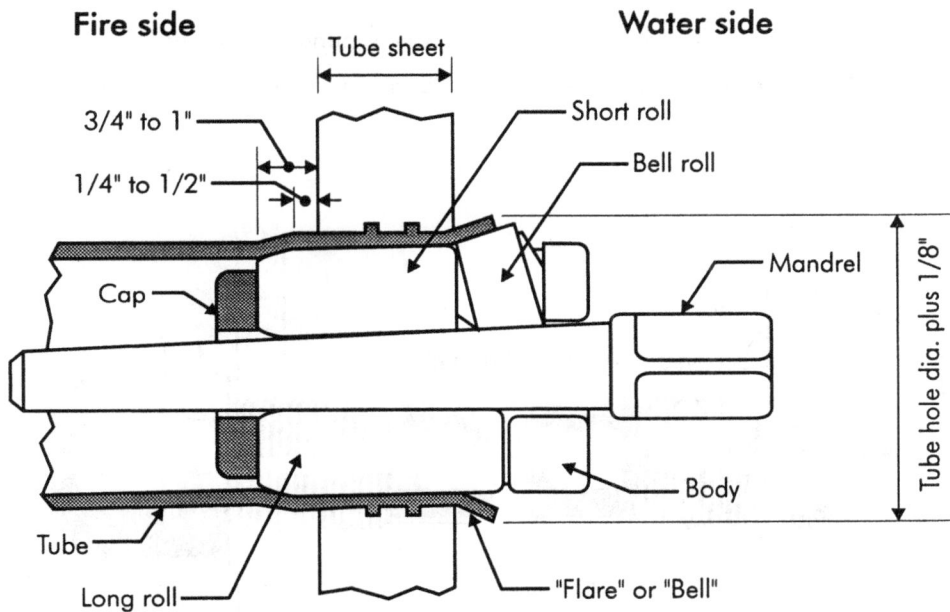

Figure 2-7. Position of the expander and mandrel after tube and mandrel are expanded into tube sheet (water-tube boiler)

Fire-tube boilers are also classified by the number of passes, or times the flue gas travels the length of the boiler. Figure 2-10 shows a simple two-pass boiler. These are the least expensive and generally the least efficient. By adding additional passes, the flue gas has additional opportunity to surrender its heat to the water in the boiler. The manufacturer has to decide on the number of passes by balancing the boiler's efficiency against the cost of production. Figure 2-11 is an example of a three-pass wet-back boiler.

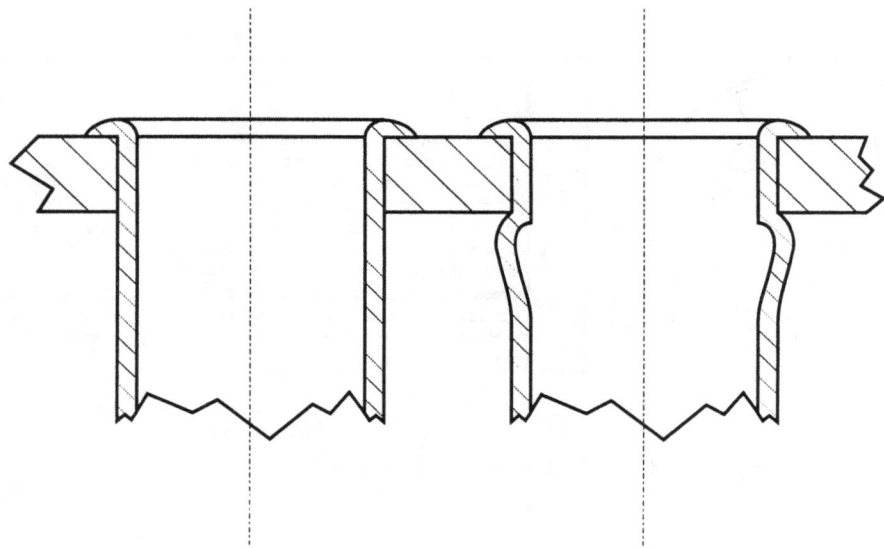

Expand roll and bead   Prosser, expand roll, and bead

Figure 2-8. Tubes expanded into and ends beaded back onto tube sheet (fire-tube boiler)

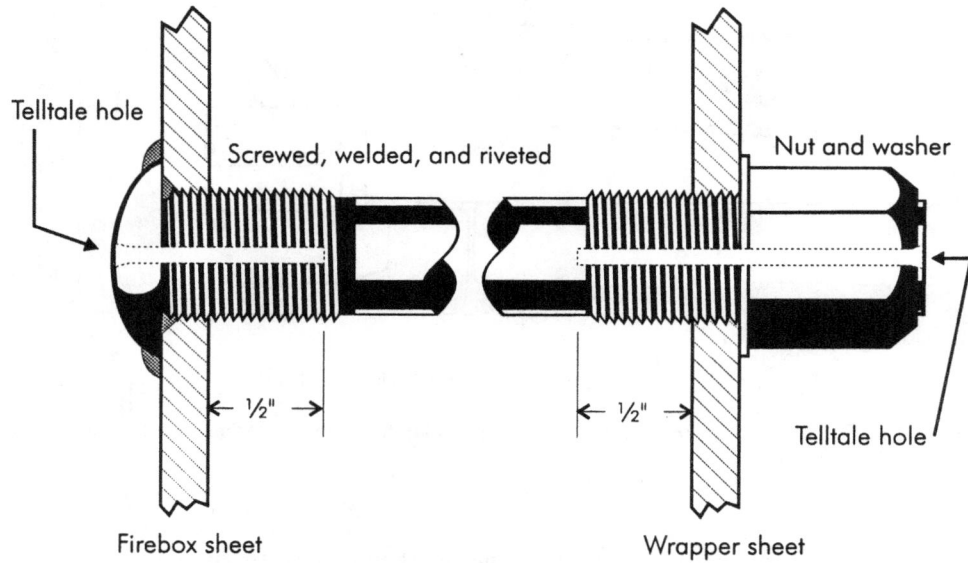

Telltale hole

Screwed, welded, and riveted   Nut and washer

½"   ½"

Telltale hole

Firebox sheet   Wrapper sheet

Figure 2-9. Stay bolt showing telltale hole

The practical economic upper limit of fire-tube boilers is 200 psi operating pressure and 30,000 pounds per hour. Beyond this, water-tube boilers become competitive.

WATER-TUBE BOILERS

Water-tube boilers have been around since the mid-1800s. One particular boiler design, created by Babcock and Wilcox, lasted into the 1930s. On this design straight tubes were rolled into vertical headers, and the front headers were connected to the steam drum. The rear header was connected to a horizontal box header, which was also attached to the steam drum. The tubes were inclined to promote water circulation. There was also a weight-and-lever safety valve on top of the drum. This type of valve is no longer permitted, because it is prone to nuisance leakage. Operators corrected this leakage problem by placing additional weight on the lever (see Figure 3-21). This increased its relieving pressure and thus made boiler explosions more likely. Also note that these furnaces were made of brick. As designs evolved, more of the furnace envelope consisted of steam-generating water tubes.

In the 1930s, manufacturing and material technology created rapid advances in higher operating pressures, temperatures, and capacities. Unlike the tubes being attached to flat tube sheets, as in fire-tube boilers, the tubes are attached to a steam drum and water or mud drum. Like fire-tube boilers, water-tube boilers come in different configurations. The most common oil- and gas-fired packaged boiler for industrial use is the "D" type.

In a water-tube boiler, fuel is fired into a water-cooled furnace under positive pressure. The *casing* (sheet metal and refractory enclosing the boiler) is seal-

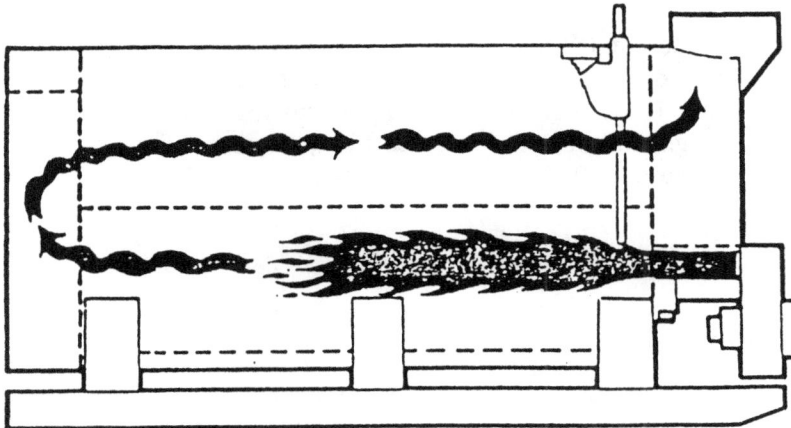

Figure 2-10. Two-pass dry-back boiler

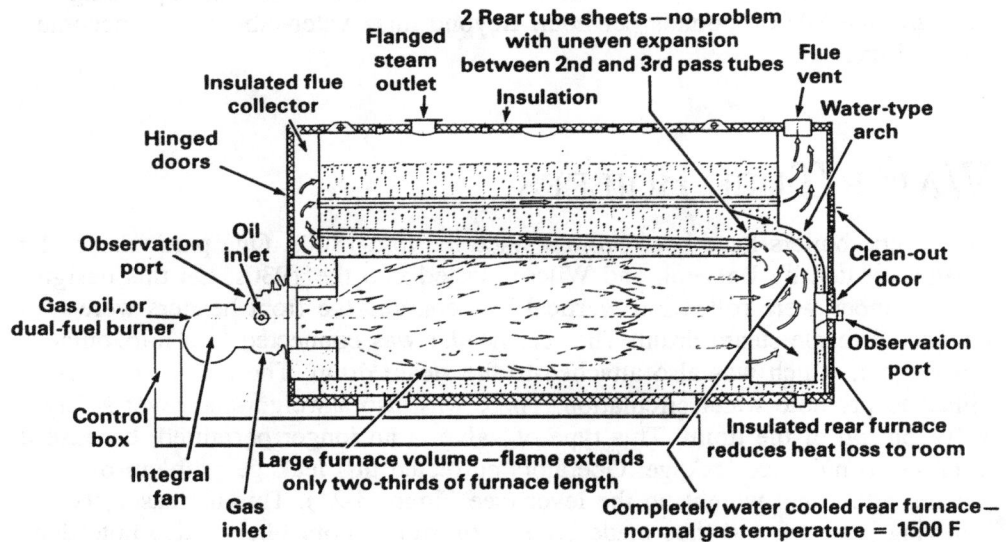

Figure 2-11. Three-pass wet-back boiler (Courtesy, North American Manufacturing Co.)

welded to form a gastight envelope. At the end of the furnace, the flue gas turns into the convection section and travels back toward the front. If superheaters are required (discussed later), they are placed between the furnace (radiant) and the convection sections.

All boilers have a *radiant* and a *convection* section. The furnace, or where the combustion takes place, makes up the radiant section. About half of the steam in commercial and industrial boilers is generated from the radiant energy of the flame. The rest of the steam is generated in the convection section, where heat is extracted from the hot flue gas before it goes up the stack. In large utility boilers, all the steam is generated in the radiant section, and the superheat is added in the convection section.

Note that the Babcock and Wilcox boiler design was made of brick. It wasn't until the 1930s that designs incorporated heat-absorbing, steam-generating tubes in the furnace area. These tubes initially had considerable space between them but were later designed so that they formed the gastight envelope called a *membrane wall*, Figure 2-12. This type of construction is often called a *water wall*.

Superheaters

Superheaters increase the amount of heat in the steam. As mentioned in the first chapter, steam temperature is determined by the steam pressure. If higher steam temperature is required, then it is run through a superheater. Locomotive-type

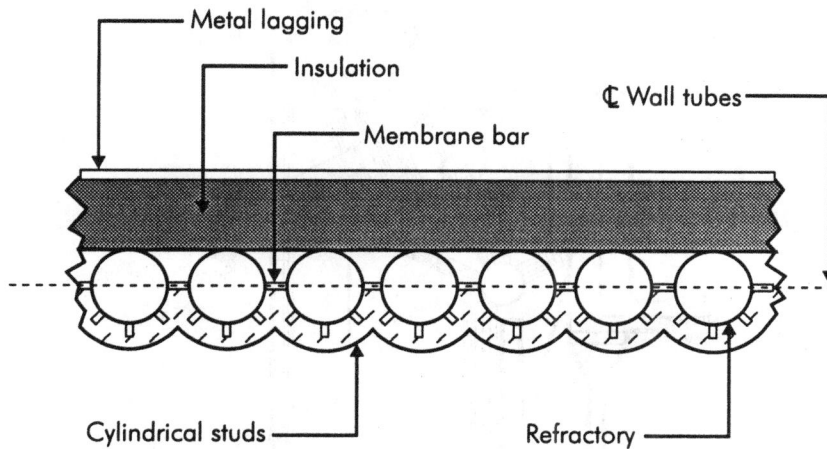

Figure 2-12. Membrane walls

boilers are equipped with superheaters, but they are rarely found in other fire-tube designs. In most industrial water-tube boilers, the superheater is placed where the flue gases make their turn from the radiant to the convection section of the boiler.

Superheaters are classified as radiant, convection, or combination. *Radiant superheaters* are placed at the rear of the furnace, where they receive the bulk of their heat by radiation from the flame, Figure 2-13. When required, *screen tubes* are placed in front of the superheater tubes to prevent damage from overheating. Screen tubes are part of the boiler water circulation system and are attached to the steam and mud drums.

A *convection superheater*, Figure 2-14, is placed in the flue gas stream where it receives most of its heat by convection.

Figure 2-15 shows that steam temperature decreases with an increase in steam production in radiant superheaters. With convection superheaters, steam temperature increases with an increase in steam production. A properly designed combination of the two yields a stable superheat temperature.

Since a constant superheat is desired over the load range of the boiler, many manufacturers place the superheater between the radiant and convection sections. With this combination superheater design, the steam temperature remains fairly constant.

Although boiler tubes are exposed to high temperatures, they remain intact because the water absorbs heat fast enough to prevent damage. Superheater tubes survive in the same environment, because the steam flowing through the tubes absorbs heat fast enough to keep the metal intact. For this reason, it is critical that steam flow through the superheater be properly maintained. Safety valves, which will be covered in the next chapter, are required on both the

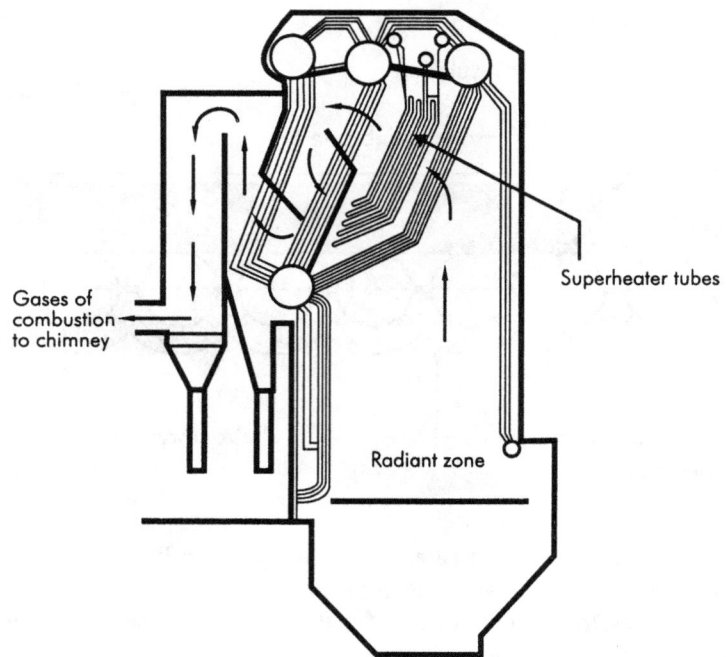

Figure 2-13. Radiant superheater

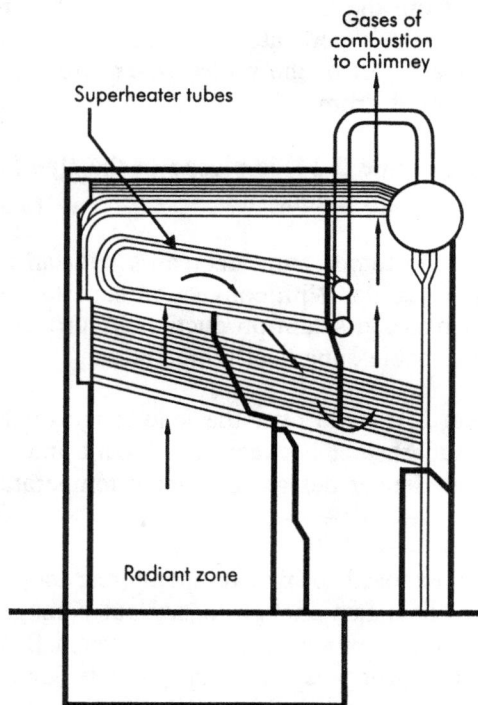

Figure 2-14. Convection superheater

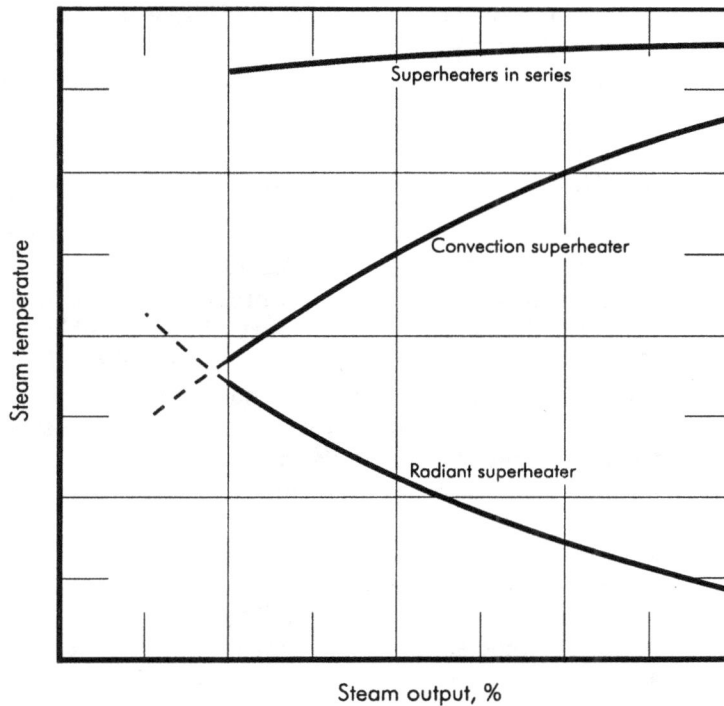

Figure 2-15. Relationships of superheaters to steam temperature and output

steam drum, where they are called the saturated safety valves, and on the superheater discharge.

To maintain steam flow through the superheater, the superheater safety valves are always set at a lower pressure than the drum safety valves. If the saturated safety valves discharged first, the reduced flow could cause the superheater to overheat.

To this point only natural-circulation boilers have been addressed. There are also controlled or forced-circulation water-tube boilers. The heating surface of a forced-circulation boiler consists of a continuous single coil and requires only a small amount of water in the boiler to produce the desired steam output. Consequently, start-up time can be as short as 15 minutes. This feature results in fuel and labor savings when the boiler is not required to operate 24 hours a day. This coil forms the furnace and convection section.

In operation of a controlled or forced-circulation boiler, water is pumped from a receiver to the accumulator. The level in the accumulator is maintained by a liquid level controller, and the bottom of the accumulator is equipped with a blowdown valve. A pump draws water from the accumulator and then forces it under pressure through the continuous heating coil. Steam generated in the coil is returned to the accumulator, where the steam and water are separated. The steam leaves through the top of the accumulator, and the water falls to the bottom and is pumped back into the coil.

Steam Separators

The water surface inside a steaming boiler is very turbulent. Under these conditions it is easy for some impure boiler water to carry over into the system with exiting steam. Steam separators are located in the steam and water drum and work on the principle that when steam changes direction, which it does with much less difficulty than water, the heavier water droplets will separate from the steam. The steam separator removes as much moisture as possible from the steam. Removal of this moisture prevents damage to the steam engine if water is carried over into the system. It also protects the system components from erosion, which may be caused by wet steam

The *dry pipe* steam separator is simple and effective for small boilers, Figure 2-16. This separator is located in the steam and water drum. The top half of a dry pipe separator consists of many small holes. The steam-water mixture must change direction in the dry pipe separator to navigate through these holes and out the steam outlet. With each change in direction, water drops out. The separated water drains out through one or two small pipes that run from the bottom of the dry pipe to under the water's surface.

Larger boilers, especially those equipped with superheaters, need more sophisticated separators. Centrifugal force, the force that tends to move objects outward from the center of rotation, is an excellent method to separate steam and water. As shown in Figure 2-17, the steam-water mixture is directed into a series of

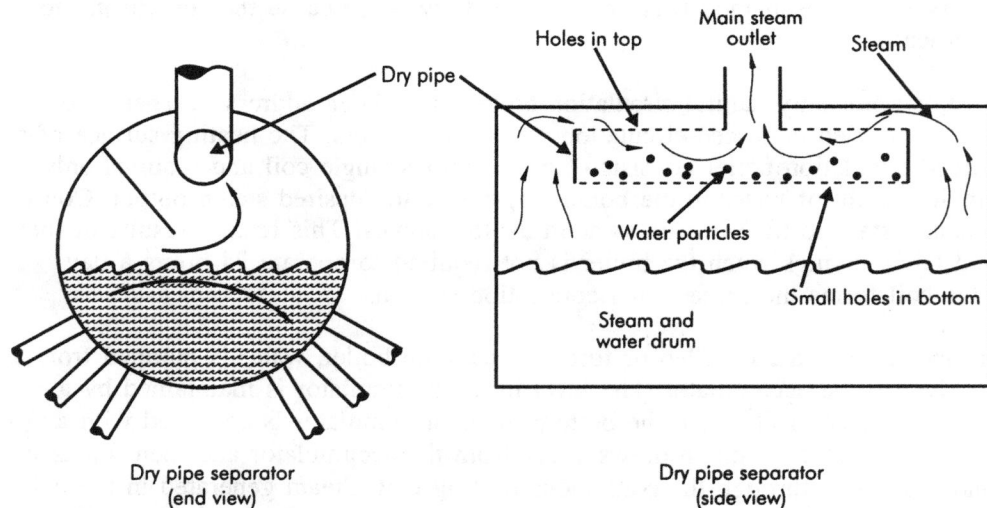

Dry pipe separator
(end view)

Dry pipe separator
(side view)

Figure 2-16. Dry pipe steam separator

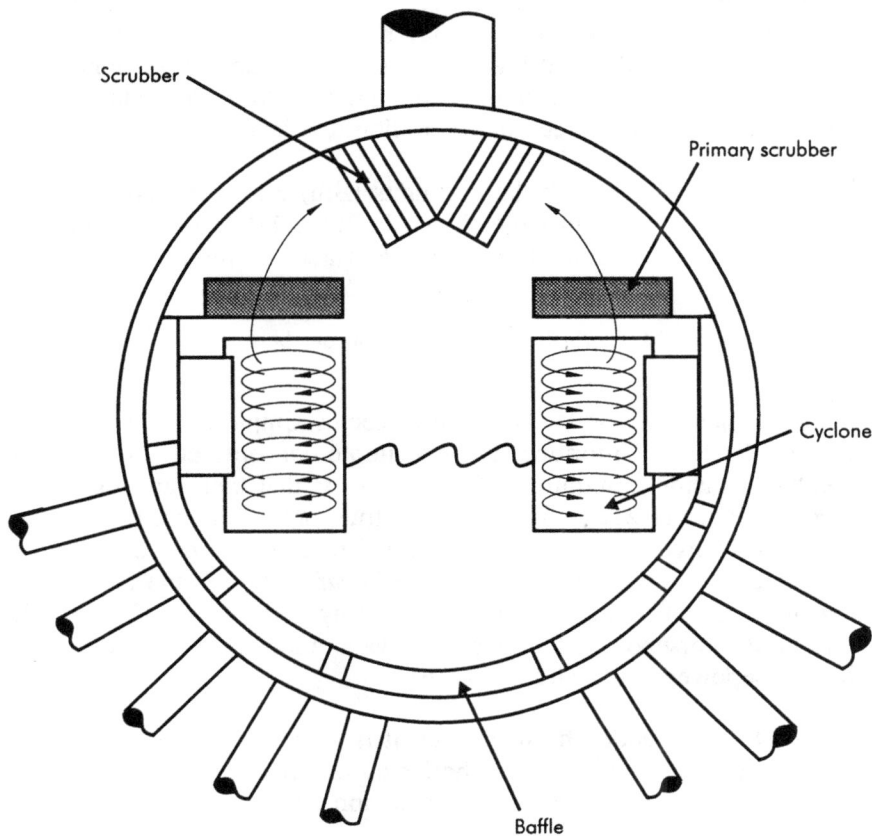

Figure 2-17. Centrifugal or cyclone separator

cyclones. Water is thrown to the side by centrifugal force and returned back to the drum, while the steam travels on to the boiler's outlet.

BOILER CAPACITY

Before discussing fire-tube versus water-tube boilers, it is necessary to understand boiler capacity. Steam boiler capacity is measured in pounds of steam per hour at operating pressure and temperature with a specified feedwater temperature. Boiler capacity used to be measured in boiler horsepower (bhp), however, that measurement is now only used to designate the size of fire-tube and cast iron boilers. *A boiler horsepower is equal to the evaporation of 34.5 lb of 212°F water per hour.* For example, consider how much steam a 600-horsepower boiler can generate:

(600 bhp) (34.5 lb/hr/bhp) = 20,700 lb/hr

Do not confuse *boiler* horsepower with *shaft* horsepower. Shaft horsepower may also be called *brake* horsepower. This may become confusing, because the abbreviation for both boiler and brake horsepower is bhp. Brake horsepower is the power needed to drive a machine at full load. It does not include the power lost in transmission of the force due to friction and heat.

To determine how many British thermal units (Btu) are contained in one boiler horsepower, refer to the steam table shown in Table 1-1. The table indicates that it takes 970.3 Btu to evaporate 1 lb of 212°F water. Therefore:

$$(970.3 \text{ Btu/lb}) \ (34.5 \text{ lb/hr/bhp}) = 33{,}475 \text{ Btu per bhp}$$

The concept of obtaining boiler horsepower seems simple enough and is sufficient most of the time. Unfortunately, this simple concept gets complicated when greater precision is required. The problem is that there are few boilers that evaporate water at 212°F. To obtain the true, or *developed*, boiler horsepower, a factor of evaporation must be applied to convert operating conditions to equivalent 212°F conditions. The *factor of evaporation* is the heat added to the water in the actual boiler in Btu/lb divided by 970.3 Btu/lb. To find the developed boiler horsepower, the factor of evaporation must be multiplied by the boiler horsepower (from and at 212°F).

To determine the heat added by the boiler, it is necessary to have a steam table and know the feedwater temperature, boiler outlet pressure, and the percentage of moisture or degree of superheat. The equation is:

$$(\text{Btu/lb})_{steam} - (\text{Feedwater temperature} - 32°F)$$

The units of measure do not work out in this equation; however, the feedwater is assumed to have a specific heat of one, therefore, one degree feedwater temperature is equivalent to one British thermal unit. The steam tables start at 32°F, so 32°F must be subtracted from the feedwater temperature.

Rated boiler horsepower is yet another method of calculating a boiler's capacity. Since rated bhp only requires boiler type and square feet of heating surface, Table 2-1, this method is frequently used to determine horsepower in the absence of reliable data.

The *percentage boiler rating* is determined by dividing the developed bhp by the rated bhp and then multiplying by 100.

It is easier to understand these concepts when a sample problem is worked out.

Type of Boiler	Surface (ft²/bhp)
Vertical fire-tube	14
Horizontal fire-tube	12
Water-tube	10

Table 2-1. Rated horsepower

Example 2-1. Find the rated boiler horsepower, heat added to the boiler, factor of evaporation, developed horsepower, and percentage boiler rating for a water-tube boiler with the following specifications:

Heating surface: 4,000 sq ft

Superheated steam produced by boiler: 30,000 lb/hr

Pressure: 140 psia at 400°F

Feedwater temperature: 230°F

Solution 2-1. First find the rated boiler horsepower (use Table 2-1):

$$\frac{4,000 \text{ ft}^2}{10 \text{ ft}^2/\text{bhp}} = 400 \text{ bhp}$$

To find the heat added to the boiler, use the following equation (the superheated steam table in Table 1-4 shows an enthalpy of 1,220.8 Btu/lb):

$$1,220.8 \text{ Btu/lb} - (230° - 32°\text{F}) = 1,022.8 \text{ Btu/lb}$$

Determine the factor of evaporation as follows:

$$\frac{1,022.8 \text{ Btu/lb}}{970.3 \text{ Btu/lb}} = 1.054$$

Then determine the developed boiler horsepower as follows:

$$\frac{(30,000 \text{ lb/hr})(1.054)}{34.5 \text{ lb/hr/bhp}} = 916.5 \text{ bhp}$$

The percentage boiler rating is as follows:

$$\left(\frac{916.5 \text{ bhp}}{400 \text{ bhp}} \right)(100) = 229\%$$

(It is very common for modern boilers to develop twice their rated horsepower.)

Heating Surface

The heating surface of a boiler consists of those areas that are in contact with heated gases on one side and water on the other. The Code states that the fire side is used when calculating heating surface. For example, consider that tube and wall thicknesses are given for a fire-tube boiler. The given dimension of boiler tubes is always the outside diameter (od). In this case, however, the fire side is the inside diameter (id); therefore, twice the wall thickness must be subtracted to obtain the correct diameter.

Example 2-2. An HRT boiler is 72 inches in diameter, 20 feet long, and contains 110 three-inch tubes each with a wall thickness of 0.135 inches. Calculate the HRT's heating surface and rated boiler horsepower.

Solution 2-2. Part of the exterior shell of an HRT boiler forms the furnace and thus is a heat-absorbing surface. The furnace is where the combustion takes place. Unless otherwise stated, to find the heating surface of an HRT, use two-thirds the area of the shell, all the tube surface area (use inside diameter), and two-thirds the area of both tube sheets minus the area of the tube holes. *Note: When taking the licensing exam, be sure to write these categories on the answer sheet so the examiner knows the basis of your calculations.*

Notice that dimensions are given in both inches and feet. Since heating surface is usually in square feet, all measurements must be converted to feet. The area of the shell equals its circumference multiplied by its length:

$$(\text{Diameter})(3.14)(\text{Length}) = \left(\frac{72 \text{ in.}}{12 \text{ in./ft}} \right)(3.14)(20 \text{ ft}) = 376.8 \text{ ft}^2$$

Note: $\pi = 3.14$

Assume two-thirds of the shell is exposed to heat:

$$\left(\frac{2}{3} \right)(376.8 \text{ ft}^2) = 251.2 \text{ ft}^2$$

For a fire-tube boiler, the heating surface of the tubes equals their id circumference multiplied by their length. To obtain the id, subtract twice the tube thickness from the od:

$$id = 3 \text{ in.} - [(2)(0.135 \text{ in.})] = 2.73 \text{ in.}$$

$$(id)(3.14)(\text{Length})(\text{No. of tubes}) = \left(\frac{2.73 \text{ in.}}{12 \text{ in./ft}}\right)^2 (3.14)(20 \text{ ft/tube})(110 \text{ tubes}) = 1{,}571.57 \text{ ft}^2$$

Now calculate the surface area of each head as follows:

$$\text{Head area} = (0.7854)(\text{Diameter})^2$$

$$\text{Head area} = (0.7854)\left(\frac{72 \text{ in.}}{12 \text{ in./ft}}\right)^2 = 28.27 \text{ ft}^2/\text{head}$$

Note: The area of a circle can be πr^2 or $0.7854 d^2$.

Assume two-thirds of each head is exposed to heat:

$$\left(\frac{2}{3}\right)(28.27 \text{ ft}^2) = 18.85 \text{ ft}^2$$

Then calculate the hole area:

$$\text{Hole area} = (110 \text{ holes})(0.7854)(3 \text{ in.})^2 = 777.55 \text{ in}^2$$

$$\frac{777.55 \text{ in}^2}{144 \text{ in}^2/\text{ft}^2} = 5.4 \text{ ft}^2$$

Therefore, the heating surface of each head is:

$$(18.85 \text{ ft}^2) - (5.4 \text{ ft}^2) = 13.45 \text{ ft}^2/\text{head}$$

There are two heads:

$$(2)(13.45 \text{ ft}^2) = 26.9 \text{ ft}^2$$

The heating surface equals the sum of the heating areas, tubes, and heads:

$$251.2 \text{ ft}^2 + 1,571.57 \text{ ft}^2 + 26.9 \text{ ft}^2 = 1,849.67 \text{ ft}^2$$

The rated boiler horsepower is as follows (use Table 2-1):

$$\frac{\text{Heating surface}}{12 \text{ ft}^2/\text{bhp}} = \frac{1,849.67 \text{ ft}^2}{12 \text{ ft}^2/\text{bhp}} = 154 \text{ bhp}$$

Example 2-3. Find the rated boiler horsepower, heat added to the boiler, factor of evaporation, developed horsepower, and percentage boiler rating for a water-tube boiler with the following specifications:

Heating surface: 18,000 ft^2

Steam produced by boiler: 75,000 lb/hr

Pressure: 150 psia

Heat contained in 150 psia steam: 1,194.1 Btu/lb

Feedwater temperature: 180°F

Solution 2-3. First find the rated boiler horsepower (use Table 2-1):

$$\frac{18,000 \text{ ft}^2}{10 \text{ ft}^2/\text{bhp}} = 1,800 \text{ bhp}$$

The heat added to the boiler is as follows:

$$1,194.1 \text{ Btu/lb} - (180°F - 32°F) = 1,046.1 \text{ Btu/lb}$$

Determine the factor of evaporation as follows:

$$\frac{1,046.1 \text{ Btu/lb}}{970.3 \text{ Btu/lb}} = 1.07$$

Then determine the developed boiler horsepower:

$$\frac{(75,000 \text{ lb/hr})(1.078)}{34.5 \text{ lb/hr/bhp}} = 2,343.5 \text{ bhp}$$

The percentage boiler rating is as follows:

$$\left(\frac{2{,}343.5 \text{ bhp}}{1{,}800 \text{ bhp}} \right)(100) = 130\%$$

FIRE-TUBE VS WATER-TUBE BOILERS

The choice between a fire-tube or a water-tube boiler comes down to how much pressure and capacity are required. If the pressure is under 150 psi and the capacity is under 25,000 lb/hr, then a fire-tube boiler works well. If higher values are required, then water-tube boilers are a better choice. Table 2-2 shows the advantages and disadvantages of both fire-tube and water-tube boilers.

Fire-Tube	Water-Tube
Lower initial cost	——
Pressures to 150 psi (some to 300 psi) but flat surfaces must be stayed	Pressures to 5,000 psi +
Most produce steam up to 25,000 lb/hr	7 million lb/hr +
——	Superheaters may be added easily
——	More flexible burning of different fuels
Larger water capacity (lb of steam per hr/lb of water storage)	——
Slower in coming up to operating pressure	Monotube boilers can be on-line in 15 minutes
Provides accumulator action and can meet load changes quickly	——

Table 2-2. Advantages and disadvantages of fire-tube and water-tube boilers

Why can water-tube boilers be built for much higher pressures than fire-tube boilers? Take a look at the formula below for the maximum allowable working pressure (MAWP):

$$P = \frac{(TS)(t)(E)}{(R)(FS)}$$

where:
    P = maximum allowable working pressure in psi inside drum or shell
    TS = tensile strength of plate, (psi - use 55,000 psi for steel)

t = thickness of plate (inches)

E = efficiency of the joint (for welded joints, use 100%, which is represented in the equation as 1.0.)

R = inside radius of drum or shell (inches)

FS = factor of safety (ultimate strength divided by allowable working stress or bursting pressure divided by safe working pressure. It can vary between four and seven depending on age, type of construction, and condition. Use five for most calculations.)

The bursting pressure is obtained by omitting the factor of safety (FS) from the above formula. There will be some examples of this equation at the end of the chapter.

When the radius of a boiler tube or shell increases for a given thickness, the maximum allowable working pressure decreases. Therefore, thin boiler tubes, with their small diameters, can stand a great deal of pressure, while the thick shells of fire-tube boilers, with their large diameters, can take only a moderate pressure. For example, a 1-3/4", 9-gauge (0.150" wall thickness) boiler tube is allowed to handle 2,020 psi, while a 5" tube of the same thickness is allowed to handle only 590 psi.

Another fact about steel structures is that they are much stronger in tension than in compression. In a fire-tube boiler, the tube is in compression (the pressure is trying to collapse the tube). The 1-3/4", 9-gauge tube that was good for 2,020 psi in a water-tube boiler is only good for 680 psi in a fire-tube boiler. Remember, boiler tubes are always measured by their outside diameter (od).

Cast Iron Boilers

Cast iron boilers have no tubes. Instead, water is routed through one set of cavities of the cast sections and the products of combustion through another set. The sections are cast from a pattern and then assembled either in the shop or in the field. Some section designs resemble pork chops, thus the nickname "pork chop boiler." Figure 2-18 shows a typical cast iron boiler. The sections are held together with tie rods. A circular gasket called a *sealing ring* keeps steam and water from leaking between the sections. A heat-resistant gasket seals in the flue gas to prevent it from leaking into the boiler room. This gasket resembles a rope and is fitted in a grove that runs around the edge of each section.

Forced draft is an arrangement in which the air required for combustion is pushed through the boiler, Figure 2-19. Since this puts the inside of the boiler under slight pressure, boilers with forced-draft firing have to be gastight to prevent flue gas from leaking into the boiler room. Forced-draft firing permits the use of a short chimney and reduces boiler room space requirements. Forced-

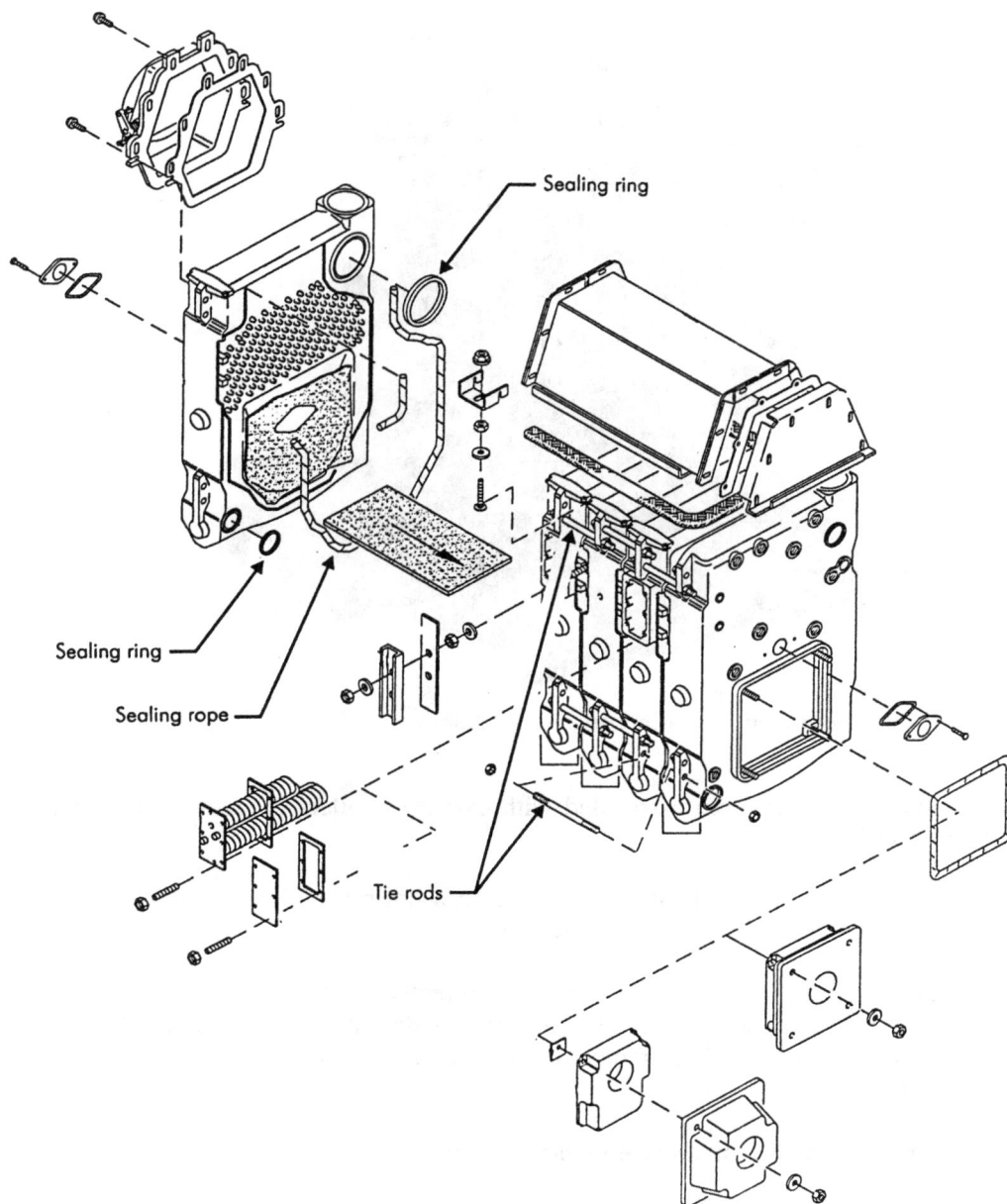

Figure 2-18. Assembly of cast iron boiler sections with sealing rings (Courtesy, Weil-McLain/A United Dominion Co.)

draft firing will be discussed in more detail in Chapter 4. The burners, controls, and trim are the same as those found on fire-tube and water-tube boilers.

Cast iron boilers are limited to 15 psi steam, 60 psi water, and are for heating service only. If used for process heating, the potentially large amount of cold make-up water would damage the cast iron sections by thermal shock.

Figure 2-19. Forced-draft firing (Courtesy, Weil-McLain/A United Dominion Co.)

The most serious problem associated with cast iron boilers is cracked sections. Cracks can be caused by:

- the rapid introduction of cold water into a hot boiler due to the operator or defective automatic feedwater device.

- the feedwater make-up line connected directly to one section instead of to the return line, where the returning hot condensate tempers the cold make-up water. (ASME Code requires make-up water to enter the return line first.)

- malfunction of temperature or pressure controls.

- low water in the boiler.

- water-side scale deposits, which cause overheating or water circulation obstruction.

- casting defects (which may become apparent only after years of service).

- poor assembly and/or improper installation (poor alignment or improper tie rod tightening).

When a section cracks, the decision is whether it should be repaired by welding or replaced. For the most part, that decision rests with the insurance boiler inspector after an evaluation of the location and length of the crack. If the crack is small and in an accessible area, a welded repair may be recommended. Otherwise the damaged section must be replaced.

Welded repairs must be done by an experienced welder. Special fusion welding electrodes are typically used and the cold-welding procedure followed. Cold welding is accomplished by minimizing the heat applied to the casting while keeping the repair area at a low temperature until the welding is completed.

Figure 2-20 shows a gravity return connection called a Hartford Loop. This piping arrangement has saved countless heating boilers from running dry. The return pipe connection is brought up to the normal water level, and an equalizer loop is arranged to prevent boiler water from being forced out below the safe water level.

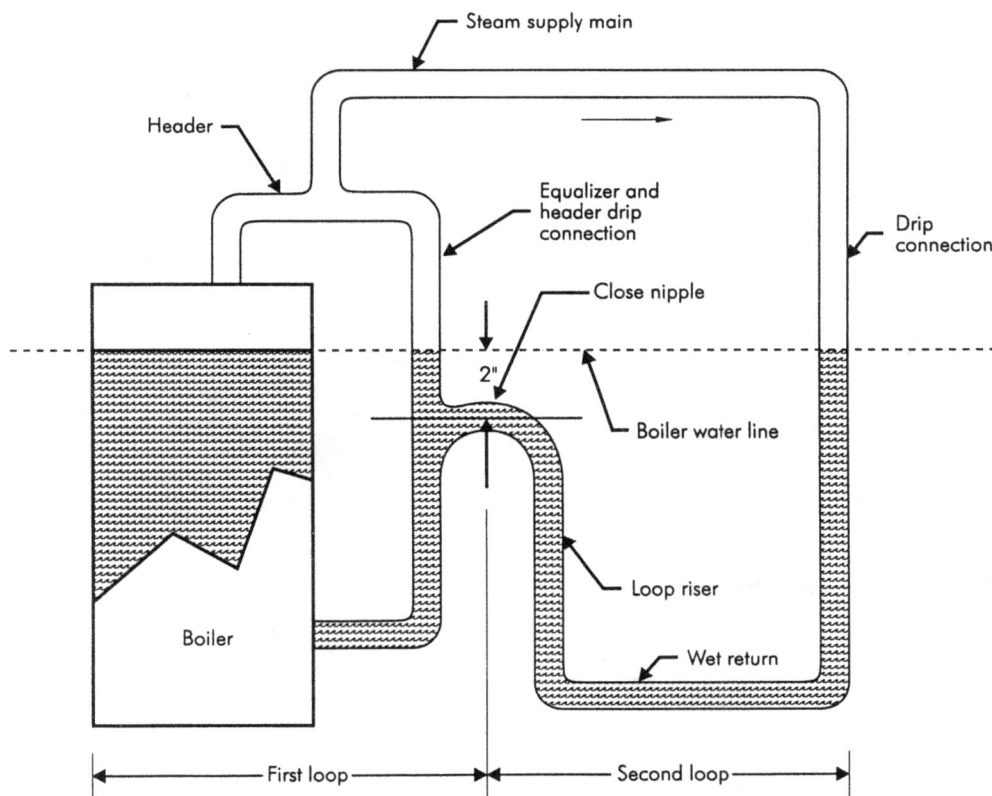

Figure 2-20. Hartford Loop

MATERIAL FOR ADVANCED EXAMINATIONS

Though exams often ask design questions, only professional engineers are permitted to design pressure vessels. Exam questions are only to determine your general knowledge of the subject. The examples that follow are for illustrative purposes only and are not intended to be used for designing actual pressure vessels.

Pressure-Vessel Bursting and Working Pressures

A common question deals with the bursting pressure of boilers. Closely associated is the calculation for the maximum allowable working pressure (MAWP), which was discussed earlier:

$$P = \frac{(TS)(t)(E)}{(R)(FS)}$$

Designers subject boilerplates to one-fourth to one-seventh of their tensile strength. Use a factor of safety of 5 unless otherwise stated in the problem.

Example 2-4. A welded boiler drum 3 ft in diameter is made of 1-inch plate steel. What is its bursting pressure?

Solution 2-4. Notice that the MAWP equation calls for the radius of the boiler drum, but the problem states the drum's diameter. Just about every problem has at least one of these pitfalls. In this case, just divide the diameter by 2 to obtain the radius. Note that before any number crunching is done, the numbers should be in compatible units of measurement.

It is first necessary to make sure that the radius is in compatible units of measurement:

$$\left(\frac{3 \text{ ft}}{2}\right)(12 \text{ in./ft}) = 18 \text{ in.}$$

Then it is possible to plug the other numbers into the MAWP equation:

$$P = \frac{(TS)(t)(E)}{(R)(FS)} = \frac{(55,000 \text{ psi})(1 \text{ in.})(1.0)}{(18 \text{ in.})(5)} = 611.11 \text{ psi}$$

The answer to the question is 611.11 psi.

Example 2-5. A 4-foot diameter riveted pressure vessel with a joint efficiency of 85% requires a MAWP of 325 psi. What plate thickness is required with a tensile strength of 55,000 psi and a factor of safety of 6?

Solution 2-5. In cases like this where the pressure is given and the problem is to find the required thickness, it is necessary to rearrange the equation:

$$t = \frac{(P)(R)(FS)}{(TS)(E)}$$

Then figure out the radius in inches:

$$\left(\frac{4 \text{ ft}}{2}\right)(12 \text{ in./ft}) = 24 \text{ in.}$$

Then plug the other numbers into the equation:

$$t = \frac{(325 \text{ psi})(24 \text{ in.})(6)}{(55,000 \text{ psi})(0.85)} = 1.0 \text{ in.}$$

Longitudinal Stress Compared to Circumferential Stress

The pressure inside a shell or drum is trying to push the ends off and rip open the cylinder lengthwise. *Circumferential*, or girth, stress is what tries to push the ends off, and *longitudinal*, or hoop, stress is what tries to rip apart a pressure vessel lengthwise. The longitudinal stress is always twice as much as the circumferential stress. Longitudinal seams have to resist twice the force as compared with circumferential seams.

$$\text{Hoop stress} = \frac{pd}{2t}$$

$$\text{Girth stress} = \frac{pd}{4t}$$

where:

p = internal pressure (lb/in^2)
d = internal diameter (in.)
t = cylinder wall thickness (in.)

Notice that the pressure vessel's length has no effect on stress, only diameter, plate thickness, and internal pressure.

Example 2-6. What is the force on the ends of a tank that is 7 ft long, 3 ft in diameter, 3/4 in. thick, with an internal pressure of 150 psi? What are the hoop and girth stresses?

Solution 2-6. First find the area of the head in square inches:

$$\text{Area} = 0.7854d^2 = (0.7854)\ (3\ \text{ft})^2\ (144\ \text{in}^2/\text{ft}^2) = 1{,}017.87\ \text{in}^2$$

Then find the force by multiplying pressure times area:

$$\text{Force} = (150\ \text{lb/in}^2)\ (1{,}017.87\ \text{in}^2) = 152{,}680.5\ \text{lb}$$

$$\text{Hoop stress} = \frac{pd}{2t} = \frac{(150\ \text{lb/in}^2)(36\ \text{in.})}{(2)(0.75\ \text{in.})} = 3{,}600\ \text{psi}$$

$$\text{Girth stress} = \frac{pd}{4t} = \frac{(150\ \text{lb/in}^2)(36\ \text{in.})}{(4)(0.75\ \text{in.})} = 1{,}800\ \text{psi}$$

Thermal Expansion

Another number that should be memorized for closed-book exams is 0.00000734 ft per ft-°F. This number shows that for every °F increase in temperature, a foot of the material will expand 0.00000734 feet. If heated, it will expand by this much, and if cooled, it will decrease by this amount. The temperature in this case is the difference between the starting and final temperatures. Sometimes the term ΔT (temperature difference) or Δt is used.

Example 2-7. Water at 250°F that is circulating through a 75-ft straight pipe is cooled to 70°F. How much is this pipe going to contract?

Solution 2-7.

$$\left(0.00000734\ \frac{\text{ft}}{\text{ft-}°\text{F}}\right)(75\ \text{ft})\ (250° - 70°\text{F}) = 0.099\ \text{ft}$$

The Code requires steel to meet certain specifications before being used in pressure vessel construction. The standard tension test determines the maximum load that a material can withstand before breaking based on yield strength, tensile strength, and ductility. The test is performed by placing a test sample of a specified length and diameter in a machine that pulls the sample apart and indicates the force used. Figure 2-21 shows the relationship between stress (load per unit area in psi) and the corresponding strain (deformation in inches) for a typical mild steel and high-tensile steel tension test.

The metal begins to stretch when the load is applied. Note the straight line from point O to the proportional limit at point A (called Hooke's law, elongation is proportional to load divided by the area). Strain is directly proportional to the stress applied. It is also called the *elastic limit*, because if the stress is released at any point along this straight line, the test sample will return to its original shape.

That dip right after the proportional limit is the *yield point* (YP). This is the stress at which a ductile material suddenly continues to elongate without further increase in load.

Now the steel becomes plastic; that is, instead of springing back to its original shape, it stays elongated. As the sample is further pulled apart, it breaks. The maximum applied load, in pounds, required to pull the sample apart divided by the area of the original sample is known as the *ultimate* or *tensile strength* (TS).

Metal ductility is determined by measuring the increase in length and the area where the sample broke. The strength of a metal decreases as the ductility increases. Since both strength and ductility are desirable, steels are mechanically worked or heat-treated to develop the best combination for a particular application.

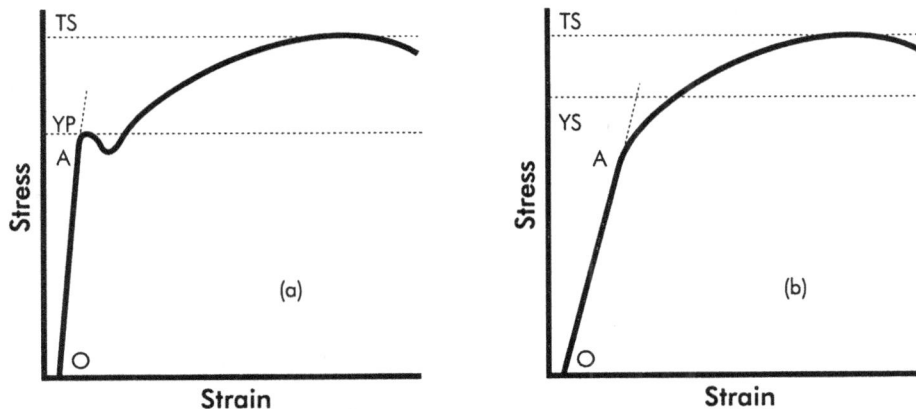

Figure 2-21. Relationship between stress and the corresponding strain for a typical mild steel (a) and high-tensile steel (b) tension test

CHAPTER THREE

Boiler Trim

The term *boiler trim* applies to just about every device connected to the boiler except the burners and their controls. Trim includes the water column, gauge glass, try cocks, steam pressure gauge, water level regulators, low-water cutoffs, safety valves, and blowdown valves.

There are many potential questions that can be asked about boiler trim. Most are easily answered once the functions of these devices are understood.

GAUGE GLASSES AND TRY COCKS

Maintaining the correct water level in a boiler is the boiler operator's most important responsibility. Improper water level results in boiler damage. That's why boilers are required to have at least two methods of determining water level. *Gauge glasses* and *try cocks* are two recognized ways to meet this requirement, Figures 3-1 and 3-2. There are several excellent remote reading devices on the market, but the gauge glass and try cock are the best.

On small boilers, the gauge glass and try cocks are sometimes attached directly to the shell, but they are usually attached to a water column. When steaming, water in the boiler is constantly turbulent. If a gauge glass were attached directly to the boiler, the water would rise and fall so rapidly that an accurate reading would be difficult to obtain. The construction and placement of a water column averages out these wild fluctuations, providing a stable water level. This stable water level allows the gauge glass and other devices to yield accurate readings.

The location of the gauge glass is explained in the ASME Code. On water-tube boilers, the bottom of the glass must be at least two inches above the lowest permissible water level. On fire-tube boilers, the bottom of the glass must be three inches above the lowest permissible water level.

Figure 3-1. High-pressure gauge glass (Courtesy, Clark-Reliance Corporation)

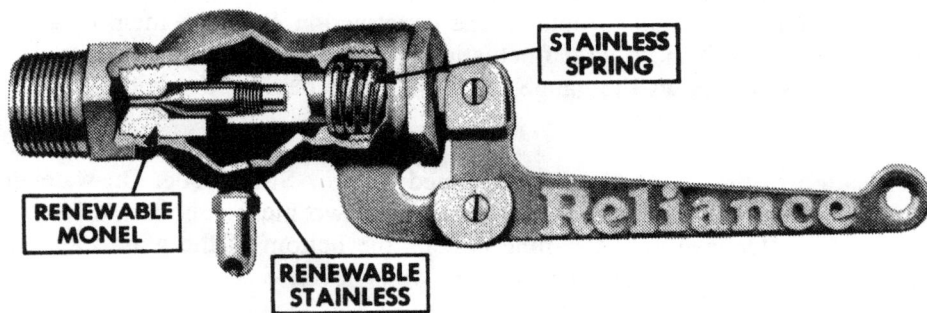

STAINLESS SPRING

RENEWABLE MONEL

RENEWABLE STAINLESS

Figure 3-2. Try cock for a boiler water column (Courtesy, Clark-Reliance Corporation)

Water columns should be mounted as close to the boiler as practical to obtain the most accurate water level reading. Long piping running between the boiler and water column causes a drop in water temperature, resulting in a lower reading. Examples of water column installations are shown in Figure 3-3.

No outlet connections (except water level recorders, feedwater regulators, drains, or steam gauges) may be connected to the water column or on the pipes connecting the water column to the boiler. A device that requires water or steam flow would create an inaccurate level reading for everything else attached to the water column. Notice that crosses with plugs or blind flanges are installed at all right angle turns. This allows for the easy inspection and cleaning of the piping when required.

The pipes connecting the boiler and water column must be at least one inch in diameter. The water column blowdown line must be at least 3/4 inch in diameter, and the gauge glass blowdown line must be at least 3/8 inch in diameter. Boiler water blowdown will be discussed later in this chapter.

If shutoff valves are used between the boiler and water column, outside screw and yoke valves must be used and arranged to prevent accidental closing. (Valves are not permitted between the boiler and water column on low-pressure boilers.) Globe valves cannot be used because sludge accumulates in their valve bodies and plugs lines. Valves will be discussed in detail later in this chapter.

Both the water column and gauge glass must have blowdown valves that should be opened daily to prevent sludge accumulations and sediment from blocking lines. Common operating procedure requires that the water column and gauge glass be blown down every time the shift changes.

Observe the level in the glass when it is blown down. When the valve is opened, the water should drop out of the glass rapidly. When the valve is closed, it should return rapidly to its previous level. If the glass water drains slowly or sluggishly, this is an indication that the column and/or gauge glass lines are clogging up. The situation must be corrected as quickly as possible, even if the boiler has to be shut down.

What if you find the gauge glass full, but the top try cock shows steam? Then, after blowing down the glass, the gauge glass refills completely again. This is a symptom of a closed or clogged gauge glass shutoff valve. Remember, for the gauge glass to work properly, both the top and bottom connections must be clear. Sometimes the problem can be cleared by: 1) closing off both gauge glass valves, 2) opening the blowdown valve, and 3) opening and closing one gauge glass valve at a time.

Why does glass fill under these circumstances? Remember that normally invisible steam occupies the space above the water level. The steam in the glass quickly gives up its heat to its surroundings and condenses. As it condenses, water takes its place until the entire glass fills. Both valves have to be clear for the glass to give an accurate reading.

REVIEW OF CODE REQUIREMENTS, WATER COLUMN CONNECTION AND OPERATIONS

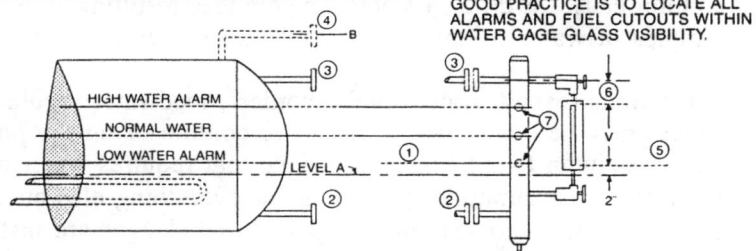

GOOD PRACTICE IS TO LOCATE ALL ALARMS AND FUEL CUTOUTS WITHIN WATER GAGE GLASS VISIBILITY.

HIGH WATER ALARM
NORMAL WATER
LOW WATER ALARM
LEVEL A

SHUTOFF VALVES BETWEEN DRUM AND COLUMN MUST BE OS&Y, OF THROUGH-FLOW DESIGN AND ORIENTATION, SHOW POSITION AS OPEN OR CLOSED, AND HAVE LOCK-OPEN CAPABILITY.

ASME WATER GAGE REQUIREMENTS
UNDER 400 PSIG
  2 Direct Reading Gages Or
  1 Direct Reading Gage And 3 Gage Cocks
400 PSIG TO 900 PSIG
  2 Direct Reading Gages
900 PSIG AND ABOVE
  2 Direct Reading Gages Or
  1 Direct Reading Gage And 2 Remote Reading Gages

SYSTEM SCHEMATIC DETAIL
1. **Lowest permissible water level** – at which level there will be no danger of overheating (Level A).
2. **Water connection for Water Column** – must be at least 1″ below low visibility point of gage glass – must be at least 1″ NPT. Line should be level or slope downward from column to drum.
3. **Steam connection for Water Column** – must be at 1″ NPT minimum and located above high visibility point of gage glass. Line should slope downward from drum to column.
4. **Steam connection may come out of top of vessel** – centerline of steam connection would be at point marked "B".
5. **The lowest visible part of water gage glass** – must be at least 2″ above the lowest permissible water level (Level A).
6. **The highest visible part of water gage glass** – must be at least 1″ below center of steam connection.
7. **Gage Cock connections** – shall not be less than ½″ pipe size and located within gage visibility range "V." Gage Cocks not required with two gages are used.

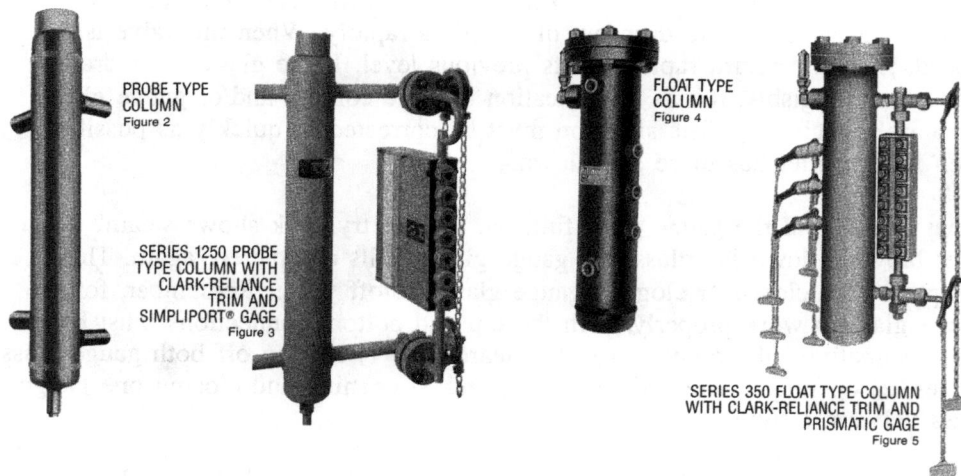

PROBE TYPE COLUMN
Figure 2

SERIES 1250 PROBE TYPE COLUMN WITH CLARK-RELIANCE TRIM AND SIMPLIPORT® GAGE
Figure 3

FLOAT TYPE COLUMN
Figure 4

SERIES 350 FLOAT TYPE COLUMN WITH CLARK-RELIANCE TRIM AND PRISMATIC GAGE
Figure 5

Figure 3-3. Water column installations (Courtesy, Clark-Reliance Corporation)

Eventually a gauge needs replacing. Either it's broken or fouled to the extent that viewing is obscured. Figure 3-4 shows the detail of a typical gauge assembly.

The following procedure is recommended for changing a gauge glass: *Note: Always wear a face shield, safety goggles, and gloves when working with a pressurized gauge glass.*

1. Close top and bottom valves, open the drain.

2. Remove the gauge glass nuts.

3. Discard the old glass, but save the flat brass packing washers.

4. Obtain or cut to length a new glass. There should be 1/4 inch of play when the glass is replaced, to allow for expansion.

5. Slide the nuts and flat brass packing washers back on the new glass.

6. Center glass between the valves and take up on the nuts until hand tight, then add a 1/4 turn with a wrench.

7. Crack open the steam (top) valve to allow the glass to warm slowly and evenly.

8. When warm, open the steam and water valves completely.

9. Close the blowdown valve.

The glass is now back in service.

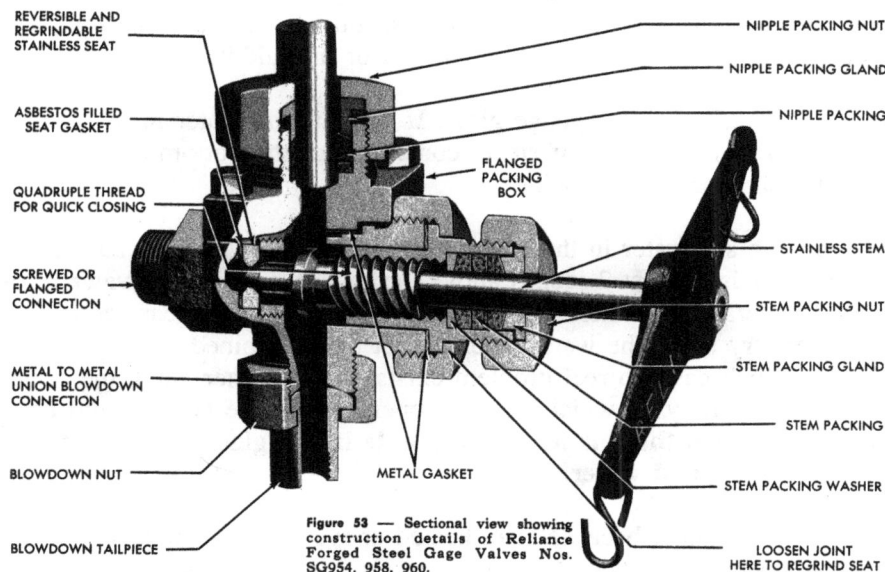

Figure 53 — Sectional view showing construction details of Reliance Forged Steel Gage Valves Nos. SG954, 958, 960.

Following steps 7 and 8 in proper sequence is important to minimize the chance of breaking the glass when putting it back in service. Although the glass is strong, it can break if subjected to thermal shock. With the blowdown valve open, just cracking open the top valve allows steam through the glass to warm it up evenly. When hot boiler water hits a cold glass, it breaks.

Although the flat brass packing washers are not absolutely necessary, they facilitate gauge glass installation. The hard, smooth surface of the flat washer eases tightening of the packing nut for leak-free operation and reduces the torsion placed on the glass.

Some gauge glasses have a lateral stripe to show the water level. The water in the glass magnifies the submerged portion of the stripe, which then looks wider than the part above the water line, Figure 3-5.

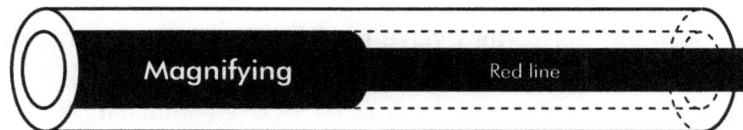

Figure 3-5. Gauge glass with lateral stripe

Try cocks are valves mounted on the water column. There are usually three on boilers that have less than 100 square feet of heating surface. At normal water level, water comes out of the bottom cock and steam comes out of the top. A combination of steam and water should come out of the middle.

Try cocks are the backup to the gauge glass. If you find no water in the gauge glass, check the bottom try cock. If steam comes out of the bottom cock, shut down the boiler because it's in trouble.

As long as you can see water in the glass, it is safe to add water, because the bottom of the glass is at least 2 inches above the lowest permissible water level for water-tube boilers. When there is no water in the glass, and steam comes out of the bottom try cock, the water level cannot be determined. It may be just out of sight, in which case no real problem exists. On the other hand, the boiler could be two-thirds empty, in which case the operator and the boiler could be in serious danger. **When there is no water visible in the glass — shut down the boiler and do not add water!**

Water should not be added, because steel heated above 700°F rapidly loses strength. When steam or hot water pressure vessels fail, they explode. The acronym used to describe this situation is BLEVE (Boiling Liquid Expanding

Vapor Explosion). *Warning: Never add water to a boiler if water is not visible in the gauge glass or if steam comes out of the bottom try cock. Shut down the fire and allow the boiler to cool slowly. When cool, dump the boiler's water and check for heat damage inside the boiler. Call your boiler inspector to get a third-party opinion.*

LOW-WATER CUTOFFS

A low-water cutoff is a device that electrically shuts off the fuel supply when the water level falls below a predetermined level. The float-type cutoff is the most common, but electrodes and differential pressure transmitters are also in use. It's common practice to install two low-water cutoffs, one positioned slightly lower than the other. The highest cutoff is the primary, the lower is the backup. The primary cutoff is usually the type that resets itself after the boiler recovers from a low-water condition. That is, it lets the boiler restart automatically without the operator's intervention. The backup is always the type that must be reset manually. When the backup low-water cutoff has to be reset, this indicates that the first cutoff failed to shut off the burner and the backup was used instead. The failure of the primary cutoff must be investigated and repaired.

Figure 3-6a shows low-water cutoffs in which a float positions a lever, which activates switches. The wet and electrical parts are separated by a packless bellows seal or are linked by a magnetized activated switch. Figure 3-6b shows the probe-type low-water cutoff, which works according to the principle that water conducts electricity much better than steam. When water covers or uncovers the tip of an electrode, a relay coil is either energized or de-energized. Depending on the make and model, the switches can sound an alarm for high- or low-water conditions, shut off the fuel during low-water conditions, and/or control a pump to maintain proper water level.

PRESSURE GAUGES

All boilers must have at least one pressure gauge connected to the steam space where it can be easily seen by the operator. The top of the water column is a convenient spot for a connection.

The piping to a gauge measuring steam should always be full of water to prevent damage from the high-temperature steam. If the piping cannot be flooded, then a *siphon loop* is used. Figure 3-7 shows various siphon loop configurations.

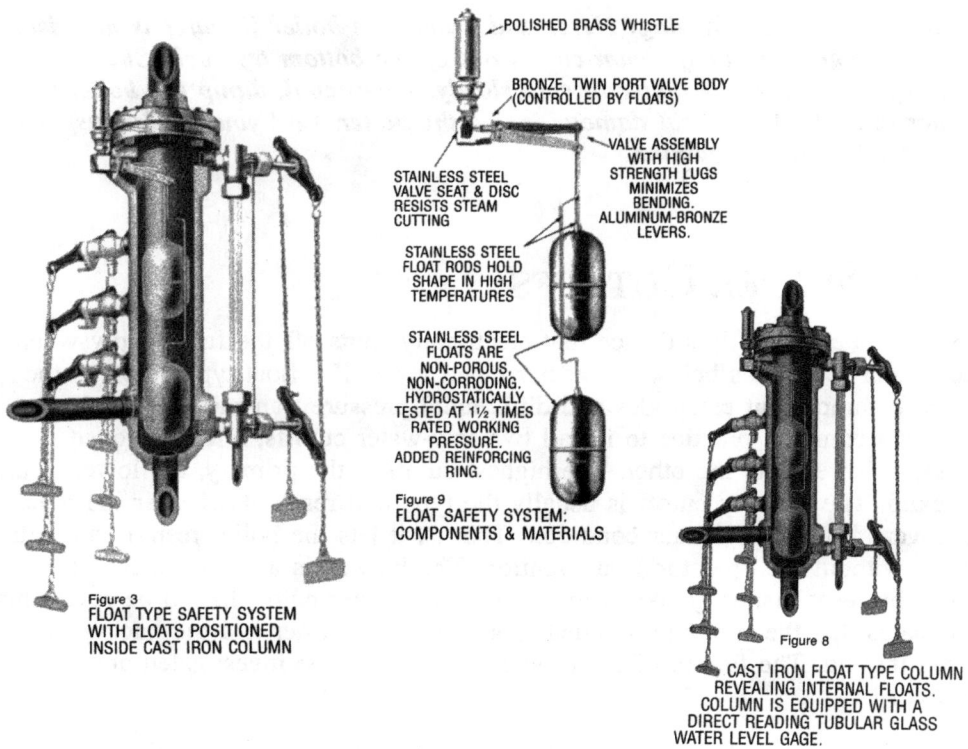

POLISHED BRASS WHISTLE

BRONZE, TWIN PORT VALVE BODY
(CONTROLLED BY FLOATS)

VALVE ASSEMBLY
WITH HIGH
STRENGTH LUGS
MINIMIZES
BENDING.
ALUMINUM-BRONZE
LEVERS.

STAINLESS STEEL
VALVE SEAT & DISC
RESISTS STEAM
CUTTING

STAINLESS STEEL
FLOAT RODS HOLD
SHAPE IN HIGH
TEMPERATURES

STAINLESS STEEL
FLOATS ARE
NON-POROUS,
NON-CORRODING,
HYDROSTATICALLY
TESTED AT 1½ TIMES
RATED WORKING
PRESSURE.
ADDED REINFORCING
RING.

Figure 9
FLOAT SAFETY SYSTEM:
COMPONENTS & MATERIALS

Figure 3
FLOAT TYPE SAFETY SYSTEM
WITH FLOATS POSITIONED
INSIDE CAST IRON COLUMN

Figure 8
CAST IRON FLOAT TYPE COLUMN
REVEALING INTERNAL FLOATS.
COLUMN IS EQUIPPED WITH A
DIRECT READING TUBULAR GLASS
WATER LEVEL GAGE.

(a)

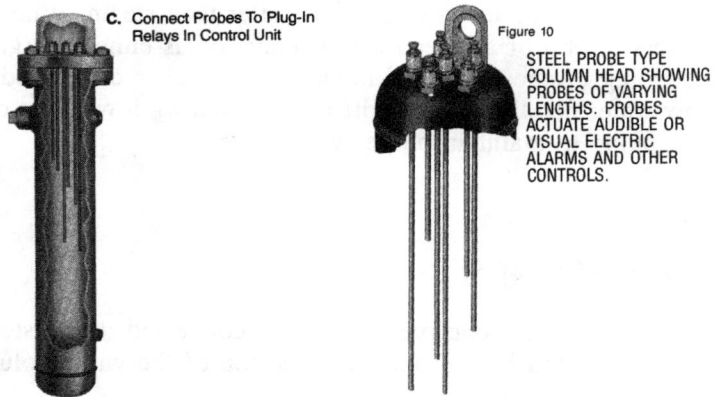

C. Connect Probes To Plug-In
Relays In Control Unit

Figure 10
STEEL PROBE TYPE
COLUMN HEAD SHOWING
PROBES OF VARYING
LENGTHS. PROBES
ACTUATE AUDIBLE OR
VISUAL ELECTRIC
ALARMS AND OTHER
CONTROLS.

(b)

Figure 3-6. Low-water cutoffs (Courtesy, Clark-Reliance Corporation)

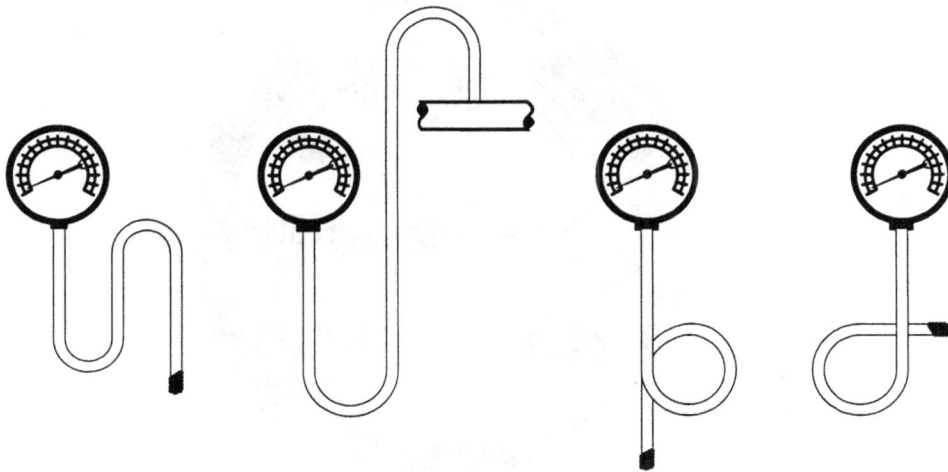

Figure 3-7. Siphon loops

A shutoff valve or cock is always a good idea in case the gauge has to be replaced. If a valve is used, lock or seal it open. If a cock is used, make sure it is the type that shows by the position of its handle whether it is open or closed.

The range of the gauge must be at least 1.5 times the maximum allowable working pressure. For operating convenience, use a gauge that covers twice the operating pressure. Then you know the pressure is correct without having to read the numbers by just by glancing at the gauge and seeing its pointer straight up.

Figure 3-8 shows a *compound gauge* that can measure both pressure and vacuum. Found on many low-pressure boilers, zero pressure is usually near the top. On a typical gauge that goes clockwise on the dial, the vacuum portion goes from 30 inches of mercury (in. Hg) to zero. Then the pressure portion starts at that zero, and in this example, goes up to 200 psi.

Most gauges use a *Bourdon tube* to sense pressure. The Bourdon tube is a flat, hollow tube shaped like a question mark, Figure 3-9. When pressure is applied to this tube, it straightens, positioning the pointer by a series of linkages and gears.

In installations where the pressure gauge is located some distance below the water column, the reading may be inaccurate due to the hydraulic head of water in the line, Figure 3-10. For example, say the gauge is 25 ft below the water column. One foot of water exerts 0.433 psi; therefore, 25 ft times 0.433 psi per foot equals 10.8 psi. The gauge reads almost 11 psi higher than the actual reading. This is corrected by resetting the gauge pointer by 10.8 psi.

Another pressure-measuring device usually found on larger boilers is the *draft gauge*. Draft is the pressure difference or force that moves air. Draft units of

Figure 3-8. Compound gauge

Figure 3-9. Bourdon tube (Courtesy, Dwyer Instruments, Inc.)

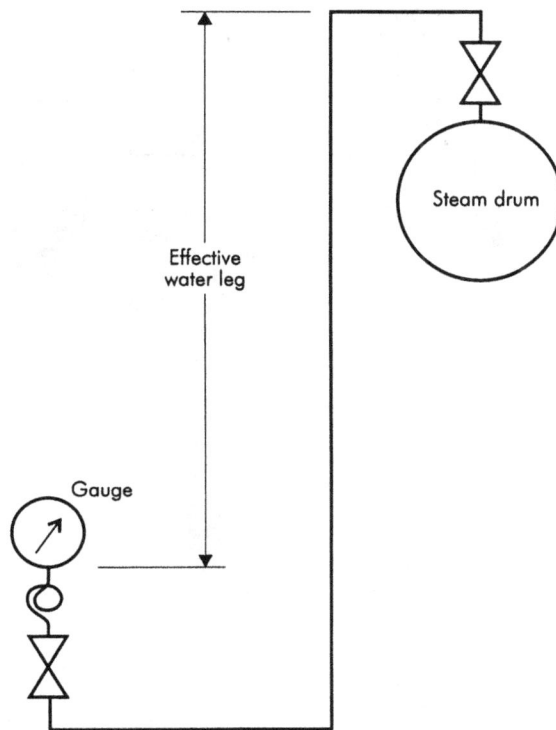

Figure 3-10. Gauge located below water column

measure are very small, usually inches of water column (in. w.c.), because it takes very little pressure to move a lot of air. One inch w.c. = 0.0361 psi.

Inches of water column came from the first instrument used for measuring draft, a simple U-tube manometer, Figure 3-11. This glass U-tube was half filled with water. One end was connected to the area to be measured by a rubber tube, the other end left open to the atmosphere. The difference in elevation between the water surfaces in the two legs of the U-tube was the draft measurement. If the water surface in the leg open to the atmosphere was lower than the other, then negative pressure was being measured. If the water surface was higher, then positive pressure was being measured. Mechanical draft gauges, Figure 3-12, have largely replaced U-tube manometers.

There are also gauges for measuring higher pressures. The status of an oil filter can be determined by measuring its pressure drop or *differential pressure*. The common unit of differential pressure is psid (pounds per square inch differential).

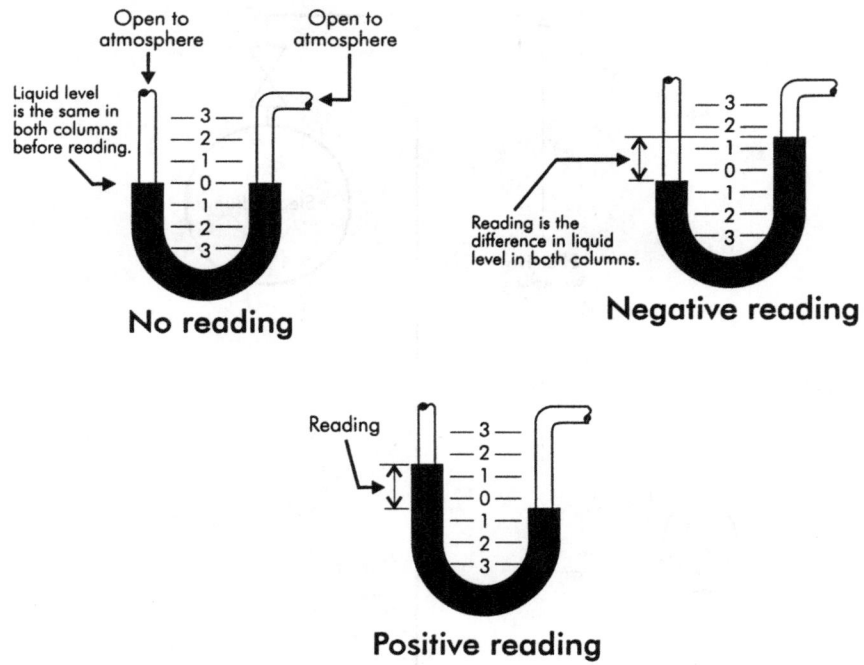

Figure 3-11. U-tube manometer

Figure 3-12. Mechanical draft gauge (Courtesy, Dwyer Instruments, Inc.)

Pressure Switches

A *pressuretrol* is the brand name of a pressure switch. In some examinations expect to see the term *pressuretrol* in place of pressure switch.

A *vaporstat* is also a pressure switch, but it responds to low pressure. It is calibrated in inches of water instead of pounds per square inch (27.7 inches of water = 1 psi). It is used in low-pressure gas applications to verify gas line and windbox pressure.

VALVES

Valves are mechanical devices used to control the flow of fluids. Some are used either to completely stop or start flow, some regulate flow, and some do both. The following discussion addresses the valves encountered most often in the boiler room.

Valves are classified by stem action (whether the stem rises out of the valve body as the valve is opened) and by internal configuration. The terms *gate*, *globe*, *check*, *ball*, and *butterfly* are used to describe the internal configuration of a valve body.

The threads that move the stem and disk in the rising-stem valve are located in the bonnet; the threads that move the disk in the nonrising stem valve are built into the disk, Figure 3-13.

The *outside screw and yoke* (OS&Y) rising-stem valve, Figure 3-14, is popular and sometimes mandatory, because an operator can tell from a distance if the valve is opened or closed. For example, the OS&Y type is required for steam stop valves over 2 inches. With a nonrising stem valve, you don't know if the valve is opened or closed. If it's stuck, it's only a guess if the valve is being turned in the right direction. Nonrising stem valves are used where a rising stem would interfere with the valve's operation, such as underground valves, and where the exposed threaded stem of an OS&Y would become corroded or excessively dirty. *Note: The term OS&Y shows up on many exams in which the acronym is not defined.*

A *gate valve*, as the name implies, consists of a gate that can be raised or lowered in a valve body. It is designed to stop flow or allow full flow. Consequently, a gate valve should be either fully open or closed. As shown in Figure 3-15, if operated partly open, or in a throttling position, its internal surfaces erode due to fluid velocity and impingement, rendering the valve useless. When fully open, this valve offers little resistance to flow.

If a valve is left in the wide-open position and is not backed off a half turn, changes in temperature cause various parts of the valve to expand at different

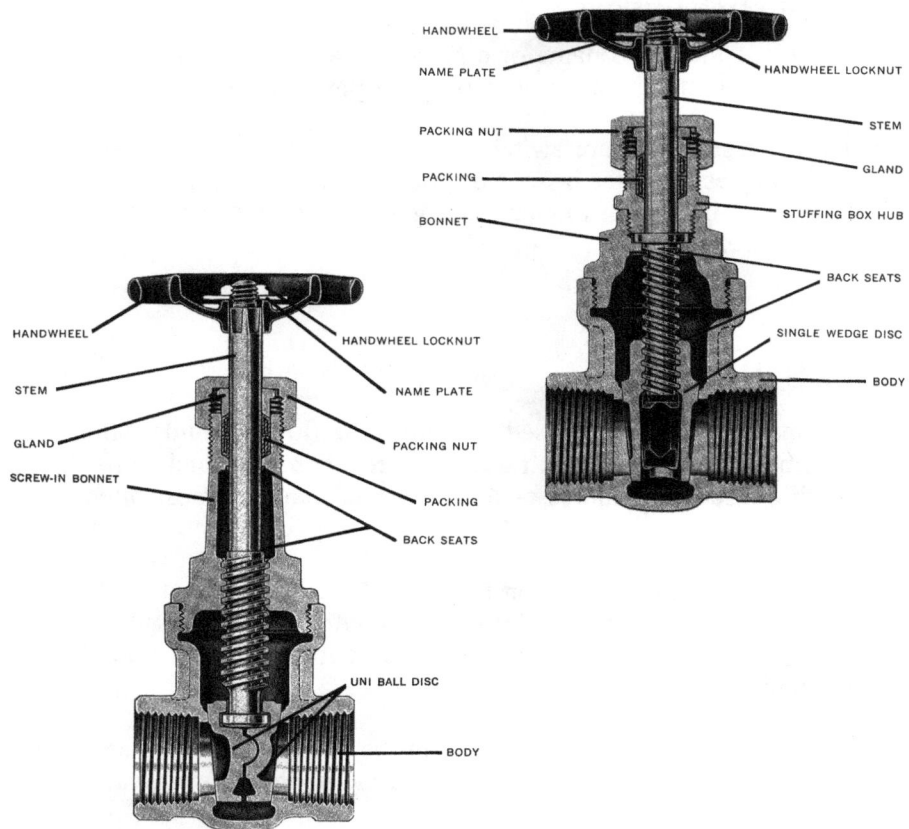

Figure 3-13. Rising-stem valve and nonrising stem valve (Courtesy, Cincinnati Valve Company)

rates. The result is often a jammed valve that refuses to turn. On large valves it might require a chain hoist and sledgehammer to correct this condition. On the other hand, it should not take much force to seat a valve closed. If you have to beat the valve to shut it, then it is time to repair or replace it.

A small *bypass valve* is used to make opening a gate valve easier. If it's opened before opening the main valve, the downstream pressure will be equalized. It also warms up the downstream line before opening the main valve.

Some valves have a back seat bushing. When the valve is wide open, this bushing forms a seal that stops leakage out of the gland. This feature only permits packing replacement while the valve is in service. It is not intended to compensate for leaky stem packing.

Globe valves regulate or throttle flow through a valve. These valves feature a Z pattern flow route and a plug or disk, which is forced into a tapered hole called a seat, Figure 3-16. Fluid faces more resistance in globe valves than in gate

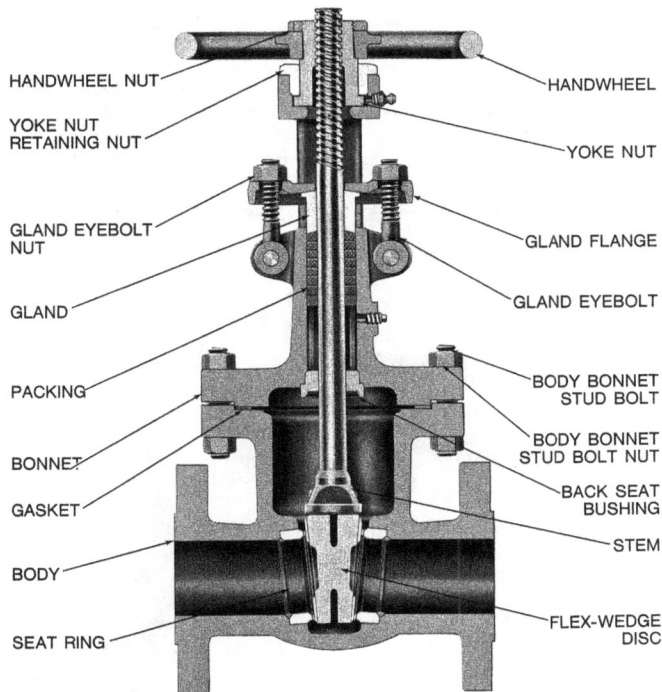

Figure 3-14. Outside screw and yoke rising-stem valve (Courtesy, Cincinnati Valve Company)

Figure 3-15. Gate valves, at right and in the middle, should be fully opened or closed to prevent erosion of their internal surfaces. Fluid encounters more resistance in a globe valve, left (Courtesy, Cincinnati Valve Company)

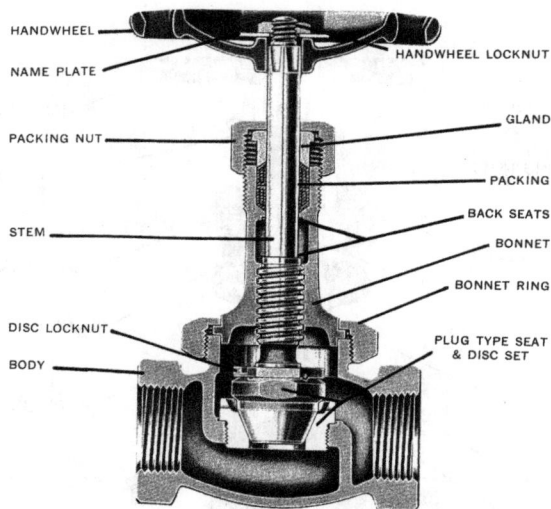

Figure 3-16. Globe valve (Courtesy, Cincinnati Valve Company)

valves, because it has to flow through the globe valve's Z pattern. Foreign matter tends to accumulate and cause blockages in the Z pattern.

Automatic control valves consist of an operator and a valve body. The valve operator takes a signal from a controller, either pneumatic or electric, and positions the valve stem accordingly. The control valves are designed so that it takes a small force to move the stem. Control valves use a balanced design similar to a globe valve with two seats and two disks on the same stem. On one seat the disk opens against pressure, and on the other seat the disk opens with pressure. This way the forces on the stem are balanced, and the valve can be opened or closed with little effort. Though good for control, it is hard to obtain a perfect shutoff with this type of valve.

When installing control valves, they should be piped with isolating valves and a globe bypass valve, as illustrated in the steam pressure reducing station shown in Figure 3-17. This way, the control valve can be serviced without taking the equipment off line.

A *check valve* is used when unidirectional flow is required. The two most common types are *swing* check valves, Figure 3-18a, and *lift* check valves, Figure 3-18b. One of the most important check valves on a boiler is installed in the feedwater line. This check valve is required by ASME and prevents boiler pressure from discharging the feedwater if the feedwater line breaks.

A *nonreturn valve* is an automatic-stop check valve used to prevent a backflow of steam into a boiler when its pressure falls below the header pressure. Essentially, it is a lift check valve — it allows steam to flow in one direction only. The valve can also be locked in the closed position by turning down the valve

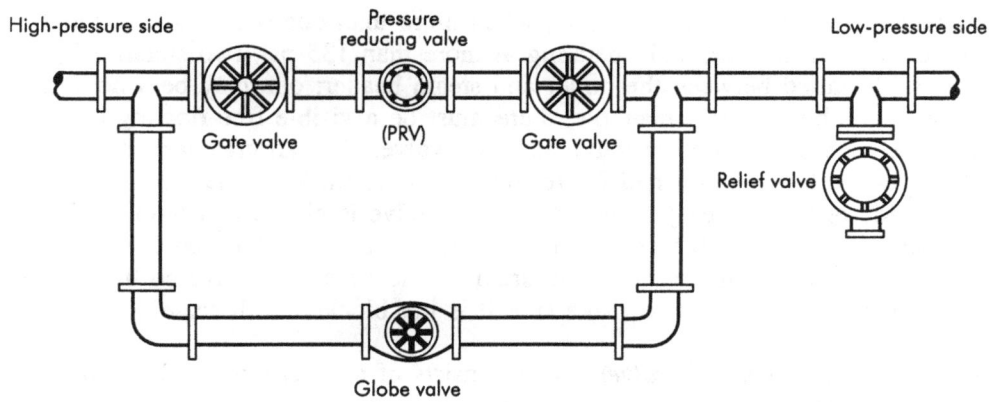

High-pressure side   Pressure reducing valve   Low-pressure side

Gate valve   (PRV)   Gate valve

Relief valve

Globe valve

Figure 3-17. Steam pressure reducing station

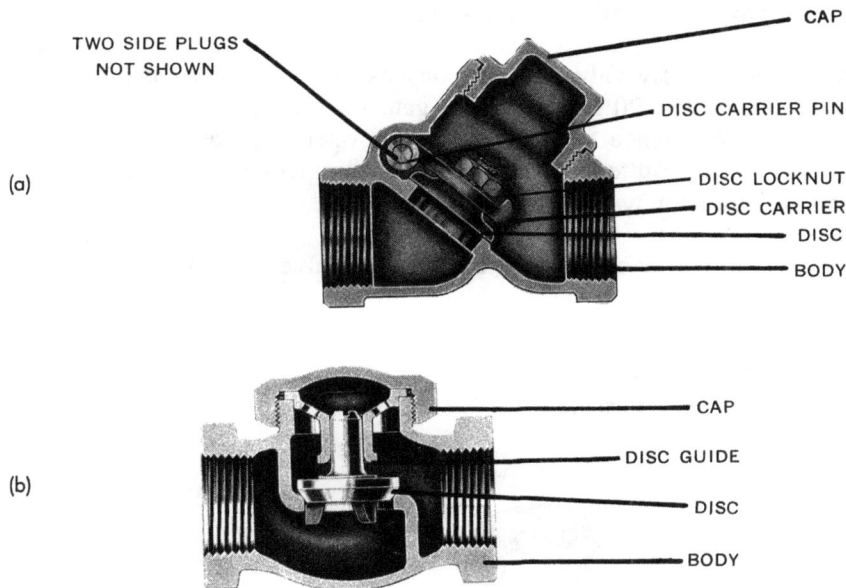

(a)

TWO SIDE PLUGS NOT SHOWN

CAP

DISC CARRIER PIN

DISC LOCKNUT
DISC CARRIER
DISC
BODY

(b)

CAP

DISC GUIDE

DISC

BODY

Figure 3-18. Swing check valve and lift check valve (Courtesy, Cincinnati Valve Company)

stem and holding the disk against the seat. Then the valve allows no flow in either direction. The nonreturn valve is located as close to the boiler steam outlet as possible. On fire-tube boilers, they are mounted directly on the steam outlet. There is also a *triple-acting nonreturn valve*, which also automatically closes when the header pressure drops by approximately 8 psi below the boiler pressure to isolate the boiler in case of a steam-line break.

When boilers are set in battery (i.e., when more than one boiler is connected to the same steam header) and operating at more than 135 psi, two steam valves must be installed between the boiler and steam header; one may be a nonreturn valve. If the boiler has a manhole, there must be a visible free-flowing drain valve between the nonreturn valve and stop valve. This arrangement is used whenever the boiler is opened for repair or inspection. With the free-flowing drain wide open, it is easy to tell if the stop valve is closed and holding. If the stop valve should fail, the nonreturn valve is still between the people inside the boiler room and the live steam. The drain should be a gate valve so that a rod can be inserted through it to make sure it is not blocked with scale.

Figure 3-19 shows a *ball valve*, which consists of a sphere with a hole through it. This sphere sits between two seats, and the valve has a handle connected to it so it can be rotated 90 degrees. Ball valves are popular substitutes for gate valves because of their good shutoff characteristics and easy operation. With the valve all the way open, there is no flow obstruction. Although commonly rated for water, oil, and gas, some ball valves can be used for steam.

Figure 3-20 shows a butterfly valve, which consists of a flat, round disk attached to a stem that rotates 90 degrees. Like gate valves, both butterfly and ball valves offer little resistance to flow when wide open. Unlike gates, they can be used for throttling. Butterfly valves are a less expensive substitute for gate valves and are used for water, oil, and gas service.

The following are common abbreviations used in the valve industry:

- SP - steam pressure
- WP - working pressure
- WOG - water, oil, gas pressure

Figure 3-19. Ball valve (Courtesy, Henry Valve Company)

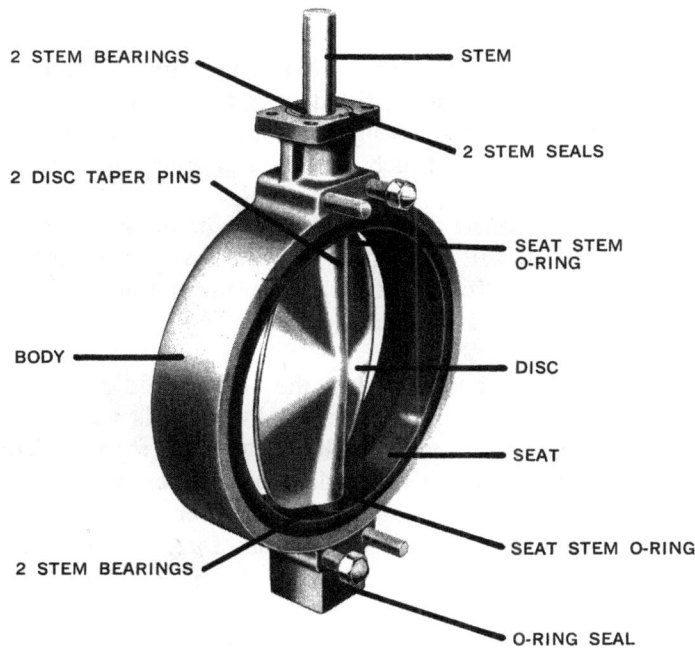

Figure 3-20. Butterfly valve (Courtesy, Cincinnati Valve Company)

- BR - bronze
- IBBM - iron body, bronze-mounted
- AI - all iron
- CS - cast steel
- SS - stainless steel
- OS&Y - outside screw & yoke
- NRS - nonrising stem
- RS - rising stem

If a catalog describes a valve as *10" gate, OS&Y, IBBM, 125 lb SP 450 F, 200 lb WOG*, the valve would have an inside diameter equal to that of 10-inch pipe; a gate internal configuration; an outside screw and yoke rising stem; a cast iron valve body with bronze trim (disk, seat rings, stem, and yoke); and a valve body strong enough to control 125 psi steam not exceeding 450°F, or 200 psi water, oil, or gas.

Why must valves have two pressure ratings (one for steam and one for water, oil, and gas)? Steam contains more energy than water, oil, or gas; therefore, the pressure for steam service is derated in most valves.

Vent Valves

Every boiler must have a valve connected to its highest point so that air can be removed while the boiler is being filled with water. This is to prevent air, or noncondensable gases, from entering the steam system when the boiler is put on line. The vent is usually a 1/2-inch to 1-inch gate or ball valve.

A vent valve is also opened when the boiler is being shut down or emptied (dumped) so a vacuum will not form. Vacuums can cause problems, especially in low-pressure systems. To detect a vacuum, many low-pressure boilers are equipped with a compound pressure gauge.

Some licensing examinations ask when the vent should be closed as a cold boiler is brought on line. The most practical answer is when the exiting steam starts making a noise, because the boiling water has forced all the air out by this point. However, it is more likely that the choices will be a) 10 lb, b) 25 lb, c) 50 lb, d) none of the above. The correct answer would be 25 lb. Most pressure gauges start indicating at 10 psi; therefore, all air should be vented at this point.

Safety Valves

The safety valve is the most important valve on the boiler. All pressure vessels are designed for a certain maximum operating pressure. If this pressure is exceeded, the danger of an explosion exists. The safety valve is the last line of defense against disaster. Safety valves are required on all boilers.

The first type of pressure-relief device used was the *lever safety valve* shown in Figure 3-21. Unfortunately, this design was prone to leakage. To stop the nuisance leaking, extra weight was added. This naturally raised the lifting pressure and boiler explosions continued. The use of lever safety valves is no longer permitted.

The *spring loaded, pop-action safety valve* is now the required type. It pops fully open when its set pressure is reached, then slams shut after the pressure has dropped a few pounds. The difference between the popping and the closing or seating pressure is called the *blowdown* or *blowback* pressure. Do not confuse this blowdown with boiler water blowdown, which will be discussed later in this chapter.

Chatter is the rapid motion of the movable parts in the valve when the disk slams repeatedly against the seat. It can sound like a machine gun. This happens because the valve's blowdown is too small so it does not know if it should be in the open or closed position. More on adjusting the blowdown later.

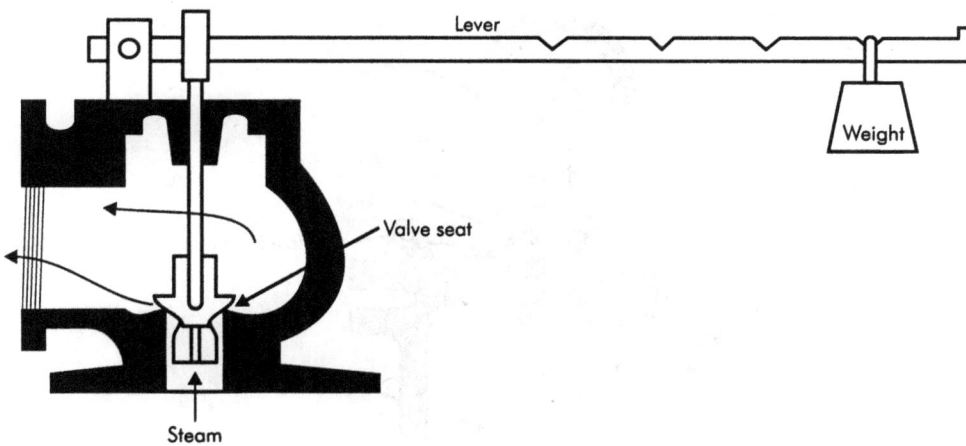

Figure 3-21. Lever safety valve

In operation, the spring of the safety valve holds the valve tightly closed against its seat until set popping pressure is reached. At first, the pressure slowly lifts the valve off its seat, allowing steam to enter the *huddling chamber*. As shown in Figure 3-22, the huddling chamber is located just above the valve's seat and provides additional area for the steam to push against. As the force against the spring increases, the popping action is produced.

The valve is held open by the steam pushing against the disk and spring. This action utilizes dynamic energy, that is, the high-velocity mass of the steam changing direction to create lift. When the pressure blowdown point is reached, the valve slams shut because there is not enough dynamic energy to hold it open.

There are two adjustments on simple safety valves: spring tension and blowdown ring. The popping pressure is changed by adjusting the spring tension. The adjustment range of any particular spring is just a few percent, so most of the time a change in popping pressure requires a different spring.

The blowdown ring changes the configuration of the huddling chamber. For example, raising the ring directs more of the escaping steam against the bottom of the valve and thus has more effect on it. This means that the system pressure must fall further below the popping pressure for the valve to close. Therefore, raising the ring increases blowdown. This is the adjustment a qualified service technician uses to cure a chattering condition. Once adjustments are made, the locking screws are often sealed.

Pop action is only produced by expandable gases such as steam and air. When used for liquids, pop safety valves act like standard relief valves. The difference between safety and relief valves is that safety valves are either fully open or fully closed, while the opening of a relief valve is proportional to pressure. Relief valve lift is in direct proportion to overpressure. To lift the disk to full

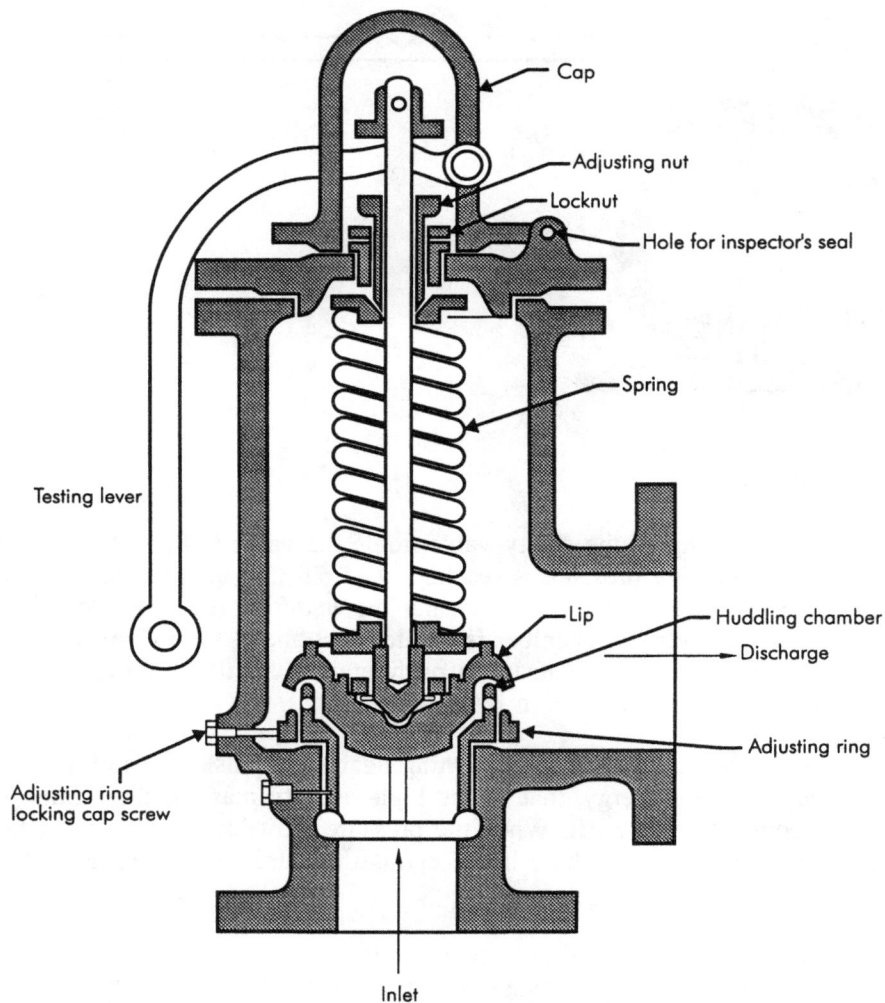

Figure 3-22. Spring loaded, pop-action safety valve

discharge capacity, 10% overpressure is required. On liquid relief valves that are not certified by code, 25% overpressure is allowed. Relief valves have only one adjustment — the spring tension.

The ASME Code goes into great detail about safety valves. Over the years, standard rules have been developed regarding their design, construction, operation, and maintenance. The following excerpts from the code illustrate basic do's and don'ts. Naturally, you are not expected to memorize the code for a test; however, the comments following each excerpt point out potential test questions. Should there be any need for specific rulings, consult your boiler inspector and the latest ASME Power Boiler Code with all the latest addenda and interpretations. The sections from the code are in bold type, and explanations (if any) follow in normal type:

- **Each boiler shall have at least one safety valve, and if it has more than 500 ft^2 of water heating surface, it shall have two or more safety valves.**

- **The safety valve capacity of each boiler shall be such that the safety valve or valves discharge all the steam that can be generated by the boiler without allowing the pressure to rise higher than the maximum allowable working pressure (MAWP) or more than 6% above the highest pressure to which any valve is set.** Safety valve capacity is the amount of steam, in pounds per hour, that a safety valve can discharge at its popping pressure. If a boiler has a steaming capacity of 50,000 lb/hr, then the combined capacity of the safety valves must exceed 50,000 lb/hr.

  An *accumulation test* determines if there is enough safety valve capacity. With the steam stop valve closed, the boiler is brought up to high fire. Under these conditions the safety valves are naturally going to lift. The boiler inspector watches the boiler's pressure gauge to make sure that there is enough safety valve capacity. Accumulation test time is always exciting. Along with the noise of fully discharging safety valves, the water level is going crazy because the steam is leaving the boiler in a way in which it was not designed to leave the boiler. Accumulation tests are not done on boilers equipped with superheaters. The superheaters would be damaged with no cooling steam flowing through them.

- **The safety valve or valves shall be connected to the boiler independent of any other steam connection, and attached as close as possible to the boiler without any unnecessary intervening pipe or fitting.** Though it doesn't come right out and say it, the main intent is to prohibit the use of any stop valves where safety valves are concerned. Is a valve allowed on the inlet or outlet of a safety valve? Absolutely not! Can the safety valve be attached to the boiler via a long pipe? Absolutely not! The standard practice is to bolt, screw, or weld the safety valve directly to the boiler.

- **The opening or connection between the boiler and the safety shall be at least as large as the area of the valve inlet. When a discharge pipe is used, the cross-sectional area shall not be less than the full area of the valve outlet or the total area of the valve outlets discharging there into and shall be as short and straight as possible and so arranged as to avoid undue stresses on the valve or valves.**

- **The vent pipe shall be large enough to accommodate the full capacity of the valve without causing steam to escape by flowing backward through the drip pan.** In other words, don't neck down, or reduce, the pipe's diameter. This is especially critical on the discharge. Restrictions decrease valve capacity due to flow friction. If each discharge has an area of 5 square inches and if there are three valves, then the discharge line must be at least 15 square inches.

- **A slip joint shall be provided between the safety valve and discharge pipe with ample clearance to allow for thermal expansion. The vent pipe shall not be allowed to exert any force on the safety valve.** Problems result when the safety valve has to support the weight of the discharge piping. Although the heavy valve seems rigid, the weight of the discharge piping causes slight distortions. The valve's seal depends on metal-to-metal contact. If the seat is distorted the slightest bit, leakage results. When installed properly, the weight of the riser, which is supported separately, is not placed on the valve.

- **Each safety valve shall have a substantial lifting device by which the valve disk may be lifted from its seat when there is at least 75% of the set pressure under the disk.** By *lifting device* they mean *lever*. When this testing lever is lifted, it shows that the valve's mechanism is moving freely. Try to move that lever against the spring when there is no pressure under the seat and you'll find that it's almost impossible. If the valve is put on a boiler that is close to the valve's popping pressure, the lever is easily lifted.

The licensing exam may have a question about this topic such as: what is the minimum pressure that should be under the valve seat before the lifting device is used? The answer is 75% of the popping pressure. At or above this pressure the test lever is easily lifted without doing damage to the valve. The lifting device has nothing to do with the popping pressure, which is determined by the spring pressure and the area of the valve seat. The lifting device only determines if the valve mechanism is not binding. How often should valves be tested using the lifting device? Remember that every time the valve is tested, there is some wear and tear and the seal is only maintained by metal-on-metal contact. Once a month is all that is required.

HYDROSTATIC TESTS

New boilers and boilers whose pressure parts have been repaired or replaced must be given a hydrostatic test before being placed back in service. This test is performed to ensure that the pressure vessel is leak free. The testing procedure is as follows:

1. Either remove the safety valves and blank off their openings, or use test clamps to gag them closed. Otherwise, the safety valves would open long before the test pressure of 1.5 times the maximum allowable working pressure would be reached.

2. Inspect the boiler to make sure all openings, except the vent valve, are closed.

3. Fill the boiler with clean water at ambient temperature, but not less than 70°F. This prevents unnecessary metal stress. Nondrainable superheaters should be filled with softened water. As the boiler is being filled, keep the vent valve open so all air is removed. Inspect for leaks as the boiler is filled.

4. When filled, close the vent and gradually apply the test pressure of 1.5 times the MAWP. Always keep the pressure under control, so that the required test pressure is never exceeded by more than 6%. Close visual inspection for leaks is not required at this stage.

5. The test pressure may then be reduced to the MAWP and maintained at this pressure while the boiler is carefully examined for leaks.

Figure 3-23 shows a *test clamp*, or *gag*, as it is commonly called. It is used to hold a safety valve closed when performing a hydrostatic test. The gag works under the same principle as a wheel puller; it clamps down on the valve stem preventing it from rising, thus holding the disk to the seat. To avoid overstressing the valve, don't screw down the gag too tight. All test clamps must be removed after the hydrostatic test.

Water Level Regulation

The most important responsibility of the operator is to maintain the proper water level. The most common method of water level regulation in fire-tube boilers is *on/off control*. With this type of control, a switch is connected to a float that senses the water level. On an operating boiler as the water reaches its

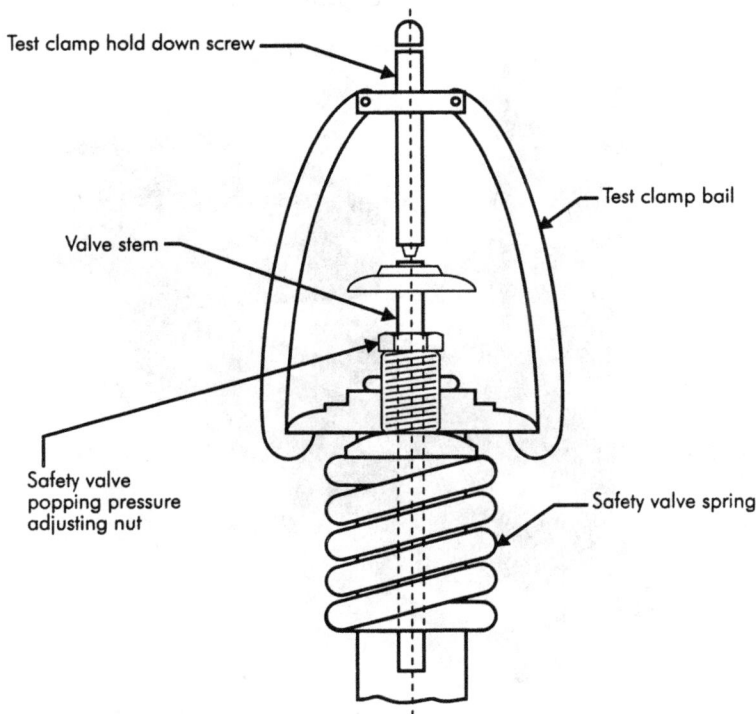

Figure 3-23. Test clamp or gag

lowest desired point, the float closes a switch, which in turn opens the feedwater valve and/or turns on the feedwater pump. (A pump is sometimes not necessary on low-pressure boilers at 15 psi or lower with a high city water pressure.) When the water reaches its highest desired point, the float opens the switch, which closes the feedwater valve and/or shuts off the pump.

Very often the same float that operates the feedwater pump switch also operates one of the low-water cutoff switches. Figure 3-24 shows a typical float type combination feedwater pump control and low-water cutoff switch.

Water-tube boilers have lower water reserve capacity than fire-tube boilers. Water reserve capacity is the ratio of the amount of water in the boiler (measured in pounds) compared to the boiler's steam output (measured in pounds per hour). Because of this reduced operating margin, simple on/off feedwater control can cause unstable water level conditions in water-tube boilers. For this reason, most water-tube boilers use a modulated feedwater control system. An example of modulated flow is a car in hilly country with cruise control. Fuel is constantly fed to the engine, but more is supplied when going uphill than downhill to maintain constant speed. It's the same in a boiler. Depending on the load, more or less feedwater is supplied to maintain a constant water level.

On some essay tests you may be asked to describe a modulated feedwater system in detail. Two such systems are the thermohydraulic (generator-diaphragm) regulator and the thermostatic expansion (Copes) regulator. Though

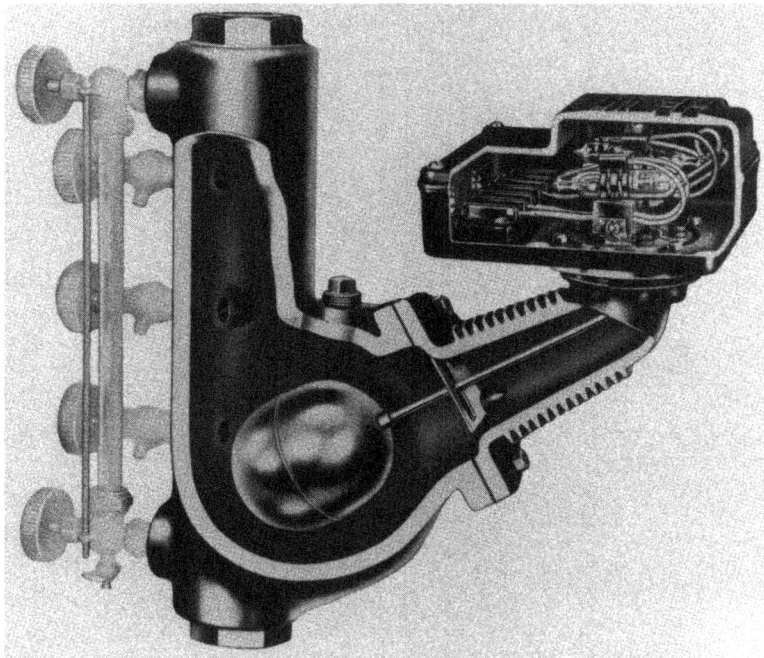

Figure 3-24. Float-type feedwater pump control and low-water cutoff switch (Courtesy, ITT McDonnell and Miller)

old designs, many are still in the field and are still often the subject of test questions.

The thermohydraulic (generator-diaphragm) regulator, shown in Figure 3-25, is often used on small water-tube boilers. This regulator consists of two concentric tubes at the normal steam drum water level. The top of the inner tube is connected to the steam space of the drum and the bottom to the water space. Thus, as in a gauge glass, the water level is allowed to rise or fall inside the inner tube. The outer tube is half full of water and has fins to dissipate heat. A small line runs from it to the diaphragm chamber of the control valve.

When the water level in the boiler drops, the inner tube fills with more steam. Since more heat is released faster from steam than from water, more heat is supplied to the confined water in the outer tube. The radiation fin surface can't get rid of all the additional heat, so the temperature and pressure of the confined water increases. This increased pressure is transmitted through the tube to

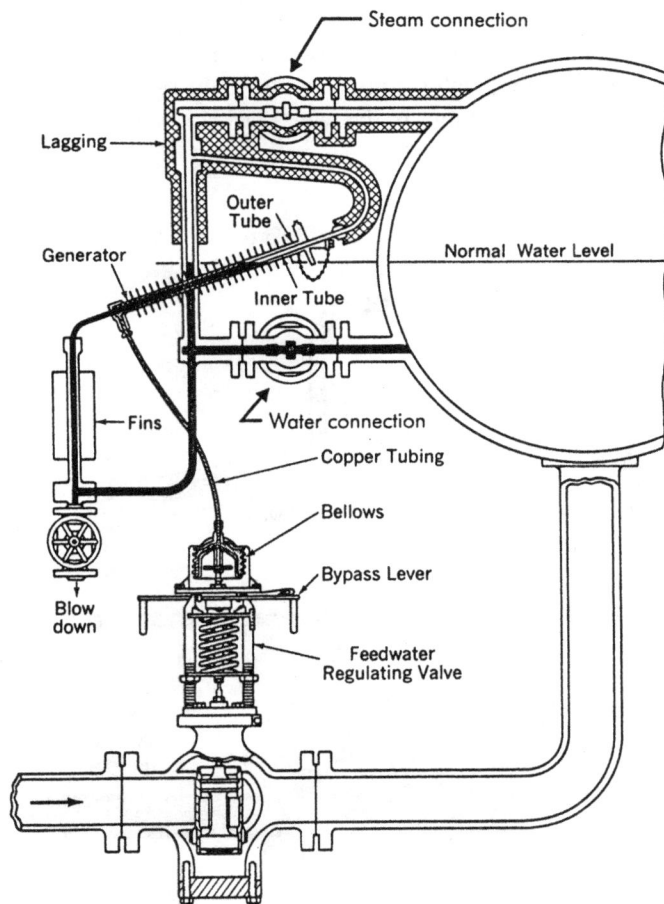

Figure 3-25. Thermohydraulic (generator-diaphragm) regulator (Courtesy, Babcock & Wilcox Company)

the feedwater valve's diaphragm, which opens the feedwater valve further. The operation is reversed when the water level increases. With more water in the inner tube, the fins dissipate more heat than is supplied. The pressure in the outer tube decreases and starts to close the feedwater valve.

This feedwater system has two main advantages: no external power is required for its operation, and since there are few moving parts, it is very reliable.

The *thermostatic expansion tube regulator* is shown in Figure 3-26. Commonly called the *Copes regulator*, its operation also depends on steam releasing more heat than water does. The Copes regulator consists of a hollow expansion tube placed at the steam drum water level. The top of the expansion tube is connected to the steam space, and the bottom is connected to the water space. Like a gauge glass, the water level in the expansion tube is allowed to rise and fall. The bottom of the tube is anchored, and the top is allowed to move as it expands and contracts. A bell crank lever attached to the top of the expansion tube amplifies the small expansions and contractions into useful movement. This movement, by way of a series of levers and linkages, opens and closes the feedwater valve. As the water level in the boiler drops, the expansion tube is filled with more steam. The tube expands as it absorbs the additional available heat from the steam. The feedwater control valve then opens more via the series of levers and linkages. Conversely, as the water level drops and the tube is

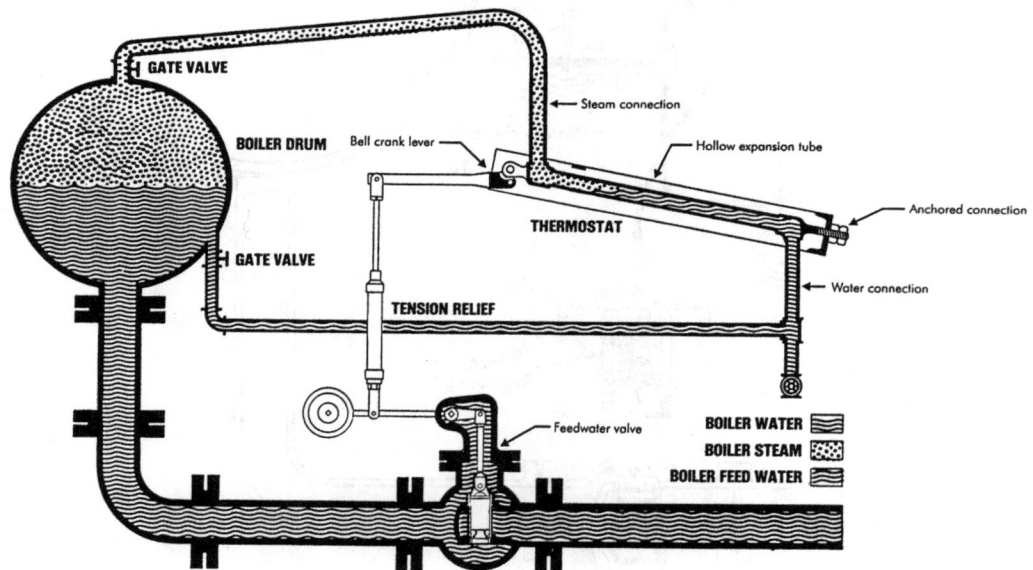

Figure 3-26. Thermostatic expansion tube regulator or Copes regulator (Courtesy, Copes-Vulcan)

filled with less steam, the tube contracts and starts to close the feedwater control valve. Like the thermohydraulic regulator, the Copes needs no external power to operate, and because it has few moving parts, it is very reliable.

Single-Element Regulators

To this point, only *single-element regulators* have been discussed, Figure 3-27. Single-element regulators control the water level by measuring the present water level. Single-element regulators are fine under steady pressure and load conditions but are unsatisfactory when the load fluctuates rapidly.

For example, consider a situation in which there is a sudden increase in demand and the steam pressure decreases before the boiler can pick up the load. This sudden decrease in pressure causes the internal steam bubbles to expand and take up more room in the boiler. This increases the water level and sends a signal to the single-element regulator to cut the feedwater rate. This of course is false information, because to maintain the proper water level, the feedwater rate must be increased. When the boiler pressure returns to normal, the size of the steam bubbles decreases, the water level drops, and the single-element regulator opens fully. This is the start of a very unstable situation that can result in wide and wild water level fluctuations.

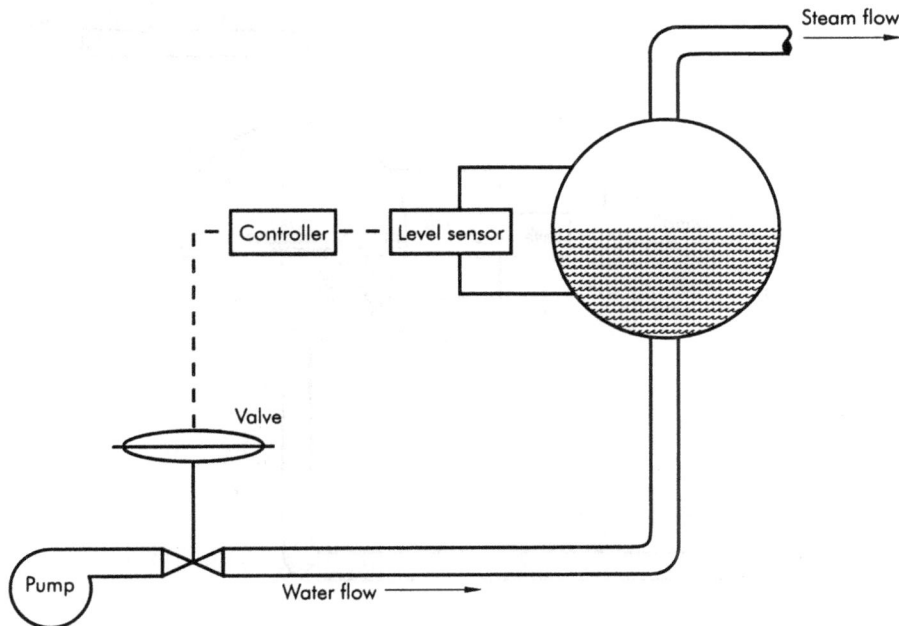

Figure 3-27. Single-element regulator

Multiple Element Regulation

Controlling the water level using both steam flow and water level is called *two-element regulation*, Figure 3-28. This method helps solve the problems associated with fluctuations caused by sudden load changes. When the controller detects a load increase, instead of using the drum level as the primary signal to control the feedwater valve, it uses the signal sent from the steam flow. Even if the drum level increases, the regulator can still sense if the load has increased and then the feedwater flow rate is increased. The steam flow measurement maintains feedwater flow in proportion to steam flow, and the drum level measurement corrects for any imbalance in water input in proportion to steam flow.

Large boilers use *three-element regulation* for precise level control, Figures 3-29 and 3-30. A three-element regulator measures steam flow, feedwater flow, and drum level. With this setup, the steam and feedwater are compared. Any difference is then used to regulate the feedwater control valve. For example, if the steam flow measures 100,000 lb/hr, the control system adjusts the incoming feedwater so that it is also flowing at 100,000 lb/hr. The drum level input keeps the level from drifting due to flowmeter errors, blowdown, or other causes.

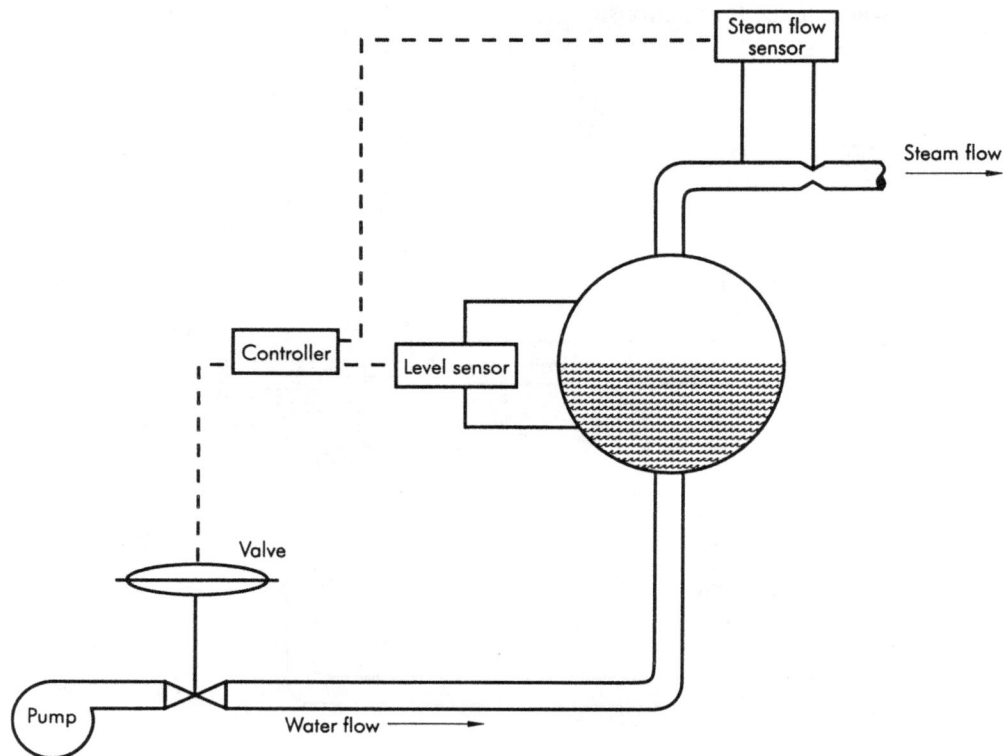

Figure 3-28. Two-element regulator

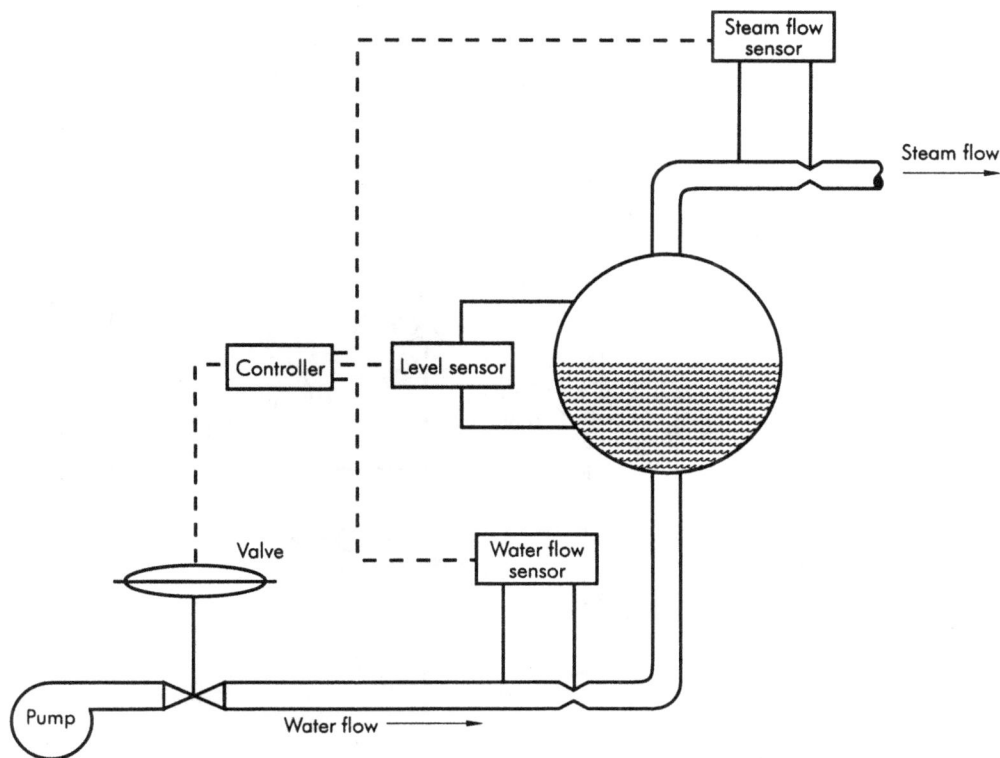

Figure 3-29. Three-element regulator

The best system is the simplest one that can do the job. The more sophisticated the system, the more it costs to install and maintain. Simple on/off regulators are satisfactory for fire-tube boilers that have large water reserve capacities. Small water-tube boilers (up to 100,000 lb/hr) that have adequate water storage and in which water level fluctuations are not critical can use the thermohydraulic (generator-diaphragm) or thermostatic expansion tube regulator (Copes regulator).

Larger boilers equipped with waterwalls tend to have small water-reserve capacities and are frequently subjected to fluctuating loads. They are typically suited for two-element controls. When load swings are large and sudden, or for very critical applications such as supply steam for large turbines, three-element regulation is required.

CONTINUOUS BLOWDOWN

Feedwater contains dissolved solids left behind when steam leaves the boiler. The preferred way to remove and control the level of these dissolved solids is to use a *continuous* or *top blowdown*. The concentration of dissolved solids is

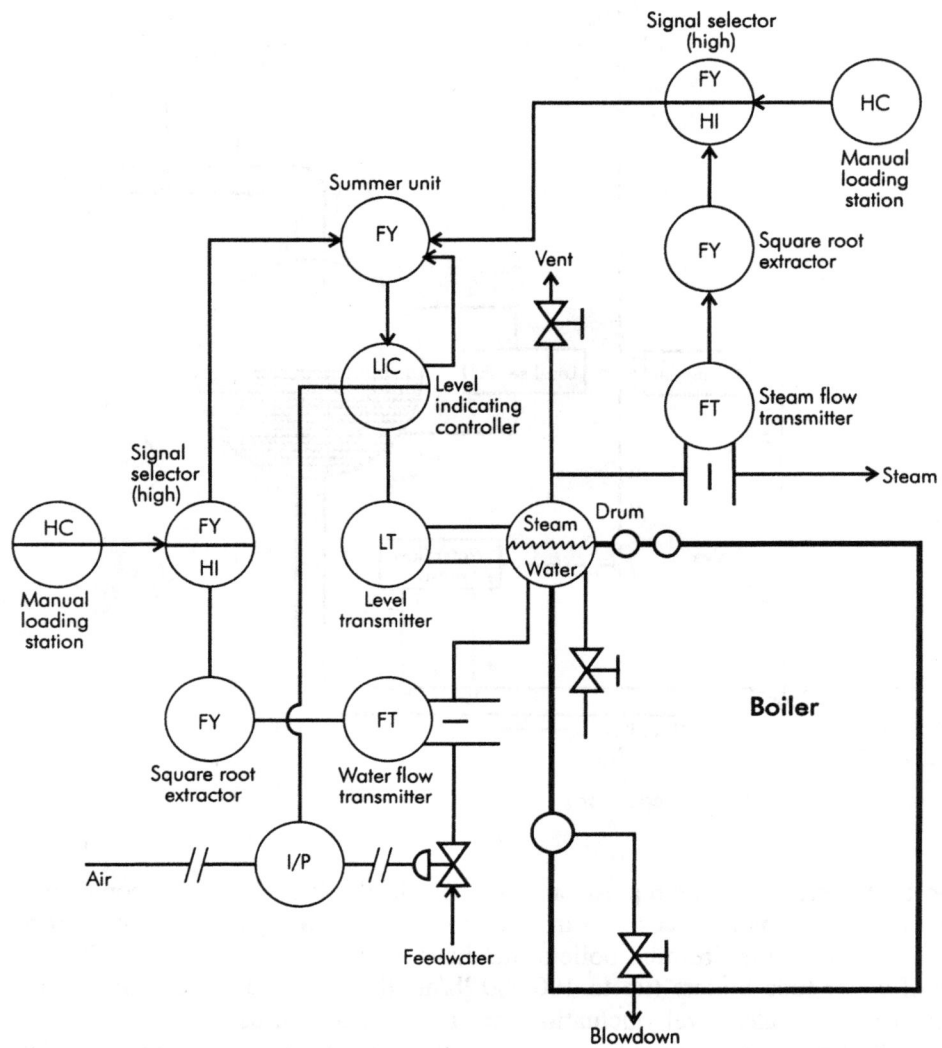

Figure 3-30. Instrument schematic using Instrument Society of America (ISA) nomenclature

highest where the steam bubbles and water separate, about two or three inches below the boiler's water level. Figure 3-31 shows various piping arrangements that manufacturers use.

As the name implies, continuous or top blowdown is done on a continuous basis. The boiler water is tested periodically for total dissolved solids (TDS) and compared to the level recommended by the boiler manufacturer or boiler chemical salesperson. If the TDS level is too high, the rate of blowdown is increased to lower the concentration of TDS. If the level is lower than recommended, the blowdown rate is decreased.

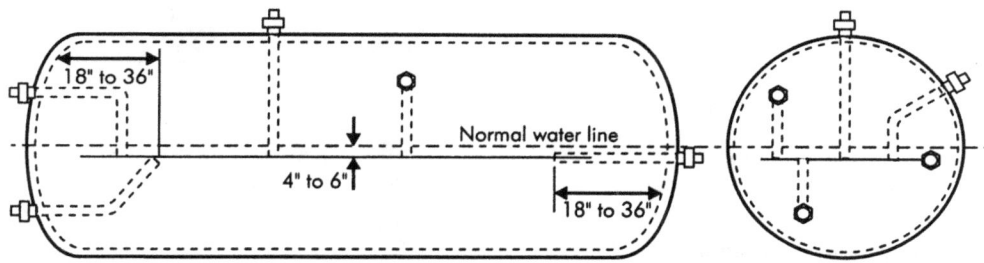

Figure 3-31. Piping arrangements for continuous blowdown lines

The blowdown rate is frequently controlled manually by an orifice-type valve with an indicator that shows how far open it is. This indicator is typically used as a guide when making adjustments. However, it is difficult to maintain a constant solids concentration in boiler water by using a manual valve, especially if the load or other conditions fluctuate. Even under the best of circumstances, several readings must be taken during each shift to maintain precise control. The continuous blowdown process can be automated by installing an electric valve and a conductivity monitor. Either a continuous or periodic conductivity reading is taken by the monitor. If the conductivity is too high, the monitor sends a signal to the electrically operated valve to open; if too low, it signals the valve to close. Figure 3-32 is a comparison of hand-controlled continuous blowdown and automatic blowdown control.

Since the top blowdown is operated continuously, a considerable amount of heat can end up down the drain. The system shown in Figure 3-33 can recover more than 90% of the heat contained in the blowdown water. The hot water is discharged into a flash tank, which is at a lower pressure than the boiler (around 10 psi in most installations). Some hot water flashes off as steam, which can be used for heating or process systems or in the deaerator. Although the steam comes from dirty water, it is still perfectly good. The remaining hot

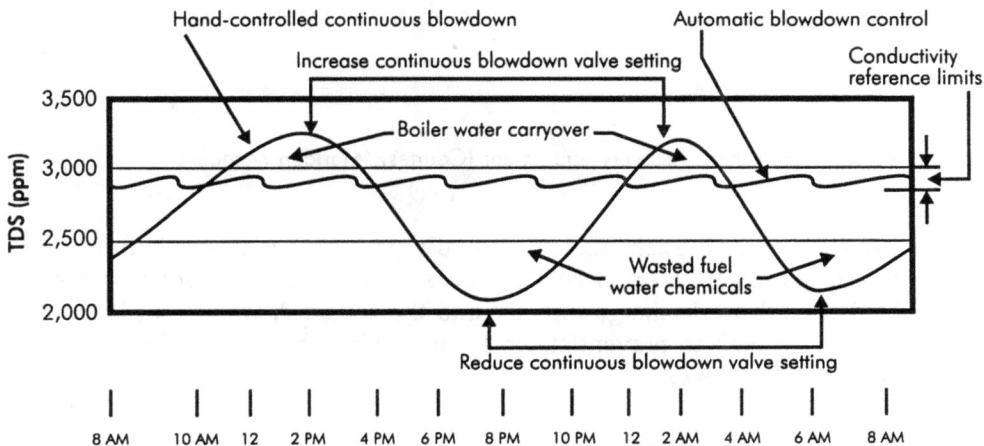

Figure 3-32. Comparison of hand and automatic blowdown control

1. ASME code welded flash tank for 150 psi for boilers rated up to 600 psi.

2. Safety pressure relief valve, normally set for 40 psi.

3. Flash steam outlet flange.

4. High water level alarm system.

5. Pressure gauge.

6. Sight gauge for water level in flash tank. *(See Cover Picture)*

7. Baffle-creates water turbulence.

8. Manifold for mounting Orifice Meter or connecting other blowdown supply pipe, one or two manifolds can handle up to six boilers.

9. Madden Orifice Meter continuous blowdown control.

10. 6" x 8" hand hole for float valve access.

11. Continuous flow balanced pressure flow float valve with 316 stainless steel float and trim is specifically designed for blowdown service.

12. Temperature gauge panel with three 3-1/2" gauges to monitor drain water, exchange water inlet and outlet temperatures, makes checking system efficiency easy. *(See Cover Picture)*

13. Heat exchanger coil, four $^7/_8$" copper tubes with bronze headers brazed to each end. Designed to provide even distribution of hot water to all four tubes to maximize thermal transfer.

14. Exchanger shell with flanged exchange water inlet and outlet provides swirling action in the water. The turbulence assures exchange water contact with the coil surfaces and improved efficiency.

15. Blowdown water discharge fitting and temperature bulb.

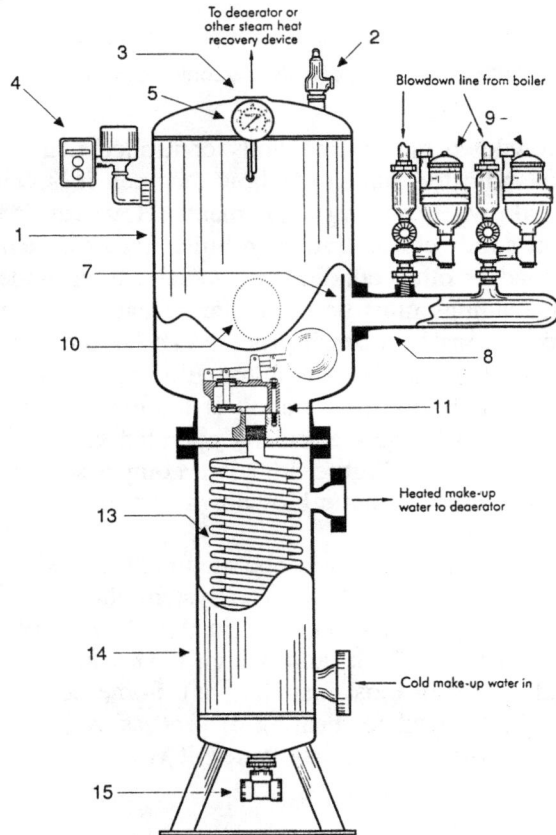

Figure 3-33. Blowdown heat recovery equipment (Courtesy, Madden Manufacturing, Inc.)

water is routed to a heat exchanger that usually preheats feedwater before it goes to the deaerator. It can also heat a process liquid. The blowdown water is usually cool enough to discharge directly into the sewer. A water seal is maintained in the flash tank to prevent steam from flowing through the heat exchanger and into the sewer.

CONCENTRATION

Every field has its own nomenclature, and water chemistry is no exception. The term parts per million (ppm) refers to the quantity (in pounds) of a particular substance contained in 1 million lb of whatever substance it is dissolved in. In this case, it is how many pounds of chlorides (or other substance) that are dissolved in 1 million lb of water.

Grains per gallon is another way to express concentration. A grain is 1/7,000 of a pound. To convert grains per gallon to parts per million, multiply by 17.1. Grains per gallon is used mostly when discussing water softening.

Cycles is yet another term used when expressing concentrations. In this case, the water in a boiler or cooling tower is compared against its make-up water. For example, if a cooling tower has a TDS level of 2,700 ppm and the city make-up water has a TDS level of 180, then the tower is concentrated 15 cycles (2,700/180 = 15).

In the example shown in Figure 3-34, the boiler water contains 2,000 ppm of solids and the make-up water contains 200 ppm. The boiler water is concentrated to 10 cycles (2,000 divided by 200 equals 10). In order to not exceed 10 cycles, it's necessary to blowdown or remove 10% of the boiler water on a continual basis. The 10% value is found by dividing the amount of blowdown water by the amount of make-up water; 200,000 ÷ 2,000,000 = 10%.

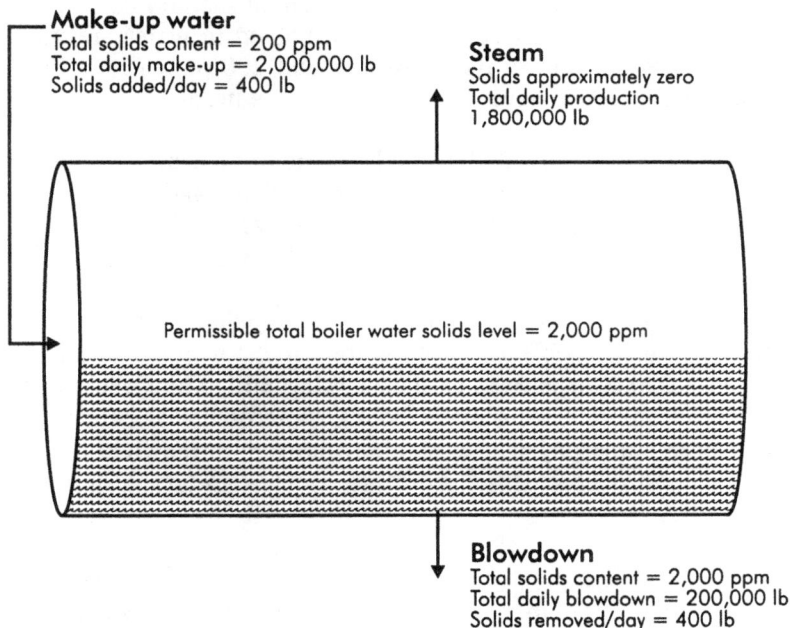

Make-up water
Total solids content = 200 ppm
Total daily make-up = 2,000,000 lb
Solids added/day = 400 lb

Steam
Solids approximately zero
Total daily production
1,800,000 lb

Permissible total boiler water solids level = 2,000 ppm

Blowdown
Total solids content = 2,000 ppm
Total daily blowdown = 200,000 lb
Solids removed/day = 400 lb

Figure 3-34. Steam/make-up water/blowdown balance of a typical operating boiler

Maintaining the recommended solids level in a boiler is very important. If the level is too high, impurities in the water may be carried over into the distribution system. If the level of solids is below the recommended level, then hot water and chemicals are being wasted by excessive blowdown.

Carryover and *priming* are two closely associated terms. Carryover describes small droplets of water leaving the boiler with the steam. A high level of TDS is the main cause of carryover. The higher the TDS level, the higher the surface tension of the water. As steam bubbles float to the surface, they must break the surface tension to break free. When the surface tension is high, water droplets are carried along with the steam.

Carryover causes problems with the steam system. Remember, the boiler water contains dissolved solids. These solids fall out and build up in places such as strainers and control valves. You can imagine that after a while, these items would require maintenance. Carryover can also contaminate a product that comes in direct contact with the steam.

Priming is an extreme case of carryover in which a slug of water enters the steam system. Serious physical damage is then possible due to water hammer. It is no problem for a gas like steam to travel 100 ft per second in a pipe. Liquids like water flow well at 5 ft per second. When a slug of water is caught up in a 100 foot per second flow, is almost acts like a solid. When a fast-moving water slug hits an obstruction, such as an elbow or pressure-reducing valve, enough force is created to cause physical damage.

Measuring the dissolved chloride levels is an excellent method of determining the cycles of concentration, because chlorides dissolved in water are very stable. If the chloride level is 20 ppm in the feedwater and 200 ppm in the boiler water, the boiler water would have 10 cycles of concentration.

Another way to measure dissolved chloride levels is to measure the electrical conductivity of the water. Pure water acts as an insulator and has very low conductivity. However, as the concentration of dissolved solids in water increases, the electrical resistance decreases and conductivity increases.

The unit of conductivity is the mho. Mho is "ohm" spelled backwards. (Ohm is the unit of electrical resistance.) The mho is too large for practical purposes, so the **micromho** (1/1,000,000) is used as the working unit. The Greek letter "μ" is the abbreviation for "micro," so in many books, "micromho" is printed "μmho." (The trend now is to name physical units after pioneers in the field of science. The new term for mho is "Siemens," so in the latest literature "μSiemens" is the unit of conductivity.)

Satisfactory condensate is typically less than 3 micromhos; city water (which varies from one source to another) runs 60 to 150 micromhos. Boiler water not exceeding 150 psi can usually operate up to 3,500 micromhos without trouble.

Although not as accurate a gauge as chloride readings for determining boiler water concentrations, the use of conductivity readings is a fast and convenient method of determining solid concentration.

BOTTOM BLOWDOWN

All boilers must be equipped with a drain connected to the lowest part in the water space so that the boiler can be emptied. This connection also doubles as the bottom blowoff. By using the bottom blowoff to blow down the boiler, accumulated sludge and sediment are removed.

Opinions differ on how to bottom-blow a boiler. To perform a short blowdown, fully open the valve, count to five, then close the valve. A short blow is more effective than a long one for several reasons. First, a localized low-pressure area forms around the blowoff opening inside the boiler as the water rushes out at high velocity. Since the water is at saturated temperature, some flashes off as steam in this lower pressure region. The steam takes up more space than the water and thus displaces the sludge and sediment away from the blowdown opening.

The second reason is the water level drops during long blowdowns. The feedwater valve opens more to compensate for this loss. Relatively cold feedwater immediately circulates to the bottom of the boiler and adds to the turbulence normally found there. This further disperses the sludge and reduces the effectiveness of the blowdown. During this time, heated water is wasted. Do not blowdown a boiler until the level in the gauge glass is lowered by 1/2 inch. This procedure merely wastes both hot water and chemicals. It is best to give the boiler a short blow once every 24 hours. If experience proves that is not often enough, then wait at least one hour between blows. This greatly increases the efficiency of the operation by allowing the sludge to settle over the blowoff opening.

If the maximum allowable working pressure of a boiler exceeds 100 psi, two blowoff valves in series are required. They may be two slow-opening valves or one slow-opening valve and one quick-opening valve. A slow-opening valve requires at least five 360-degree turns to change from fully closed to fully open.

To ensure a positive seal, the quick-opening valve is always installed closest to the boiler between the boiler and the slow-opening valve. When blowing down, first open the quick-opening valve, then the slow-opening valve. After counting to five, close the slow-opening valve before shutting off the quick-opening valve. This method puts all the wear and tear caused by the rapidly moving boiler water on the outside valve and leaves the valve next to the boiler in good condition for a positive shutoff. The quick-opening valve is used for sealing, while the slow-opening valve is used for blowing.

For tandem blowoff valves (where a one-piece block serves as a common body for both the sealing and blowing valve), the second valve from the boiler is opened first and closed last. Some manufacturers of these high-performance valves have the opening and closing sequence instructions on them.

The blowdown from a high-pressure boiler cannot be discharged directly to a sewer, because the stem and hot water could damage the sewer or cause personal injury. A blowdown tank must be placed between the boiler and the sewer. The purpose of the tank is to reduce the temperature and pressure of the discharged water to safe levels (150°F and 5 psi maximum).

In the atmospheric flash tank shown in Figure 3-35, blowdown water enters at a point above a constantly maintained water line. The flashed steam is discharged out through a vent large enough to avoid backpressure. The vent is usually capped with an exhaust hood that captures and prevents hot water from being sprayed around the vicinity. The hot water mixes with the cooler water in the tank before being discharged into the sewer.

In another variation, shown in Figure 3-36, cold water is automatically added to decrease the temperature to a safe discharge level.

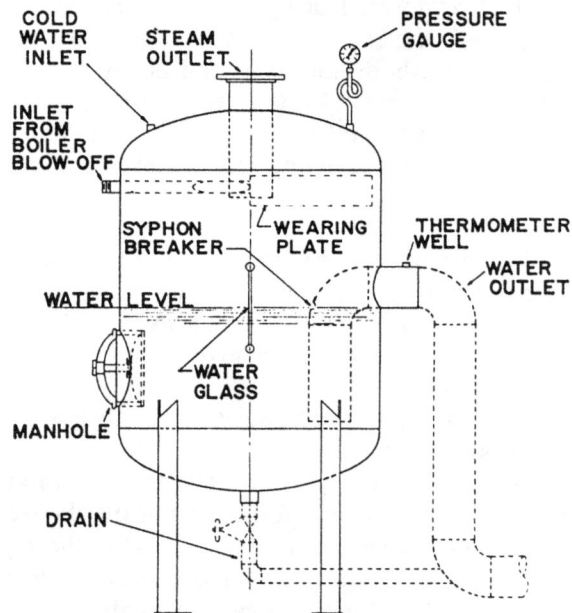

Figure 3-35. Atmospheric flash tank (Courtesy, Penn Separator Corporation)

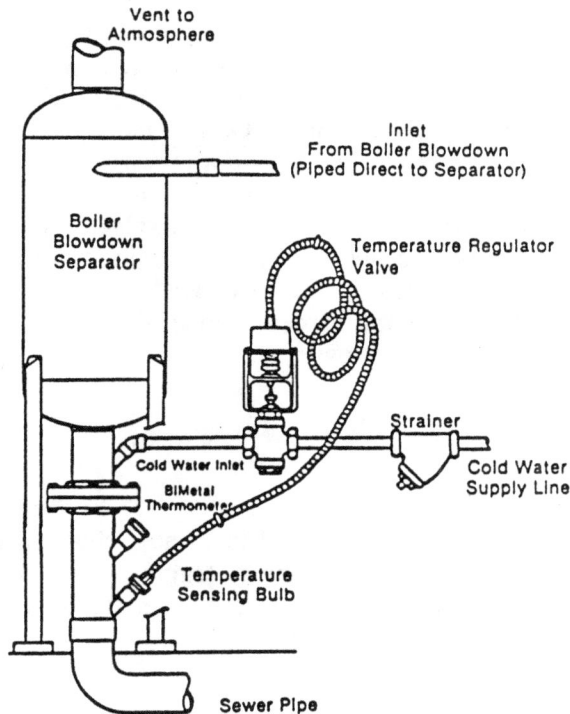

Figure 3-36. Blowdown separator with automatic waste water cooling (Courtesy, Penn Separator Corporation)

BOILER ENTRY

A valve or switch is locked and tagged when its operation could lead to injury or mechanical damage. With valves, this means chaining and locking them closed. With electrical switches, it means locking them in the open or de-energized position. In both cases, a tag includes the following information: 1) who tagged the device; 2) why was it taken out of service; and 3) when was it taken out of service.

There are seven systems that must be locked and tagged before entering the boiler: steam, feedwater, bottom blowdown, top blowdown, chemical feed, fuel, and soot blowers. The first five are on the water side, and the last two are on the fire side. *Warning: For your own safety, personally verify that each system is locked and tagged before entering the boiler. Do not take anyone's word that a system is safe.*

Steam

It is desirable to have all steam valves chained closed and tagged before entering the boiler. All boilers with manholes connected to a common steam header (boilers in battery) must have two stop valves with a free-blowing gate drain valve between them. Both stop valves can be OS&Y gate types, but most installations have a gate valve at the header and a nonreturn valve at the boiler. The purpose of the free-blowing drain is to indicate that the header stop valve is closed and holding. If it should fail, the nonreturn valve is still between the operator and disaster. Just opening the free-blowing drain is not enough. Take a length of stiff wire and poke it through the valve and into the line. Sometimes material accumulates in the valve and hardens. Under these conditions, even with the valve open, nothing would flow.

Feedwater

Since most multiple boiler plants have heated feedwater, shut off any pressurized hot water entering the boiler. In addition to tagging and locking the valves closed, it is recommended practice to have a drain after the bypass valve to indicate that all feedwater valves are holding.

Bottom Blowdown

The bottom blowdown valves must be opened to drain the boiler of washdown water. The problem is that most multiple boiler installations have a common bottom blowdown system. When an operator bottom blows the on-line boilers, the odds are very good that some very hot water will find its way into the opened boiler. To avoid this, the valve next to the boiler should be removed and the valve connected to the common drain header should be closed, chained, and tagged. Some installations remove both valves and block off the header connection.

Top Blowdown

All of the top blowdown lines are probably connected to the same flash tank. If the top blowdown valve is open on the opened boiler, steam and hot water would enter from an operating boiler. The top blowdown valve must be closed, chained, and tagged.

Chemical Feed

Some plants have common piping from their chemical feed systems. Though the volume of liquid is small, it is usually considered toxic. To avoid putting your hand in a puddle of caustic soda, make sure the chemical feed line is secured and tagged.

Fuel

It goes without saying that you don't want to be in the furnace while an operator is trying to light a burner. Though this is not likely to happen, make sure all fuel systems are secured and tagged before entering the furnace.

Soot Blowers

Be sure that an absent-minded operator doesn't use a soot blower while you are working on the boiler's fire side. Automated systems are especially dangerous since it takes only the press of a button to start operation. Lock and tag both the steam or high-pressure air and the operating mechanism.

Manholes and Handholes

It is necessary to provide access to the interior of boilers for construction, repair, and maintenance. When it is necessary to enter the boiler, manholes, sometimes called manways, are used. Handholes are used where access to tubes is necessary or when an operator just needs to get a hand in the boiler.

As shown in Figure 3-37, the cover plate is placed inside the pressure vessel. This way the pressure forces the cover plate against its seat. Gaskets are used to form a good seal. A yoke, also called a crab or dog, is used to hold the cover plate in place. Most manholes and handholes are elliptical in shape to make it possible to remove the cover plate through its own opening. The two most common sizes for manholes are 11" x 15" and 12" x 16".

The procedure used when installing a manhole or handhole is as follows:

1. Remove the old gasket, and thoroughly clean the surfaces on the boiler and cover plate. Sometimes it's necessary to buff each surface.

2. Place a new gasket on the cover plate, and verify that it's the proper size and shape.

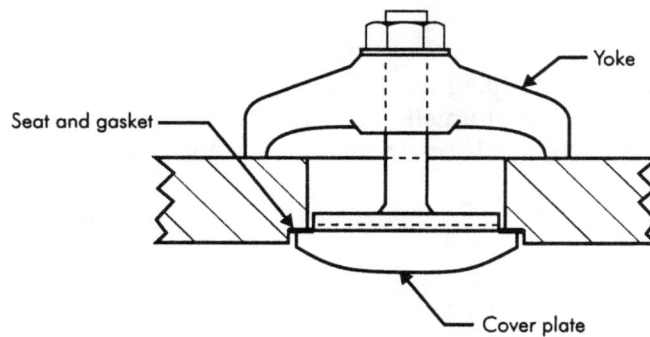

Figure 3-37. Typical handhole cover

3. Place the cover plate on its seat, set the yoke, finger-tighten the nut, then snug up a quarter turn with a wrench. If the gasket leaks, only tighten enough to stop the leakage.

4. As pressure builds up in the boiler, the yoke will loosen. It takes a short time for the gasket to reach the compression necessary to conform to the boiler pressure, so snug up the nut during this period.

ASME AND THE NATIONAL BOARD

The majority of pressure vessels in the United States are built under the guidelines set by the American Society of Mechanical Engineers' Boiler and Pressure Vessel Code. It is called the ASME Code or simply, the Code. The National Board of Boiler and Pressure Vessel Inspectors is the enforcement agency empowered to ensure adherence to provisions of the ASME Code. All states use either the Code or a similar version to define their pressure vessel construction requirements.

When manufacturers or contractors build or repair equipment that must comply with the Code, they are also required to stamp that equipment with one of the following symbols:

- S - Power boilers
- M - Miniature boilers
- L - Locomotive
- A - Boiler assembly
- H - Heating boiler
- PP - Pressure piping
- U - Unfired pressure vessel

- V - Safety valve
- R - Repair

Every code-compliant pressure vessel has a nameplate on which is stamped the manufacturer, maximum allowable working pressure (MAWP), square feet of heating surface, rate of steaming capacity, serial number, year built, and a national board registration number. This national board registration number (abbreviated NAT'L BD) refers to a permanent record that contains, among other things, the composition of materials used, wall thickness, joint details, and size and type of tubes.

Boiler inspectors must pass a national board examination and a test on state rules and regulations before they are authorized to evaluate pressure vessels. They are employed by legal jurisdictions designated to enforce the ASME Code and by insurance companies that underwrite boilers and pressure vessels.

Even if a boiler inspector works for an insurance company, he or she is still an agent of the legal jurisdiction regulating pressure vessels in a state and is required to enforce all rules and regulations. However, national board-commissioned inspectors cannot be employed by boiler or pressure vessel manufacturers or installers. This is necessary to preserve their status as independent third parties in questions of public safety. The inspectors at manufacturing facilities typically work for insurance companies who are contracted by the manufacturer.

CHAPTER FOUR

Combustion

Combustion is the rapid recombination of oxygen with a fuel that results in the release of heat. It can also be thought of as a rapid oxygenation process of fuel that results in a release of heat.

Perfect combustion (stoichiometric combustion) is obtained by mixing and burning precisely the right amounts of fuel and oxygen so that no products are left once combustion is complete. If too much oxygen (excess air) is supplied, the flame is cooler, short, and clear; this is *complete combustion* but not perfect combustion. The excess air takes no part in the combustion process, except to absorb heat from the flame and discharge it out the stack.

If there is too much fuel and not enough oxygen supplied, the flame is longer and sometimes smoky. This is *incomplete combustion* and results in unburned fuel (carbon monoxide, hydrogen, unburned hydrocarbons, and free carbon) blowing out of the stack. Hydrocarbons include compounds composed of carbon and hydrogen atoms (e.g., fuel oils, natural gas, propane, and coal).

It is important to remember the *three T's of combustion*: time, temperature, and turbulence. *Time* because all chemical processes take time to complete; *temperature* because this process must take place at elevated temperatures; and *turbulence* because there must be thorough mixing of fuel and air.

DRAFT

Combustion air is supplied by either natural or mechanical means. Draft is the difference in pressure that produces the flow of gas, measured in inches of water column (in. w.c.). Draft is produced in a stack by the difference of the hot flue gas inside and the cooler, denser, ambient air outside. The equation for natural draft is:

$$D = (0.52)(H)\left[P \left(\frac{1}{T_a} - \frac{1}{T_f} \right) \right]$$

where:
- 0.52 = constant used to convert to inches of water
- D = draft produced (inches of water)
- H = height of stack (feet)
- P = atmospheric pressure (pounds per square inch)
- T_a = absolute ambient temperature
- T_f = absolute flue gas temperature

The draft from the stack pulls the flue gas from the boiler. This is called *induced draft*. If this proves insufficient, a fan is used to help. This induced draft fan is located in the breeching, which is the connection between the boiler and stack. Packaged boilers use a forced draft fan, which pushes air into the boiler.

Special consideration must be given to large boilers, because the pressure produced by the forced draft fan forces the furnace walls outward. To compensate for this, large boilers employ balanced draft, in which both forced and induced fans are used, Figure 4-1. The control system keeps the furnace at or just above atmospheric pressure. One fan controls furnace pressure and draft, while the other controls the air/fuel ratio.

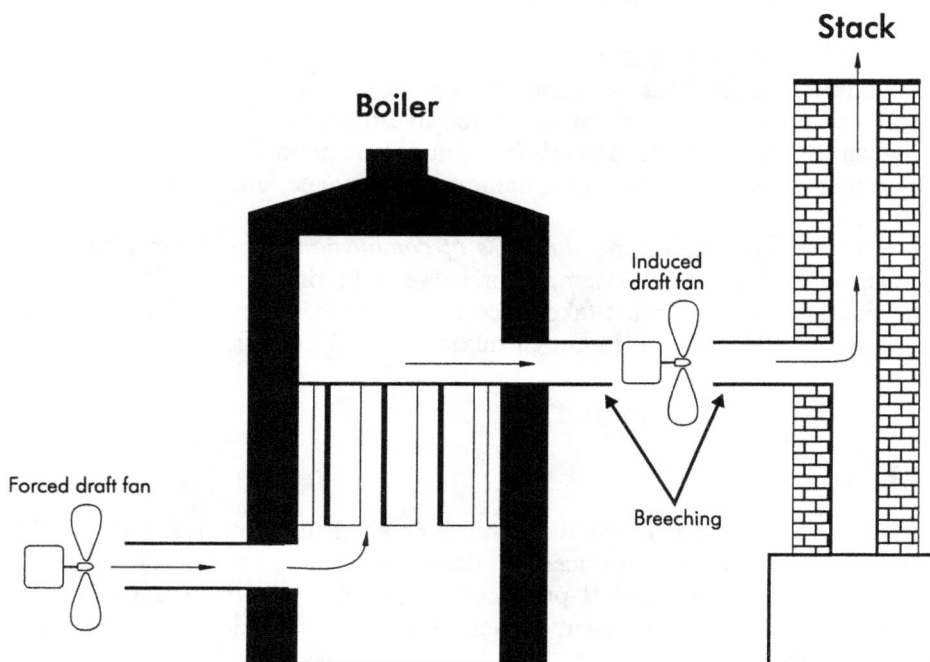

Figure 4-1. Balanced draft system

Dampers, variable inlet guide vanes, variable speed, and variable pitch are other methods used to control the output of a fan. Variable speed and variable pitch are more efficient than inlet vanes, and inlet vanes are more efficient than dampers.

Besides combustion air, furnace volume also limits how much fuel can be burned in a boiler. For example, 10 ft^3 of furnace volume is required to burn one gallon of oil per hour.

When a hydrocarbon is burned, the result is heat plus water vapor (H_2O) and carbon dioxide (CO_2). If there is not enough oxygen, carbon will produce carbon monoxide instead. Carbon monoxide can be used as a fuel. When carbon monoxide is burned, it changes to carbon dioxide.

The left column of the heat loss graph in Figure 4-2 demonstrates that operating costs increase rapidly when unburned fuel goes up the stack. The right side shows that useful heat is lost up the stack when too much air is supplied. The middle part of the graph is where the losses are at a minimum. The extent of the overlap between unburned fuel and excess air primarily depends on the efficiency of the combustion equipment.

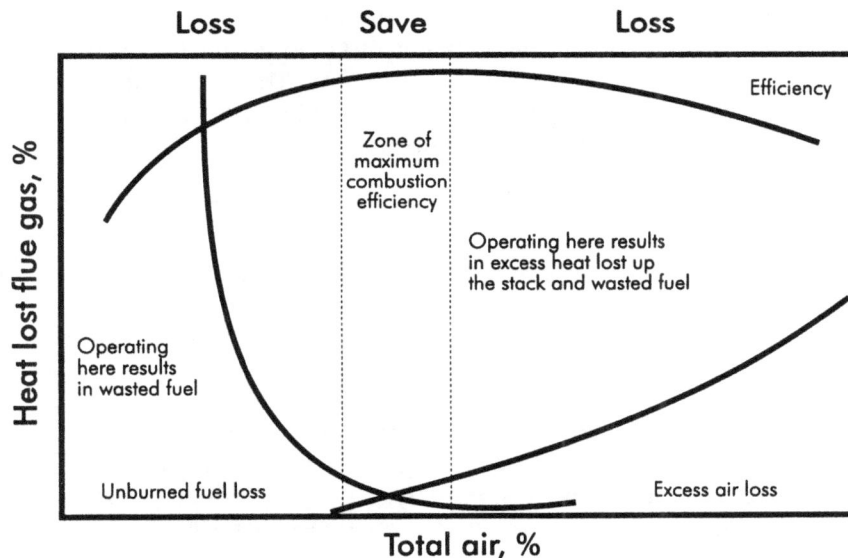

Figure 4-2. Heat loss

FLUE GAS ANALYSIS

The percentage of carbon dioxide (CO_2), and sometimes carbon monoxide (CO), in flue gas is used to calculate combustion efficiency. Oxygen (O_2) or carbon dioxide is used to calculate the amount of excess air.

Perfect combustion is attained when the flue gas analysis shows no carbon monoxide or oxygen. That means that every available fuel molecule and every available oxygen molecule came in contact with one another. Such perfect mixing is not possible, even with the most advanced burner.

Figure 4-3 shows the relationship between oxygen, carbon dioxide, and excess air. Notice that it does not really matter what fuel is used when measuring oxygen; but it matters a great deal when measuring carbon dioxide. The reason is that the O_2 reading measures how much oxygen is left over after combustion. Carbon dioxide is a product of combustion and thus depends on the ratio of carbon to hydrogen in the fuel. Coal has a higher percentage of carbon than does natural gas; therefore you would expect a higher percentage of carbon dioxide from coal than from natural gas.

Boiler firing conditions are usually either based on stack O_2 or CO_2. Both are equally valid for determining excess air, but many prefer O_2 for the following reasons:

- As mentioned before, the relationship between O_2 and excess air is not highly affected by the fuel burned. The O_2 curves for natural gas, oil, and coal are nearly coincident, while the CO_2 curves show a large variation.

- When firing at low excess air, the measurement of CO_2 requires much greater precision than the measurement of O_2 to obtain the same accuracy. For example, suppose you want to burn oil at 20% excess air with 2% accuracy. CO_2 would need to be measured within 1.5% accuracy (13.0% ± 0.2%), while O_2 would only require 8% accuracy (3.7% ± 0.3%).

- CO_2 is a product of combustion, while O_2 is associated directly with excess air.

- The instrumentation is less expensive and more reliable for O_2 than for CO_2. A zirconium oxide oxygen probe can be placed directly in the stack to give continuous analyses. Someday, however, the cost of continuously monitoring CO_2 will also become cost-effective.

Also notice in Figure 4-3 that more excess air is required for coal than for oil and gas. For example, the ideal level of O_2, CO_2, and excess air when burning oil is 2.75%, 13.25%, and 15%, respectively. Examiners often ask what is the ideal level of excess air for various fuels. Remember, it's 10% for gas, 15% for oil, and 30% for coal.

Figure 4-4 shows two other reasons why boiler firing conditions are usually based on oxygen instead of carbon dioxide. Notice how flat the carbon dioxide curve is near the zero excess air line compared to the oxygen line. A change in excess air is much easier to detect on the oxygen line. Also, when making adjustments with carbon dioxide, you don't know if the change put you in the excess fuel or excess air region without having someone look for smoke. Remember, if too much fuel and not enough oxygen are supplied, the flame is sometimes smoky.

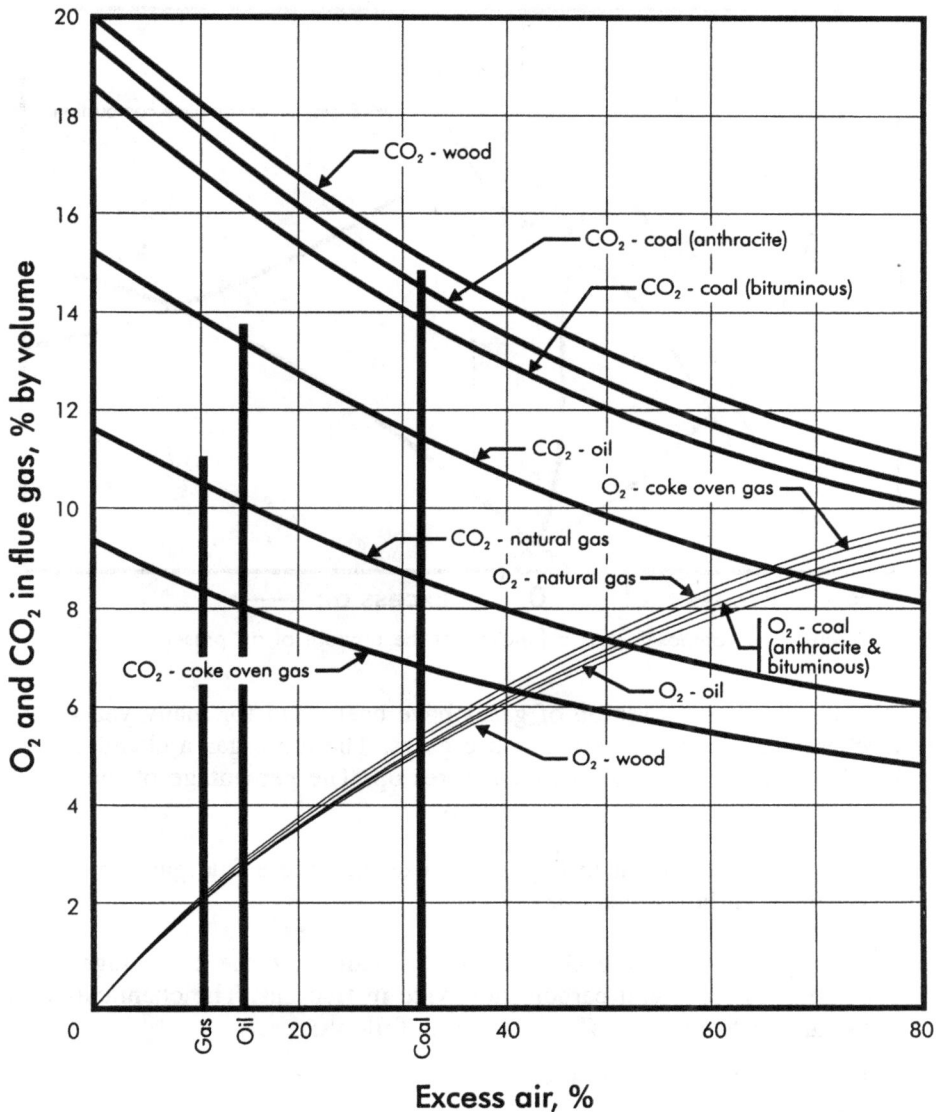

Figure 4-3. Ideal levels of oxygen and carbon dioxide for boilers burning gas, oil, and coal

Figure 4-4 also shows that the carbon monoxide content of the flue gas jumps up dramatically in the excess fuel region, because carbon monoxide is a fuel.

Remember that high carbon dioxide levels are desirable, because the higher the percentage of CO_2 in the products of combustion, the higher the combustion efficiency. Higher levels of oxygen mean there is excess air, which means low combustion efficiency.

Many types of instruments are available for analyzing flue gases and indicating the volume percentage of the various components. Instruments that use the

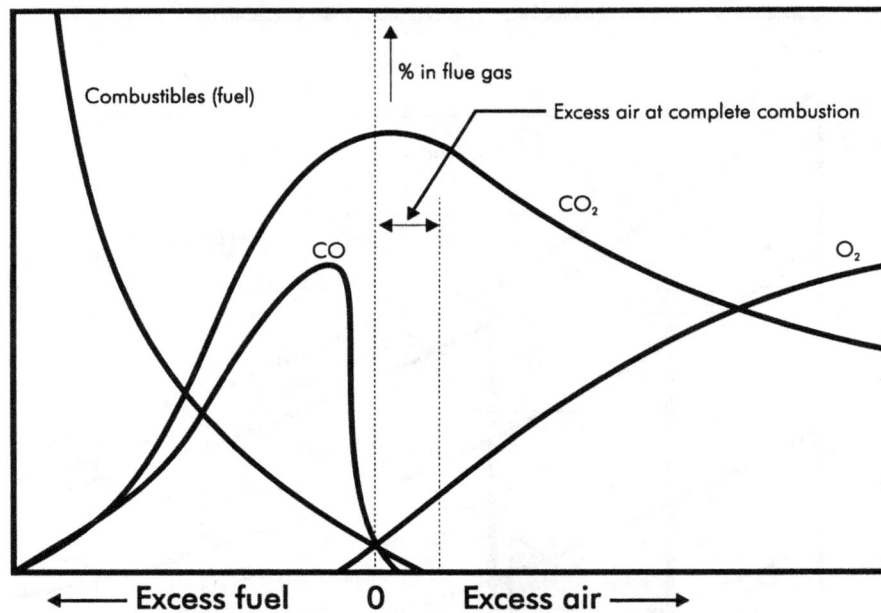

Figure 4-4. Flue gas composition as a function of the amount of air present

principle of selective absorption of gases have been used for many years. Certain chemicals readily absorb specific gases. The more gas a chemical absorbs, the more volume the chemical takes up. The percentage of the gas is read on a graduated scale on the apparatus.

The flue gas analyzer shown in Figure 4-5 can measure a flue gas sample for either oxygen or carbon monoxide.

Another flue gas analyzer, the Orsat apparatus, can show the percentages of carbon dioxide, oxygen, and carbon monoxide in flue gas. The chemicals used in this instrument to measure flue gas are as follows:

- Caustic potash for carbon dioxide
- Alkaline solution of progallol for oxygen
- Acid solution of cuprous chloride for carbon monoxide

The Orsat apparatus takes one sample at a time. There are several instruments that can take continuous samples. One popular oxygen analyzer uses zirconium oxide. A galvanic action proportional to the amount of oxygen is translated to a reading. Since this is continuous and without time delay, the output signal can be used for automatic air/fuel ratio control.

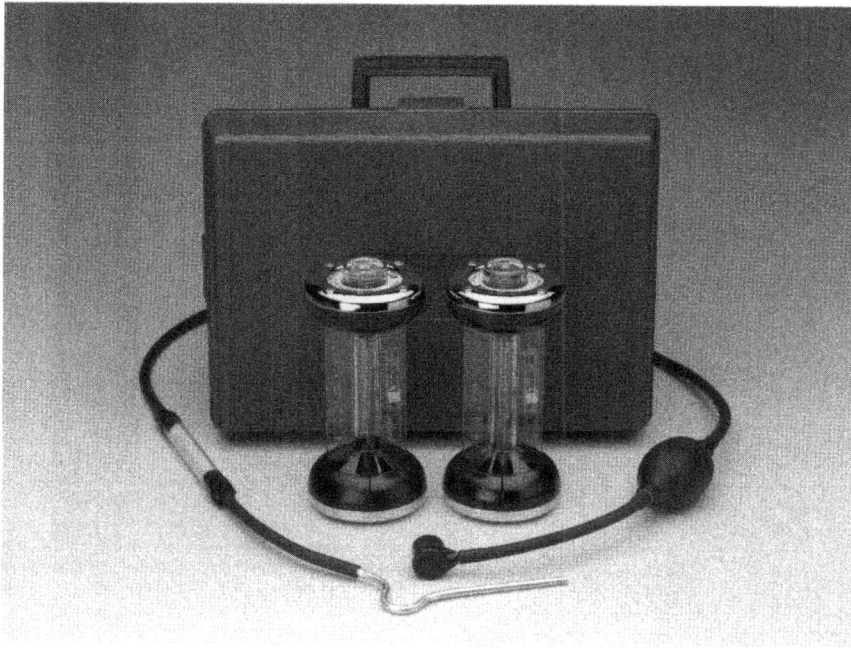

Figure 4-5. Flue gas analyzer (Courtesy, Bacharach, Inc.)

SMOKE DENSITY

A smoke test is a good way to check combustion efficiency. A sample of the smoke is taken and compared to the *Ringelmann Scale*, which consists of the four charts shown in Figure 4-6. These charts are compared to the smoke sample, and a rating is given for smoke density.

Most stacks cannot be observed while operating the boiler, so some installations have smoke density meters such as the one shown in Figure 4-7. They can be equipped with alarms to alert operators of a smoke condition.

BOILER EFFICIENCY

Boiler efficiency is directly related to the amount of energy released in the combustion process. There are two methods used to determine boiler efficiency: the *indirect* method and the *direct* method.

In the indirect method, boiler efficiency is calculated by dividing energy released in combustion by the heating value of the fuel. Complex calculations are required to obtain the answer, but fortunately tables are available to make the job simple. When the amounts of combustibles and carbon monoxide in the flue

Ringelmann charts for estimating smoke densities. The charts below are proportional reductions of standard charts issued by the U.S. Bureau of Mines. These charts are used in the following manner: Make observation from a point between 100 and 1,300 ft from the smoke. The observer's line of sight should be perpendicular to the direction of smoke travel. Place the below charts approximately 15 ft in front of the observer and as close as possible to his line of sight. (Standard ASME or U.S. Bureau of Mines charts should be placed 50 ft from the observer.) Open sky makes the best background for observations. Compare the smoke density with the charts (which, at 15 ft, are shades of gray instead of individual lines) and classify the smoke according to the Ringelmann chart number. Ringelmann Nos. 0 and 5 are 0% and 100% black, respectively. Charts for these are solid white and black (not shown).

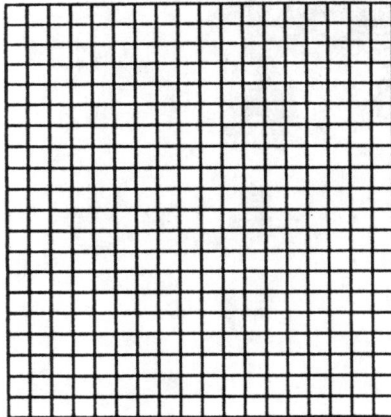

1. Equivalent to 20 percent black.

2. Equivalent to 40 percent black.

3. Equivalent to 60 percent black.

4. Equivalent to 80 percent black.

Figure 4-6. Ringelmann Scale for estimating smoke density (Courtesy, North American Manufacturing Company)

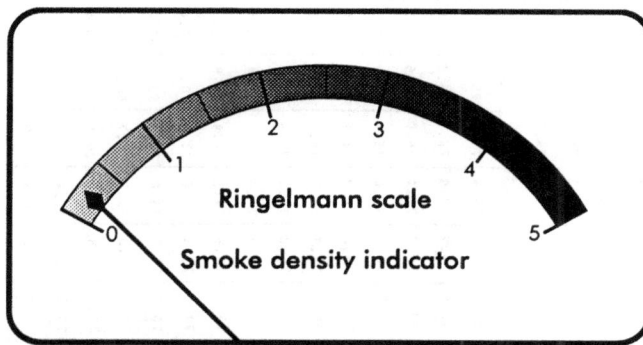

Figure 4-7. Smoke density indicator

gas are negligible, it becomes necessary to obtain both the percentage of oxygen or carbon dioxide in the flue gas and the flue gas temperature.

Note that the tables in Figure 4-8 refer to *net stack temperature*. Net stack temperature is the temperature of the flue gas minus the temperature of the combustion air (the air in the boiler room). In a well-designed, properly maintained boiler, the flue gas temperature should not be higher than 50° to 100°F above the steam temperature.

For example, consider a boiler that burns oil and has a flue gas analysis reading of 10% CO_2. The conversion chart in Figure 4-3 shows that 10% CO_2 is equal to 50% excess air. The flue gas is 550°F, and the air in the boiler room is 100°F. The net flue gas temperature is therefore 450°F. Figure 4-8 shows that a boiler is 80.7% efficient if there is 50% excess air and a net stack temperature of 450°F.

The direct method of determining boiler efficiency involves dividing the net amount of Btu in the steam leaving the boiler by the amount of Btu being put into the boiler by the burning of fuel:

$$\text{Efficiency} = \frac{(\text{Steam flow})(\text{Net heat value of steam})}{(\text{Fuel flow})(\text{Heating value of fuel})}$$

where:

Net heat value of steam = enthalpy of steam minus enthalpy of feedwater (Btu/lb)

Fuel and steam flow   = rate of flow (gal/hr, lb/hr, cfm/hr, etc.)

Heating value of fuel   = heat content (Btu/gal, Btu/lb, Btu/cu ft, etc.)

Multiply the above answer by 100 to convert the decimal to a percentage. The units used in this equation must be consistent; do not, for example, multiply gal/hr by Btu/lb.

FUEL OIL EFFICIENCY TABLE

Percent Excess Air (%)	Net Stack Temperature (°F)											
	200.	250.	300.	350.	400.	450.	500.	550.	600.	650.	700.	750.
5.	89.8	88.7	87.6	86.5	85.5	84.4	83.3	82.2	81.2	80.1	79.0	77.9
10.	89.6	88.5	87.3	86.2	85.1	84.0	82.9	81.7	80.6	79.5	78.4	77.2
15.	89.4	88.2	87.1	85.9	84.7	83.6	82.4	81.2	80.1	78.9	77.7	76.6
20.	89.2	88.0	86.8	85.6	84.4	83.2	81.9	80.7	79.5	78.3	77.1	75.9
25.	89.0	87.8	86.5	85.3	84.0	82.8	81.5	80.2	79.0	77.7	76.5	75.2
30.	88.9	87.6	86.3	85.0	83.6	82.3	81.0	79.7	78.4	77.1	75.8	74.5
35.	88.7	87.3	86.0	84.6	83.3	81.9	80.6	79.2	77.9	76.5	75.2	73.9
40.	88.5	87.1	85.7	84.3	82.9	81.5	80.1	78.7	77.4	76.0	74.6	73.2
45.	88.3	86.9	85.4	84.0	82.6	81.1	79.7	78.2	76.8	75.4	73.9	72.5
50.	88.1	86.7	85.2	83.7	82.2	80.7	79.2	77.8	76.3	74.8	73.3	71.8
55.	88.0	86.4	84.9	83.4	81.8	80.3	78.8	77.3	75.7	74.2	72.7	71.1
60.	87.8	86.2	84.6	83.1	81.5	79.9	78.3	76.8	75.2	73.6	72.0	70.5
65.	87.6	86.0	84.4	82.7	81.1	79.5	77.9	76.3	74.6	73.0	71.4	69.8
70.	87.4	85.7	84.1	82.4	80.8	79.1	77.4	75.8	74.1	72.4	70.8	69.1
75.	87.2	85.5	83.8	82.1	80.4	78.7	77.0	75.3	73.6	71.8	70.1	68.4
80.	87.0	85.3	83.5	81.8	80.0	78.3	76.5	74.8	73.0	71.3	69.5	67.7
85.	86.9	85.1	83.3	81.5	79.7	77.9	76.1	74.3	72.5	70.7	68.9	67.1
90.	86.7	84.8	83.0	81.2	79.3	77.5	75.6	73.8	71.9	70.1	68.2	66.4
95.	86.5	84.6	82.7	80.8	78.9	77.1	75.2	73.3	71.4	69.5	67.6	65.7
100.	86.3	84.4	82.5	80.5	78.6	76.6	74.7	72.8	70.8	68.9	67.0	65.0

NOTE: May be used to calculate savings from changes in efficiency for No. 2, No. 6 Low Sulfur, and No. 6 (1.5% S) fuel oils. Based on higher heating value (HHV) and total flue gas loss, including latent and sensible losses due to water formed from fuel and sensible loss due to moisture in air. CAUTION: Do not calculate savings due to changes in efficiency using efficiency values from two different sources. The various slide rules, graphs, tables and nomographs available may incorporate different assumptions.

NATURAL GAS EFFICIENCY TABLE

Percent Excess Air (%)	Net Stack Temperature (°F)											
	200.	250.	300.	350.	400.	450.	500.	550.	600.	650.	700.	750.
5.	86.1	85.0	84.0	82.9	81.8	80.7	79.6	78.6	77.5	76.4	75.3	74.3
10.	85.9	84.8	83.7	82.6	81.5	80.3	79.2	78.1	77.0	75.8	74.7	73.6
15.	85.8	84.6	83.4	82.3	81.1	79.9	78.8	77.6	76.4	75.3	74.1	72.9
20.	85.6	84.4	83.2	82.0	80.8	79.5	78.3	77.1	75.9	74.7	73.5	72.3
25.	85.4	84.2	82.9	81.7	80.4	79.1	77.9	76.6	75.4	74.1	72.9	71.6
30.	85.2	83.9	82.6	81.3	80.0	78.7	77.5	76.2	74.9	73.6	72.3	71.0
35.	85.1	83.7	82.4	81.0	79.7	78.4	77.0	75.7	74.3	73.0	71.6	70.3
40.	84.9	83.5	82.1	80.7	79.3	78.0	76.6	75.2	73.8	72.4	71.0	69.6
45.	84.7	83.3	81.9	80.4	79.0	77.6	76.1	74.7	73.3	71.8	70.4	69.0
50.	84.5	83.1	81.6	80.1	78.6	77.2	75.7	74.2	72.7	71.3	69.8	68.3
55.	84.4	82.8	81.3	79.8	78.3	76.8	75.3	73.7	72.2	70.7	69.2	67.7
60.	84.2	82.6	81.1	79.5	77.9	76.4	74.8	73.3	71.7	70.1	68.6	67.0
65.	84.0	82.4	80.8	79.2	77.6	76.0	74.4	72.8	71.2	69.6	68.0	66.3
70.	83.8	82.2	80.5	78.9	77.2	75.6	73.9	72.3	70.6	69.0	67.3	65.7
75.	83.7	82.0	80.3	78.6	76.9	75.2	73.5	71.8	70.1	68.4	66.7	65.0
80.	83.5	81.7	80.0	78.3	76.5	74.8	73.1	71.3	69.6	67.8	66.1	64.4
85.	83.3	81.5	79.7	78.0	76.2	74.4	72.6	70.8	69.1	67.3	65.5	63.7
90.	83.1	81.3	79.5	77.7	75.8	74.0	72.2	70.4	68.5	66.7	64.9	63.1
95.	83.0	81.1	79.2	77.3	75.5	73.6	71.7	69.9	68.0	66.1	64.3	62.4
100.	82.8	80.9	79.0	77.0	75.1	73.2	71.3	69.4	67.5	65.6	63.6	61.7

NOTE: Based on higher heating value (HHV) and total flue gas loss, including latent and sensible losses due to water formed from fuel and sensible loss due to moisture in air. CAUTION: Do not calculate savings due to changes in efficiency using efficiency values from two different sources. The various slide rules, graphs, tables, and nomographs available may incorporate different assumptions.

Figure 4-8. Efficiency tables for natural gas and fuel oil

Example 4-1. A boiler produces 25,000 lb/hr of 150 psi dry saturated steam and consumes 218 gal/hr of low-sulfur No. 6 oil over a one-hour period. The feedwater temperature is 210°F, and the fuel oil supplier states that the oil contains 145,955 Btu/gal. What is the efficiency?

Solution 4-1. Refer to the steam tables in the appendix. Notice that 150 psi saturated steam contains 1,194.1 Btu/lb, and 210°F feedwater contains 178.15 Btu/lb. Plug these numbers into the equation:

$$\text{Efficiency} = \left(\frac{(25{,}000 \text{ lb/hr})(1{,}194.1 \text{ Btu/lb} - 178.15 \text{ Btu/lb})}{(218 \text{ gal/hr})(145{,}955 \text{ Btu/gal})} \right)(100) = 80\%$$

FUEL OILS

Fuel oils contain from 83% to 88% carbon and 6% to 12% hydrogen. Because all fuel oils are so similar in chemical analyses and because the physical properties of oil have a far greater effect on the operation of fuel-burning equipment, the physical properties are commonly measured and specified by producers and users. The physical properties are as follows:

- Specific gravity — the weight of a given volume of fuel oil divided by the weight of an equal volume of water with both measured at the same temperature.

- Degrees API — the industry-wide standard for measuring the specific gravity of fuel oil.

- Viscosity — the measure of the resistance to flow exhibited by a liquid.

- Pour point — the temperature at which an oil starts flowing.

- Flash point — the temperature at which a fuel oil gives off enough vapor to produce a momentary flash when flame is brought near the surface of the oil. The test is performed in a special standard apparatus and under certain definite conditions.

- Fire point — the temperature at which a fuel oil gives off enough vapor to burn continuously.

Specific Gravity

Gravity, like temperature, can be measured on different scales. Before 1921, the gravity of oil was read on the Baumé scale. The API (American Petroleum Institute) scale was then adopted because it is a linear scale on a hydrometer. A hydrometer is an instrument used for measuring specific gravity.

To convert specific gravity to degrees API, use the following formula:

$$\text{Degrees API} = \frac{141.5}{\text{SpGr } 60°/60°F} - 131.5$$

SpGr is the abbreviation for specific gravity. The number 60°/60°F means that both the liquid under evaluation and the reference water were at 60°F.

Degrees API

There following facts can be obtained from the API number:

- The lower the API gravity of an oil the higher the viscosity, carbon content, ash content, weight in pounds per gallon, and amount of heat liberated per gallon when burned. It also contains more sulfur, if the sulfur was not refined out.

- The higher the API gravity, the thinner the oil, lower the sulfur, lower the percentage of carbon, and higher the percentage of hydrogen (resulting in a lower heating valve). The oil will also burn cleaner and will be easier to atomize. Atomization is the process of breaking up a liquid into a large number of tiny droplets.

- The gross heating value of fuel oil can be approximated as follows:

$$\text{Btu/lb} = 17,687 + [(57.7)(\text{API})]$$

Commercial standards have been established for fuel oil. Grades No. 1 and 2 are light and medium domestic fuel oils. Grade No. 4 is commonly used in commercial office buildings and large schools. Grade No. 5 is not commonly used. Grade No. 6, sometimes known as Bunker C oil, is a heavy industrial fuel often used in large central heating plants and power stations. Table 4-1 shows the characteristics of the common grades of oil.

Viscosity

Viscosity is one of the most important characteristics of fuel oil. Unless the viscosity is properly controlled, efficient combustion is impossible. If the oil is too thick, it doesn't atomize properly and therefore doesn't burn efficiently. If it is too thin, the oil may carbonize the fuel oil heater and burner tip, cause sparking in the furnace, and the flame may become unsteady and blow away from the burner tip.

Fuel Grade	Description	Btu/gal	API Gravity	Pour Point
No. 1	Distillate oil intended for vaporizing pot-type burners and other burners requiring this grade	137,207 to 140,521	44-36	Below 0°F
No. 2	Distillate oil for general purpose heating for use in burners not requiring No. 1	141,400 to 143,223	34-30	Below 0°F
No. 4	Preheating not usually required for handling or burning	144,168 to 146,132	28-24	10°F
No. 5	Preheating may be required for burning, and in cold climates may be required for handling	147,168 to 149,275	22-18	30°F
No. 6	Preheating required for burning and handling (over 0.3% sulfur)	150,380 to 152,681	16-12	65°F
No. 6	Low sulfur (0.3%)	About 146,000	About 22.5	90°F

Table 4-1. Characteristics of common oils

The petroleum industry uses the Saybolt viscosimeter to measure viscosity. Again, like temperature, there is more than one scale for measuring viscosity. The two scales used most often in the United States are the Saybolt Seconds Universal for low viscosity oil such as lubricating oil and Saybolt Seconds Furol for heavier fuel oils.

The viscosity test uses a viscosimeter, Figure 4-9. When the sample is heated to the specified temperature, the cork is pulled from the bottom of the tube. Light oils are heated to 100°F. Heavier oils use a larger orifice and are heated to 122°F. The time it takes to fill a 60-milliliter container is recorded as either Saybolt Seconds Universal (SSU) or Saybolt Seconds Furol (SSF).

As the temperature of oil is increased, its viscosity is reduced and it flows more easily. Heavy oils must be heated before being pumped and heated further until

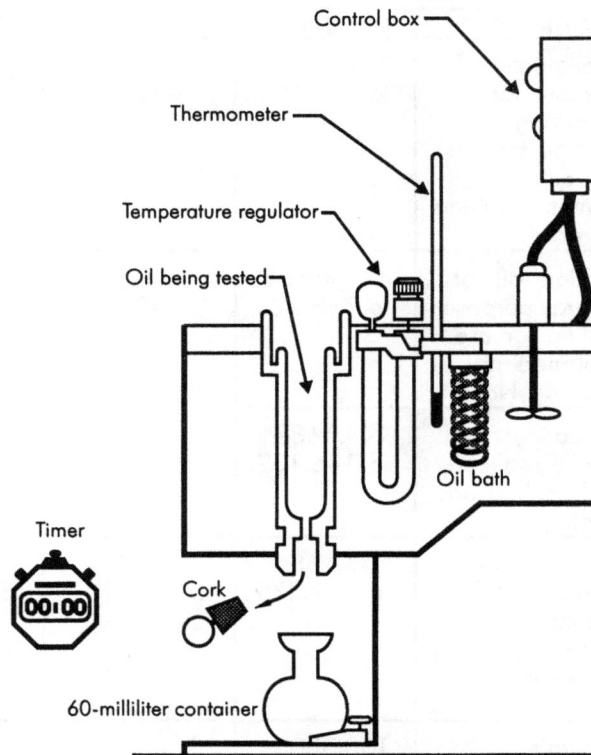

Figure 4-9. Saybolt viscosimeter

the correct viscosity is reached before they are atomized and burned. For easy pumping, oil should not be above 5,000 SSU. Most burner manufacturers recommend an atomization range between 100 and 200 SSU. Consult the burner manufacturer's manual for the correct atomization viscosity. Upon request, fuel oil suppliers can provide the specifications of the product being used. Included in those specifications is the viscosity measured in either SSU or SSF.

The chart in Figure 4-10 provides a method to determine the temperatures for easy pumping and proper atomization.

Pour Point

The pour point test is run by placing a sample of oil in a small corked jar with a thermometer in the top. The sample is heated, then cooled. As the oil cools, the fluidity of the sample is observed at 5°F intervals. When the oil appears to be solid, the temperature is noted and the pour point recorded as 5°F higher than the observed temperature. Oil from different sources has different pour points. Oils derived from an asphalt-based crude may have pour points between 30° and 40°F, while oils from paraffin-based stocks may have pour points of 90°F or higher. (Low sulfur oil is usually paraffin-based.)

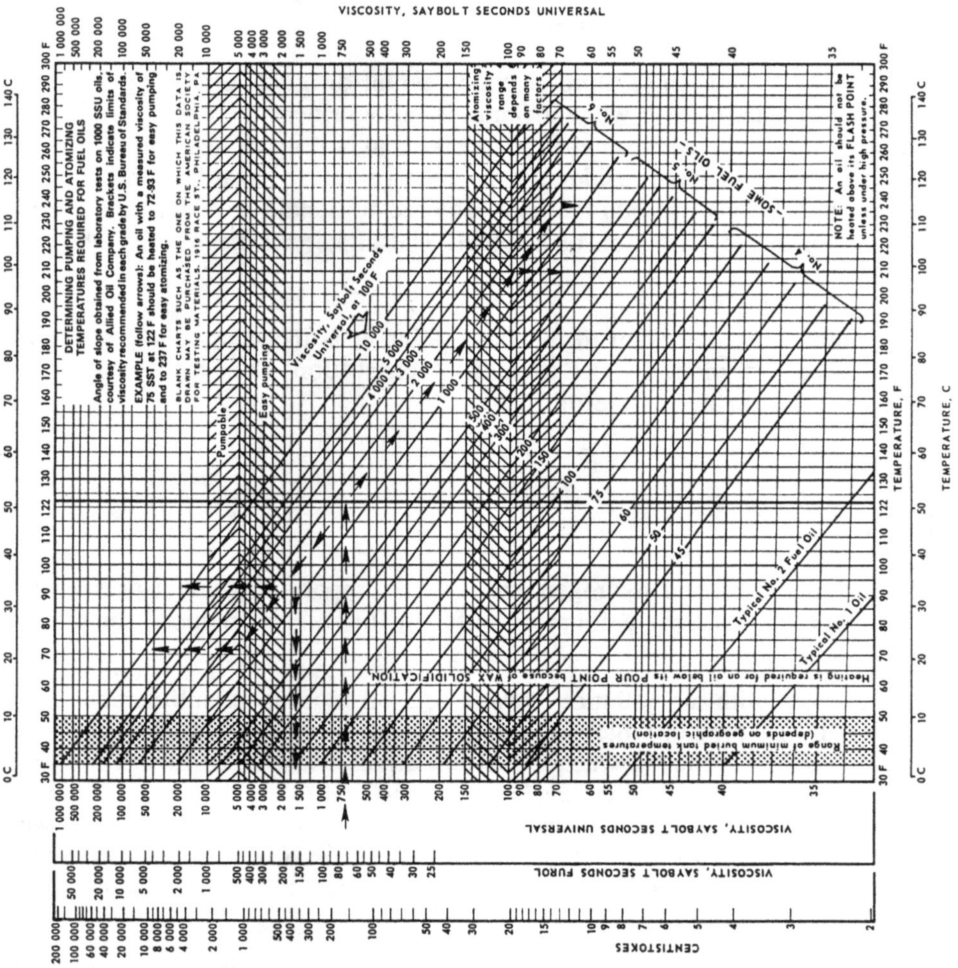

Figure 4-10. Pumping and atomizing temperatures for fuel oils (Courtesy, North American Manufacturing Company)

If the oil temperature falls below the pour point, it generally can't be pumped. This clogs strainers and/or fuel lines. Tanks should be equipped with steam coils, bayonet heaters, or line heaters to keep the oil warm. The storage temperature should be at least 20°F above the pour point. Lines should be insulated and steam- or electrically-traced so the oil can always be pumped. If oil is allowed to cool to its pour point, startup becomes very difficult.

Flash Point and Fire Point

All oils evaporate while standing. At normal temperatures, evaporation usually occurs so slowly that an explosive mixture does not occur. By adding heat, the evaporation rate of the sample can be accelerated to a point where the hydrocarbon-to-air ratio supports combustion.

The flash point is the temperature at which a fuel oil gives off enough vapor to produce a momentary flash when a flame is brought near the surface of the oil. The flash point test is performed in a special standard apparatus and under strict conditions. The test serves to identify the relative safety hazards of different products. Flash points indicate the care required to handle and store the product and to control the safe ignition of the fuel. The higher the flash point, the safer the product.

The fire point is an extension of the flash point. It is the temperature at which a fuel oil gives off enough vapor to sustain combustion when a flame is brought near the surface of the oil.

Sulfur

In addition to carbon and hydrogen, sulfur is an important element present in oil. It is important that the amount of sulfur in fuel oils be reduced for two reasons:

- To comply with air pollution laws, a fuel oil must meet sulfur requirements in the area in which it is used. Health hazards are attributed to the sulfur oxides produced when a fuel is burned. Most areas of the country prescribe maximum sulfur levels in fuels to control sulfur oxide emissions. Normally, the higher the industry and population concentration, the lower the allowable sulfur content. For example, in the New York metropolitan area, the maximum amount of sulfur allowed in fuel oil is 0.3% by weight.

- The sulfur oxides and moisture produced during combustion can combine and form sulfuric acid. The resulting corrosive attack on tubes and other metal surfaces causes serious damage. The lower the sulfur level in a fuel, the fewer corrosion problems.

Sediment and Water

Sediment and water are serious contaminants that occur in fuel oil systems. Often called BS&W (bottom sediment and water), they are found in varying amounts and compositions. Some of the causes of BS&W are as follows:

- Improperly sealed manholes, gauges, or vent lines that allow water to seep in or permit outside contaminants to enter
- Chemical reactions in the oil that form sludge

- Excessive tank heating that hastens sludge-forming chemical reactions
- Condensation of water
- Leaking steam coils that form rust and sludge
- Improper mixing of different grades and types of oils that cause separation or precipitation of materials
- Settling of the small amount of BS&W inherent in the oil

These are some difficulties caused by BS&W:

- Shutdown of the boiler due to blockage of burner tips and strainers
- Erratic and unsteady combustion, including:

    - sparking and spitting of the flame

    - flashback of the flame

    - carbon buildup

    - clinker buildup (formation of a mass of unburned carbon) in the combustion chamber

- Loss of heat release

It is good practice to take a bottom sample of storage tanks annually. If excessive BS&W is apparent, determine its source and have a reputable contractor remove the sludge. If the contractor gets caught dumping the sludge in the nearest sewer, you can be held liable.

GAS

Natural gas is easy to burn for three reasons:

- It already exists in a gaseous state and does not have to be atomized and vaporized.
- Storage is not a problem, because a utility delivers it by pipe.
- It has few or no contaminants (e.g., sulfur).

Lighter than air and odorless, natural gas is mostly methane (CH_6) with smaller quantities of other hydrocarbons such as ethane (C_2H_6). Combustible odorless gases are very dangerous because leaks can go undetected. Therefore, an odorant is added for safety.

Since natural gas has a higher ratio of hydrogen to carbon than oil, its efficiency is lower because of the increased amount of moisture produced by the burning of hydrogen. During cold days you might have noticed white "smoke" coming from a power plant's stack. This is water vapor condensing from the burning of hydrogen.

Natural gas is metered by the cubic foot and billed by the therm. A therm is 100,000 Btu. The gross heating value of natural gas is usually just over 1,000 Btu/cu ft. Propane, otherwise known as LP gas or liquefied petroleum gas, is a member of the same hydrocarbon family as natural gas. It is colorless and nontoxic, and like natural gas, it has an odorant added for safety. Unlike natural gas, LP gas is heavier than air, and instead of dispersing, it accumulates in low areas. These accumulations have been known to explode.

The main reason to liquefy propane is to reduce the volume required for storage. One cubic foot of liquid propane yields over 270 cu ft of gas. Note that the boiling point of propane is 44°F at atmospheric pressure. To keep propane liquid at room temperature (70°F), it must be pressurized to at least 120 psi.

COAL

Coal is our nation's greatest hydrocarbon source. Because the composition of coal varies so much (even samples from the same mine), it is often necessary to sample and analyze it often to make adjustments for satisfactory combustion. The two common methods of analysis are proximate and ultimate.

Proximate analysis is done on site and is used to measure coal quality. It determines the percentages of moisture, volatile matter, fixed carbon, ash, heating value, and sometimes sulfur. *Ultimate analysis* is required for heat balance calculations and the determination of theoretical air requirements. It provides the chemical composition by determining the amount of carbon, hydrogen, sulfur, oxygen, nitrogen, ash, and moisture in the fuel. The test is usually done in a chemical laboratory. Perhaps the acronym **NO CASH** will help you remember what the ultimate analysis consists of: **n**itrogen, **o**xygen, **c**arbon, **s**ulfur, and **h**ydrogen.

Coal is ranked to provide data for specifying burning and handling equipment. The major classifications are anthracite, bituminous, subbituminous, and lignite:

- Anthracite (hard coal) burns with a short blue flame (primarily used for heating homes). It contains about 84% fixed carbon and 13,300 Btu/lb.

- Bituminous (soft coal) contains about 78% fixed carbon and 13,900 Btu/lb.

- Subbituminous contains about 55% fixed carbon and 9,300 Btu/lb.

- Lignite contains about 42% fixed carbon and 7,000 Btu/lb. It is not worthwhile to transport lignite far, so plants that use it are typically near the mine.

The ash-softening temperature is also an important characteristic of coal. When heated, ash becomes soft and sticky, then eventually becomes fluid. If the ash softens and fuses at low temperatures, clinkers form in the fuel bed and slag is deposited on boiler tubes. The ash-softening temperature is determined by placing a cone made of the ash in a small furnace. As the furnace temperature is increased, the ash cone is observed. The ash-softening temperature is defined as the temperature at which the cone is fused down to a spherical lump.

POLLUTION

The Clean Air Act sets maximum levels for six pollutants: carbon monoxide (CO), lead (Pb), nitrogen oxides (NO_x), ozone (O_3), particulate matter (PM), and sulfur oxides (SO_x). Boiler operators must worry most about NO_x.

One way to control NO_x emissions is to lower the flame temperature. This can be done with low NO_x burners and flue gas recirculation. Low NO_x burners reduce temperature by enlarging the flame in a staged combustion process so either fuel or air is introduced at more than one point. This reduces available oxygen and creates partial combustion in several zones over a larger area. A large, long furnace is required to accommodate the larger flame. This method can reduce NO_x emissions by 25% to 50%.

Flue gas recirculation reduces flame temperature by returning products of combustion back to the burner and flame. It also reduces the available oxygen sufficiently to lower NO_x levels. This method can reduce NO_x by up to 75%. Problems occur if the flue gas gets too cool, because SO_x and NO_x can condense into corrosive sulfuric and nitric acids.

A more costly but more effective way to reduce NO_x is by selective noncatalytic reduction (SNCR). Especially good for oil, a chemical reagent (ammonia or urea) is injected into the boiler at critical points where the temperature makes the chemical reaction most efficient. Sophisticated automated control is required for consistent results. Capital investment is low, but the cost of the injected chemical is high. The NO_x removal rate can be as high as 90%, but 50% to 70% is more common.

Then there is selective catalytic reduction (SCR), which is used for both gas and oil. Ammonia is vaporized and mixed with the flue gas, then passed through a catalyst bed. The chemical reaction converts the NO_x to harmless nitrogen and oxygen. Its typical removal rate is 80% to 90%, but the control must be even more precise than the selective noncatalytic method. This is the most costly way to control NO_x, because the expensive catalyst has to be replaced periodically.

CHAPTER FIVE

Fuel-Burning Equipment

A practical fuel-burning system must perform the following functions:

- Start and maintain ignition
- Mix and proportion the fuel and air
- Vaporize solid and liquid fuels
- Safely supply the proper amount of heat required for the process
- Position flames for useful heat release

All fuel systems must comply with the rules and regulations of the National Board of Fire Underwriters and with local codes. Consult your local fire or code officials to make sure your installation is in compliance. In addition, all tanks and fill stations must comply with spill- and leak-containment rules. For example, underground tanks must have double walls or be installed in concrete vaults, and aboveground tanks must be surrounded by a berm to contain spills.

GAS BURNERS

Gas burners are classified either as premix or nozzle mix according to the manner in which the gas and air are brought together. Most boiler burners are the nozzle-mix type.

In premix gas systems, air and gas are mixed at some point upstream from the burner port. The burner itself serves only to maintain the flame in the desired location.

The inspirator type of premix system is used on most domestic gas-burning appliances. It uses the energy in the gas flow to draw air into the burner. On the other hand, the aspirator type of premix system uses a blower to force air through a venturi so that low pressure at the throat of the venturi draws gas into the air stream in proportion to the airflow.

Most boilers have nozzle-mix burners for the following reasons:

- Premix burners usually have limited turndown.
- Premix burners cannot be adapted to dual-fuel configurations as easily as nozzle-mix types.
- Nozzle-mix burners can be designed for a great variety of flame shapes.
- Large manifolds of premixed air and gas pose a serious danger.

The term *turndown* is used when describing the characteristics of a burner. It is the ratio of maximum to minimum firing rates. For example, if a burner has a maximum output of 1,000,000 Btu at high fire and a minimum output of 200,000 Btu at low fire, it would have a 5 to 1 turndown ratio.

Like oil burners, there is no mixing of gas and air in nozzle burners until the gas leaves the nozzle orifices. Boiler burners are designed to inject a stream of hot incandescent combustion gases into a furnace space where heat is transferred to the load by means of radiation. After the hot combustion gases leave the furnace area, the heat that remains in the gases is transferred by convection.

OIL BURNERS

Liquid fuels must be vaporized before they can be burned. Some small capacity burners such as blowtorches, gasoline stoves, and wick-type kerosene burners vaporize liquid fuel in a single step by direct heating. Larger oil burners must employ a two-step process — atomization and vaporization.

As explained in the previous chapter, atomization is the process of breaking up a liquid into many tiny droplets. *Vaporization* is the process of converting a liquid or solid into a gas, usually by the application of heat.

By first atomizing the liquid fuel, a large surface area consisting of millions of tiny droplets is exposed to air and heat. This allows oil to be vaporized at a very high rate. Satisfactory vaporization requires the following conditions:

- A large volume of air must be thoroughly mixed with oil particles.
- Airflow must be turbulent (varying at a given point in speed and direction) and moving at high speeds.
- Heat from the flame must be radiated into the incoming spray (a function of the burner tile).

The burner brick protects the burner from heat damage, while providing a source of radiated heat for vaporization. The amount of heat can be appreciated by observing that the burner tile refractory is so hot it is incandescent just after a burner is shut down. The first two requirements listed above are met by blowing air through oil or by passing oil through air.

Atomization is successfully accomplished by using one of several different methods. Mechanical atomizing burners use the pressure of the fuel itself. When oil is forced through a small orifice, it breaks into a spray of fine droplets. This type of burner is used for home heating furnaces that produce 100,000 Btuh at 75 psi, to large electric utility boilers that produce 180 million Btuh at 1,000 psi. (Btuh is an abbreviation for British thermal units per hour.) An example of a mechanical burner is shown in Figure 5-1.

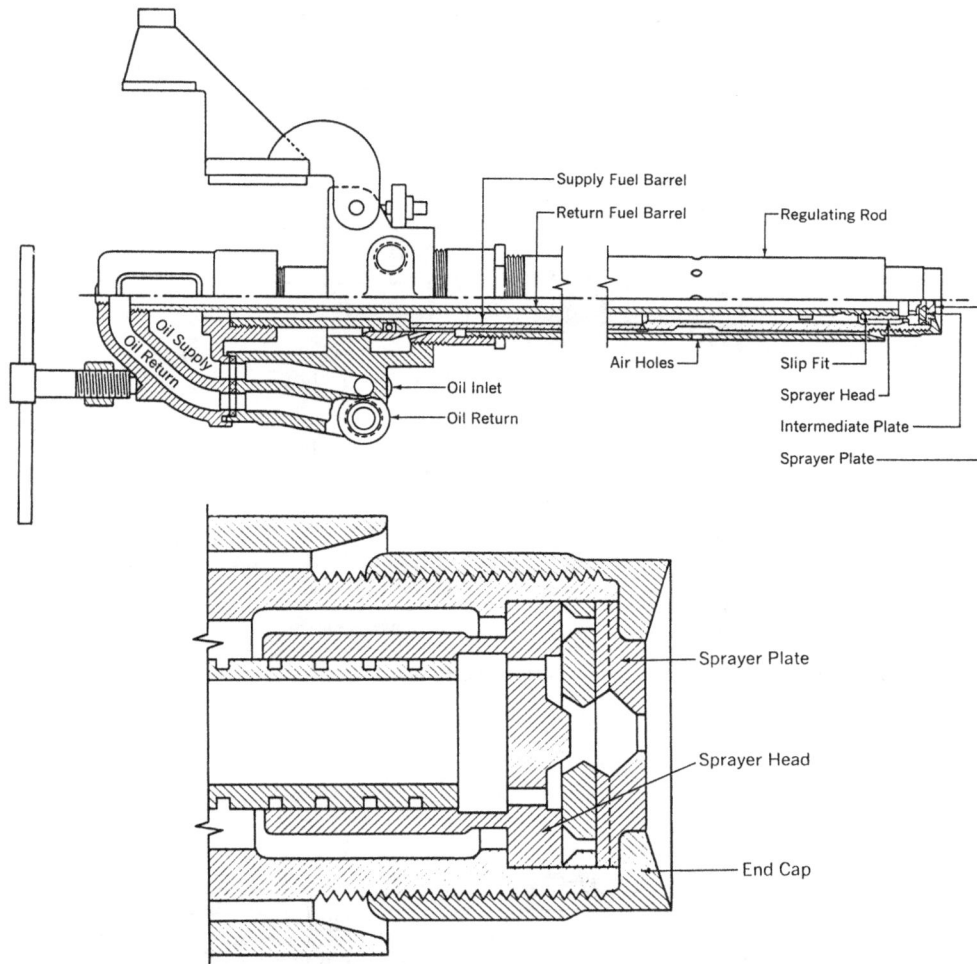

Figure 5-1. Mechanical return-flow oil atomizer assembly and detail (Courtesy, Babcock & Wilcox Company)

The high pressure required to make mechanical atomizing work can be a problem. In one installation, fuel sprayed into the boiler room because of a defective oil-heater gasket. Naturally this oil found a source of ignition, and the resulting fire shut down the facility for days. All equipment such as hoses and fittings have to be kept in top-notch shape.

Low-pressure air atomizing burners are the most popular on packaged boilers, Figure 5-2. The 1- or 2-psi atomizing air is supplied by a blower driven either from its own motor or the forced draft fan motor at a rate of 150 cu ft per gallon of oil. These burners have no trouble burning light or heavy oils as long as the viscosity is between 100 and 150 SSU.

High-pressure air or steam atomizing oil burners use steam or compressed air to tear droplets from the oil stream and propel them into the combustion space, Figure 5-3. The high velocity of the oil particles (relative to the air) produces the action required for quick vaporization. These burners can atomize light to very heavy oils, sludge, or pitch and are used for the incineration of liquid wastes. They are available in sizes that produce up to 300 million Btuh.

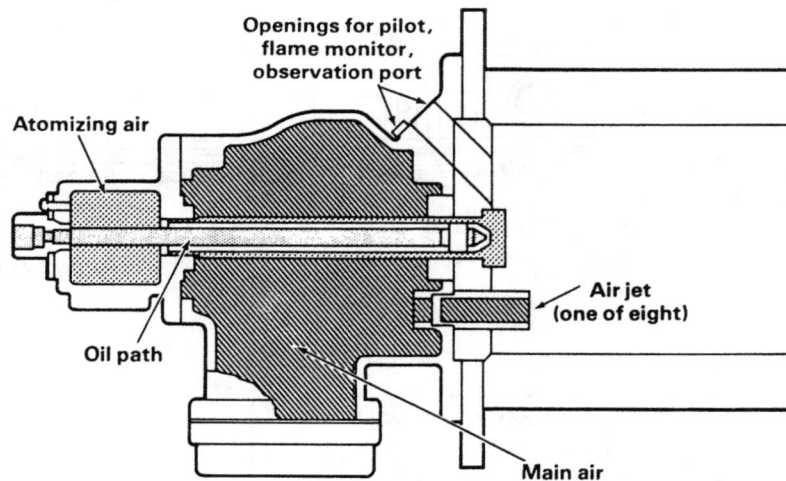

Figure 5-2. Low-pressure air atomizing burner (Courtesy, North American Manufacturing Company)

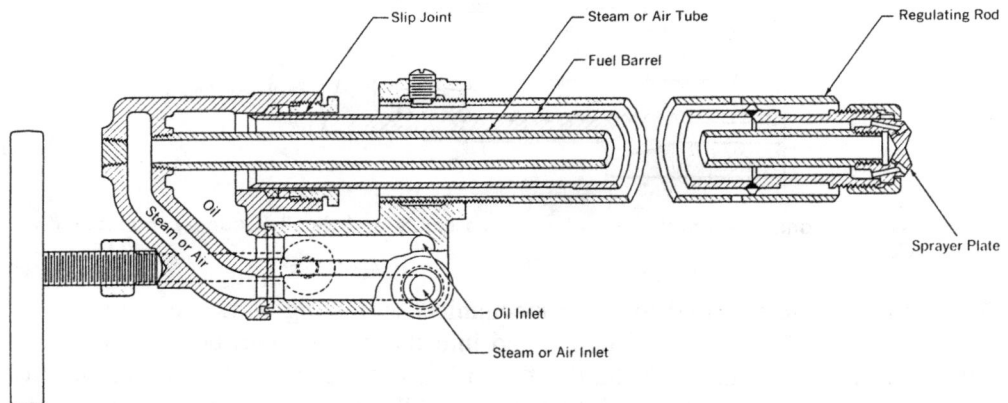

Figure 5-3. High-pressure air or steam atomizing oil burner (Courtesy, Babcock & Wilcox Company)

The required oil pressure is much lower for high-pressure air or steam atomizing burners, which have a maximum of 150 psi (mechanical atomizers have a maximum of 300 psi). A significant advantage of the air or steam design is that it can be turned down to 20% of its rated capacity. A disadvantage of the steam atomizer is its steam consumption. Even at 0.02 lb of steam per pound of oil burned, it represents a sizable loss. Despite this loss, the cost of using compressed air as a substitute is not always justifiable.

Though rotary cup burners are obsolete, they may still appear on licensing exams. As shown in Figure 5-4, a hollow, rotating shaft runs the length of the burner. Along this shaft are a high-speed motor, a gear to drive the oil pump, the atomizing air fan, and the atomizing cup. Oil is forced through the hollow shaft to the inside of the cup. Atomizing air is administered through the angular space around the rotating cup. Centrifugal force throws oil from the lip of the rotating cup in the shape of a conical sheet. This quickly breaks into a spray when hit with the atomizing air.

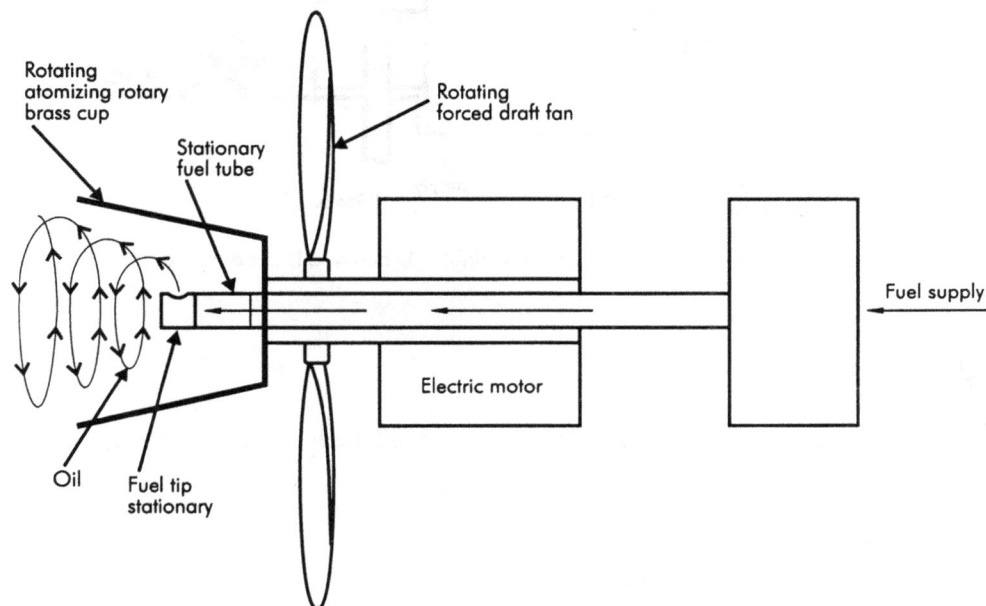

Figure 5-4. Direct drive rotary cup oil burner

Advantages of this design are that only a modest pressure is required to supply oil to the cup, and no high-pressure steam or air is required for atomization. This is a great benefit to low-pressure steam plants. However, since this burner has an electric motor and moving mechanical parts close to the furnace, they have lost favor due to the maintenance required to keep them operational. Newer designs are more efficient. Low-pressure air atomizing burners have replaced the rotary cup for small packaged boilers.

COMBINATION BURNERS

Burners can be designed to burn more than one type of fuel so that the lowest-priced or most-available fuel can be used. Figure 5-5 shows a combination burner that can burn light oil, heavy oil, propane, or gas.

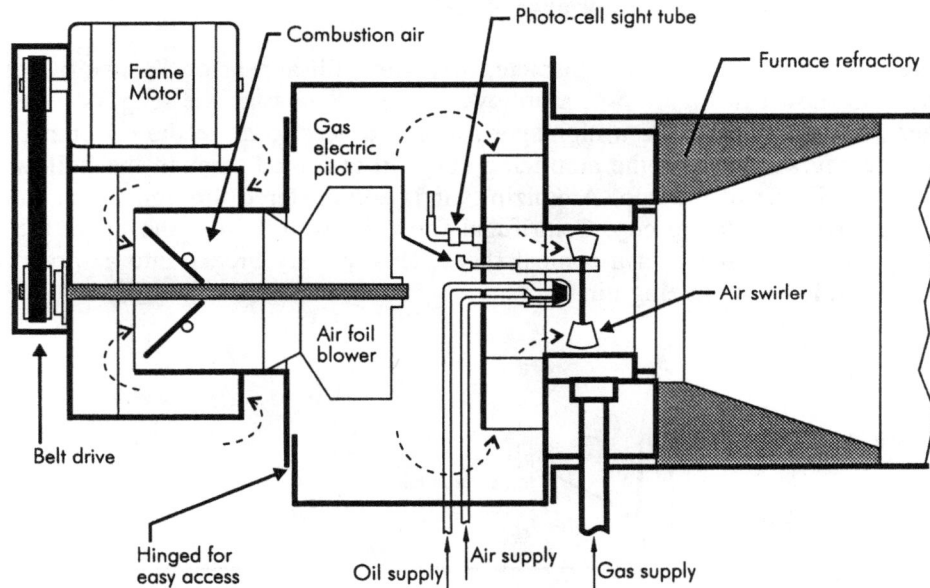

Figure 5-5. Combination burner that can burn light oil, heavy oil, propane, or gas

STOKERS

Stokers are mechanical devices used to burn solid fuels. They are designed to:

- feed fuel (continuously or intermittent);
- sustain combustion;
- supply adequate air for combustion;
- allow free passage of flue gas;
- allow the removal of noncombustible material (ash) from the furnace.

The three types of automatic stokers are underfeed, crossfeed, and overfeed.

Underfeed

The underfeed stoker consists of a trough, also called a retort, into which coal is pushed either by a ram or feed screw, Figure 5-6. Combustion air is introduced under the fuel bed through tuyeres or grate bars. The movement of the fuel caused by the feeding action discourages the formation of large clinkers. The volatile gas distilled from the coal burns above the incandescent fuel bed. The fixed carbon part of the coal is burned on the grate bars. As new coal is fed into the retort, ash is pushed aside and falls into the ash pit.

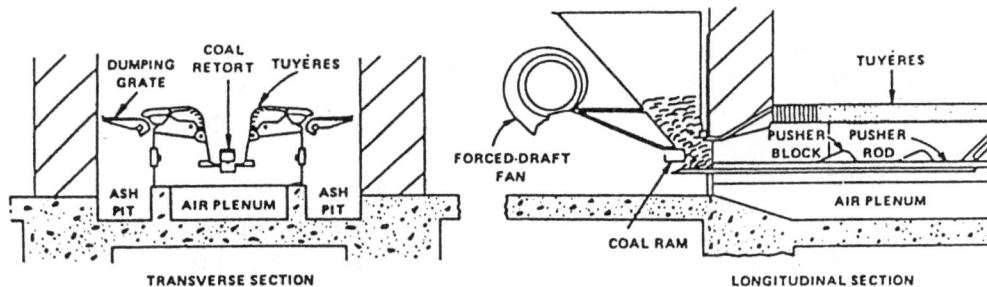

Figure 5-6. Underfeed stoker (Reprinted courtesy American Society of Heating, Refrigerating and Air-Conditioning Engineers)

The underfeed stoker is a good choice for small installations. It is simple to operate (all adjustments are made from the stoker front), and most of the fuel bed is visible and accessible. The ram was formerly driven by a steam cylinder and piston, but on newer designs, electric and hydraulic drives are also used.

Crossfeed

The crossfeed (traveling grate) stoker can burn many solid fuels, including garbage and municipal refuse. The typical traveling grate is comparable to a wide bulldozer drive track. Coal or other solid fuel is fed by a hopper in front of the stoker. The firing rate is regulated by the depth of the fuel on the grate, the speed of the grate, and the amount of combustion air supplied.

The function of the traveling grate is to carry the fuel from the gravity hopper, support the fuel during combustion, and then dump the ash into the ash pit. There are two types of grates:

- Chain grates, which consist of a series of links strung on rods in a staggered arrangement, Figure 5-7

- Bar grates, which consist of rows of bars

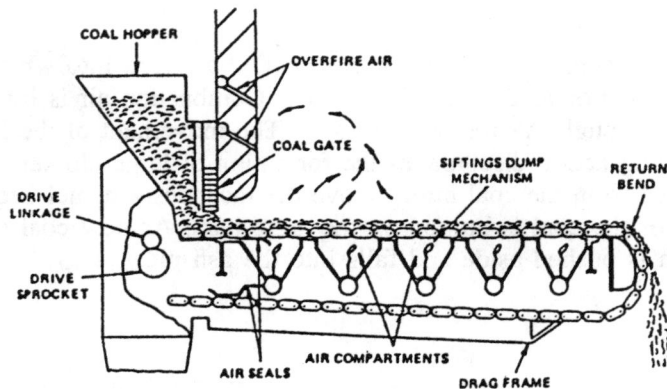

Figure 5-7. Crossfeed (traveling grate) stoker (Reprinted courtesy American Society of Heating, Refrigerating and Air-Conditioning Engineers)

Overfeed

The overfeed (spreader) stoker looks like a traveling grate stoker, but instead of feeding the fuel from a gravity hopper, a mechanism is used to throw the coal into the furnace and onto the grate, Figure 5-8. Part of the volatile matter and fine particles are burned in suspension, and the remainder is burned on the grate. Practically all types of coal can be burned on a spreader stoker.

The overfeed stoker is most commonly found in industrial installations, because it is very flexible in responding to load fluctuations. Also, the firing rate of the fuel is almost instantaneous.

Pulverized Coal Firing

Coal can be pulverized to the consistency of talcum powder and blown into the furnace. Like oil, all the fuel is burned in suspension. The resulting coal flame even looks like an oil flame.

The advantage of pulverized firing is that rapid combustion occurs when a large surface area of coal is exposed to oxygen and heat, Figure 5-9. Since the development of pulverized coal firing in the 1920s, boiler capacity has increased from a few hundred thousand pounds of steam per hour to 6.5 million lb of steam per hour.

Pulverized coal firing is limited to large installations due to its high capital and operating costs. Although coal is the least expensive fossil fuel, storage and handling equipment are expensive. The problem of minimizing air pollution is also costly. Particulate emissions must be captured by mechanical dust collectors, electrostatic precipitators, or bag houses. Sulfur dioxide (SO_2) discharges

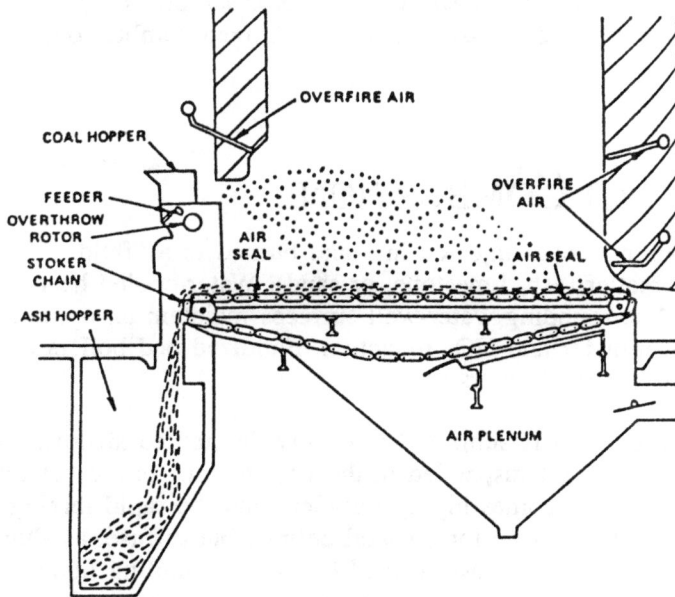

Figure 5-8. Overfeed (spreader) stoker (Reprinted courtesy American Society of Heating, Refrigerating and Air-Conditioning Engineers)

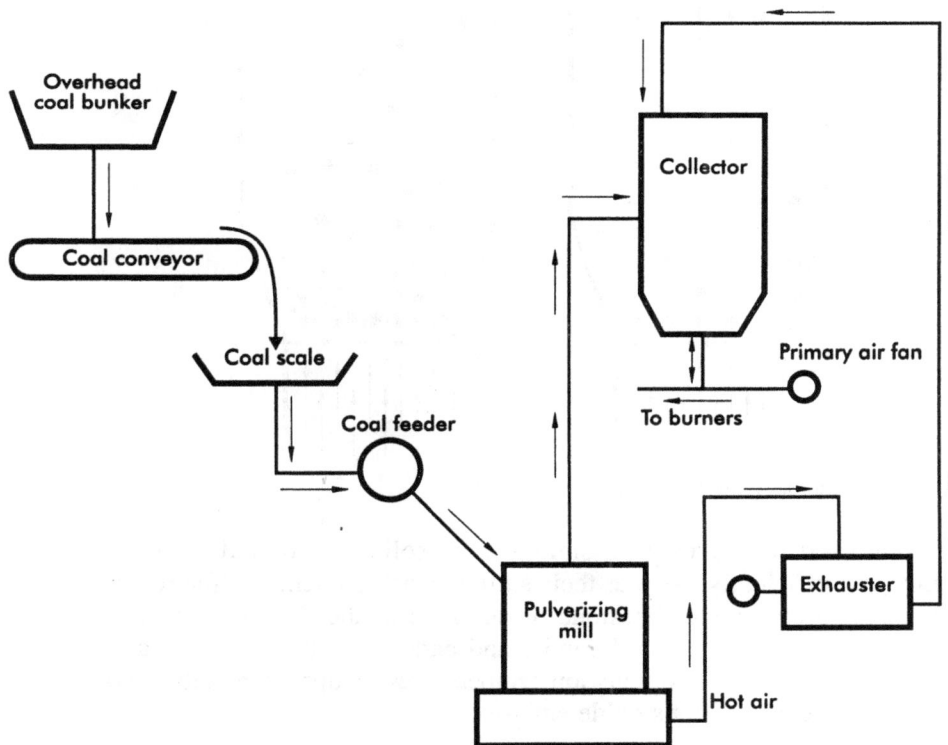

Figure 5-9. Pulverized coal firing

must be kept in check by using low sulfur fuel and/or scrubbers. Nitrogen oxides (NO_x) must be reduced by using a catalyst, staged combustion, or fluidized bed combustion.

FLUIDIZED BED COMBUSTION

In fluidized combustion, fuel is injected into and burned in a "fluidized bed" of particles of mineral matter (ash, limestone, or sand). Water-cooled heat-transfer tubes in the bed and surrounding water-wall surfaces maintain an optimum temperature for heat transfer and NO_x reduction. Fluidized bed boilers are made in both fire-tube and water-tube versions.

When a bed of inert material is subjected to evenly distributed airflow, the material is forced upward and suspended in the gas stream. As the velocity of the gas increases, the bed becomes highly turbulent and the rapid mixing of particles occurs. The surface is no longer well defined but diffuse, as shown in Figure 5-10. Bubbles, similar to those formed by gas in a boiling liquid, rise through the bed. A bed in this state is said to be fluidized; that is, it behaves like a liquid, finding its own level and possessing hydrostatic head.

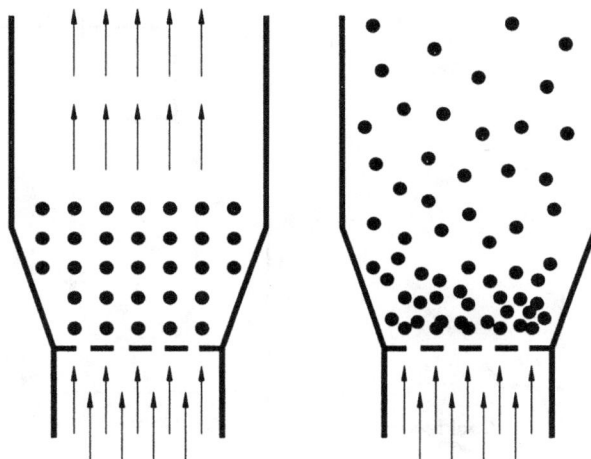

Figure 5-10. Fluidized bed combustion

This characteristic makes it possible to burn solid, liquid, and gaseous fuels cleanly and efficiently, despite their sulfur or ash content. Sulfur reduction is accomplished by adding limestone or dolomite to the fluid bed. The sulfur reacts with the limestone or dolomite and ends up in the ash instead of leaving with the flue gas. Low combustion temperatures of approximately 1,600°F significantly reduce nitrous oxide emissions.

OIL STORAGE TANKS

Suction line difficulties outnumber all other oil system problems. To avoid drawing water, dirt, and sediment into the piping system, the suction line inlet should be at least four inches above the bottom of the tank for heavy oil and two inches above the bottom for light oil.

A gauge or measuring well is usually provided to sound the tank (sounding is the process of measuring the depth of the tank's contents), Figure 5-11. Directly below the gauge well opening on the tank bottom, a doubling plate should be installed to protect the shell against the repeated impact of the sounding tape plumb bob. Though innocent looking, these plumb bobs have been known to punch holes in tanks.

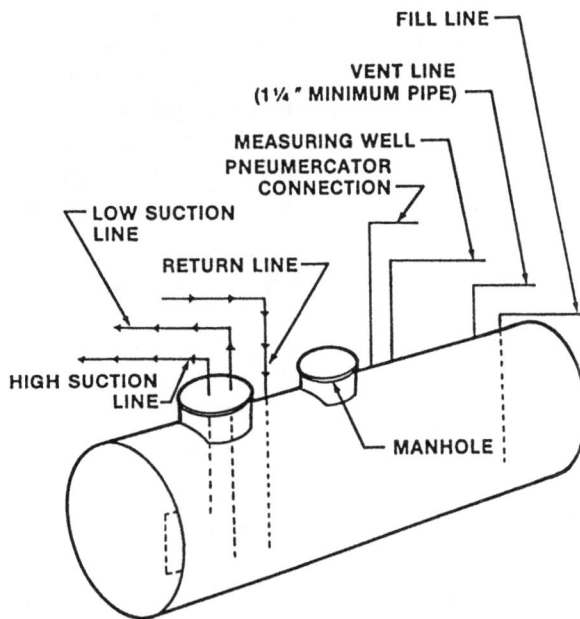

Figure 5-11. Fuel oil storage tank (Courtesy, Babcock & Wilcox Company)

Sounding a tank is never a clean job, because oil sticks to the tape, especially large tanks filled with heavy oil. To make this easier, measure the tank's *ullage* instead of the liquid depth. Ullage is the amount of liquid removed from a tank, or the amount of oil that a container is lacking. In this case, the ullage is measured from the top of the tank to oil's surface. Capacity tables are available or can be calculated to yield a tank's contents from the ullage measurement.

Remote tank level indicators are also available and are quite accurate. Through the years bubbler systems have become very popular, and good electronic systems are also on the market. However, if any doubt exists, sound the tank with a tape. Before each fill, determine the exact tank level; trying to add 6,000 gal to a tank with an available capacity of only 5,000 gal is courting disaster.

Other requirements for oil storage tanks include the following:

- The vent line should be visible from the fill station. If the tank is filled from a pump, the vent line must be as large as the pump's discharge. The lower end on the vent line should not extend more than one inch below the uppermost point of the tank and should not be connected with any other line.

- Manholes are required for inspection and repair access on tanks over 1,000 gal. Manhole covers are bolted and gasketed and should not be used for sounding or for the connection of vent or return lines.

- Depending on the grade of oil and the climate, tank heaters might be required to keep the oil temperature about 20°F above its pour point. The additional heat keeps the oil at a temperature where it can be pumped, prevents the separation of blends, and retards sludge formation. However, heating the oil above 150°F may result in the distillation of light ends such as gasoline and kerosene. Figure 5-12 shows an electric suction heater. With this arrangement, oil returned to the bell or open bottom shortens warmup time and minimizes oil heater energy consumption.

- The purpose of the return line is to route unused circulating oil back to the tank. It should have a liquid seal trap so it can't act as a vent for oil vapors when the oil level is low.

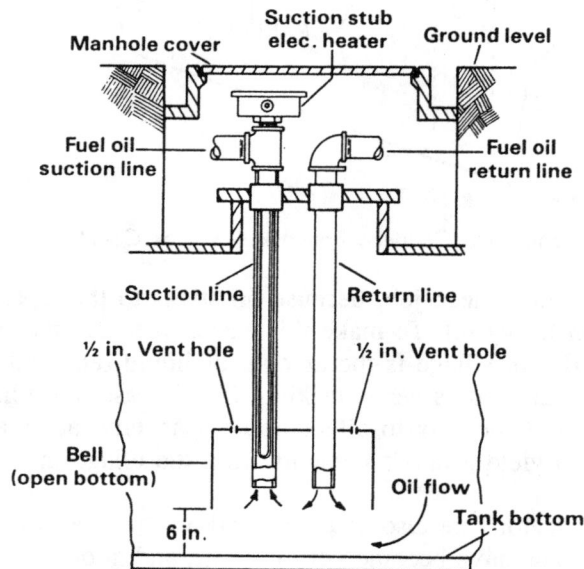

Figure 5-12. Electric suction heater (Courtesy, North American Manufacturing Co.)

- The fill line should either extend below the level of the suction inlet, or it should contain a trap so that it can't act as a vent. Fill terminals should not be located inside buildings and should never be connected to any other line.

Notice the placement of the temperature and pressure gauges in the fuel oil system in Figure 5-13. The thermometer on the suction line monitors the tank heater's operation. The two vacuum gauges on either side of the oil filters alert the operator if there is a problem in the suction line, such as a plugged tank vent.

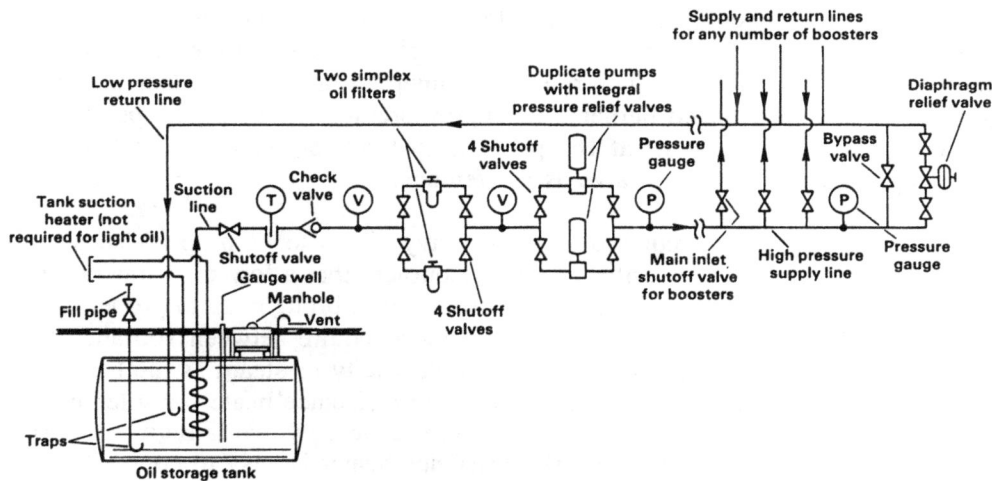

Figure 5-13. Fuel oil system (Courtesy, North American Manufacturing Co.)

The strainers prevent dirt and scale from entering the system. A 20- or 30-mesh basket is usually used. A suction line may have two strainers valved in parallel. With this arrangement, one strainer can be taken out of service and cleaned without shutting down the system. Duplex strainers have the two strainers and isolating valves built into one body. They have one inlet and outlet, and compared to two separate strainers are easier to pipe and take up less room.

Fuel oil pumps are also usually installed in pairs so that the installed standby is ready to go in case the on-line pump fails or needs maintenance. It is common for the control circuit to include a pressure switch so the installed standby will kick on automatically if the on-line pump fails.

As a rule, each pump is sized to 1.5 times the maximum fuel consumption rate. When pumps are installed in parallel, each discharge line must have a check valve to prevent backflow through the out-of-service unit. The pump discharge pressure gauge monitors the pump's performance.

Fuel oil pumps are positive-displacement pumps that deliver a constant volume of fuel regardless of demand or pressure. Therefore, all positive-displacement pumps must be equipped with relief valves to prevent excess pressure and system damage. These spring-loaded relief valves can either be built into the pump or piped between the pump's discharge and return lines. The built-in type is easy to install but cannot protect the pump from damage over a long period.

The temperature of the oil increases slightly as it passes through a pump. As this oil is circulated again and again through the relief valve, the oil eventually gets hot enough to damage the pump. External relief valves do not pose this problem, because they must be connected directly back to the return line with no shutoff valves in between.

Pump relief valves should never be used to regulate the supply-line pressure. System pressure is regulated by the relief valve placed between the supply and return lines at the end of the pipe run. In commercial or industrial installations using air or steam-atomized burners, the supply pressure is set at 135 psi, and the pump relief valves are set at 165 psi. The system's pressure is adjusted using the gauge by the relief valve as a reference.

Some fuel oil heaters are required to bring heavy oil up to its proper viscosity at the boiler. Viscosity is so critical on some burners that a low oil temperature switch is installed to prevent operation until the oil is brought up to proper temperature. (Consult the burner manual, but most require between 100 and 150 SSU.) The heater in this type of system uses electricity or steam as the heat source. The thermostat either controls an electric resistance heater or solenoid steam valve to regulate temperature. A pressure-reducing valve is required when using high-pressure steam. The electric resistance heater is easy to control, but steam is cheaper to operate.

Do not return the condensate from oil heaters to the feedwater system. Discharge the condensate in a waste line. As valuable as condensate is, it is not worth the risk of having oil-contaminated feedwater.

Gas Systems

Figure 5-14 shows a typical gas train. A train refers to all the valves, gauges, and controls arranged to perform a specific task. This train controls the flow of gas to the burner.

Both the main burner and pilot gas lines are equipped with a block-and-bleed arrangement. There are two normally closed (NC) shutoff valves in each line with a normally open (NO) solenoid vent valve between them. The idea of this arrangement is to have two valves in series to ensure good shutoff. Any gas that gets past the first valve would escape to the atmosphere through the normally open vent valve. This safeguard prevents gas from accumulating in the

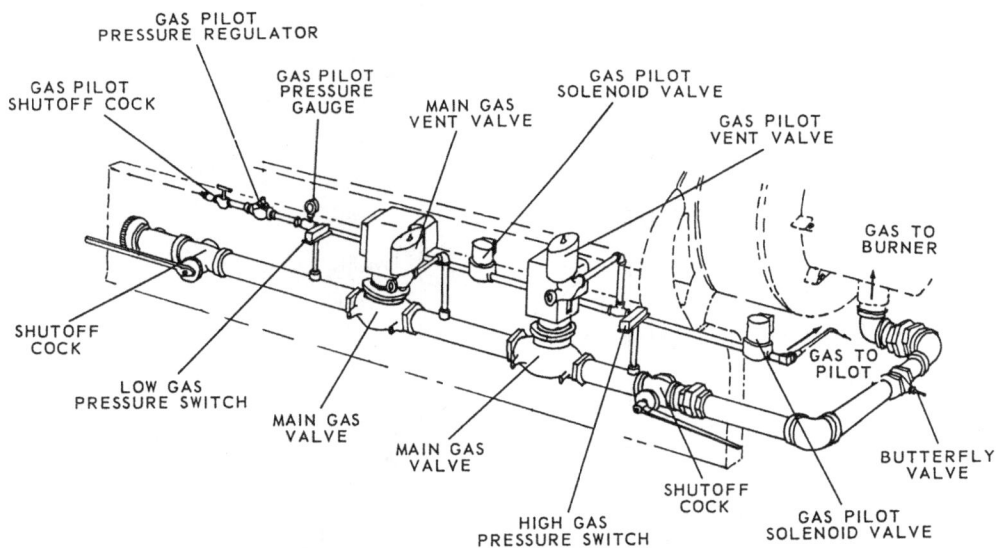

Figure 5-14. Train controlling gas flow to burner (Courtesy, Cleaver Brooks)

boiler and boiler room. When gas is called for, the two shutoff valves open and the vent closes.

The gas pilot pressure regulator reduces the incoming gas pressure to between 5 and 10 in. w.c. The low and high gas pressure switches either prevent the burner from starting, or if the burner is operating, they shut the burner off when the supply gas pressure is not within allowable limits.

The test cocks and second manual shutoff cock are used to test the tightness of the main gas valves and the solenoid vent valve. The butterfly valve is driven by the firing rate modulating motor and regulates the amount of fuel going to the burner.

If interruptible gas is used or the contract with your gas supplier requires an alternate fuel during periods of high gas demand, the gas pilot line must be connected to the uninterruptible gas meter. This is because pilot gas is still required, and any reading on the interruptible meter during the peak periods might result in a severe penalty.

COAL SYSTEMS

In industrial plants, coal is delivered by truck or rail. The system shown in Figure 5-15 can be used for both systems. A bucket elevator unloads the hopper, which is the funnel-shaped receptacle for delivering coal. The silo is arranged with an internal shelf that provides maximum use of storage space. As the coal is removed from above the shelf, reserve storage can be reclaimed from

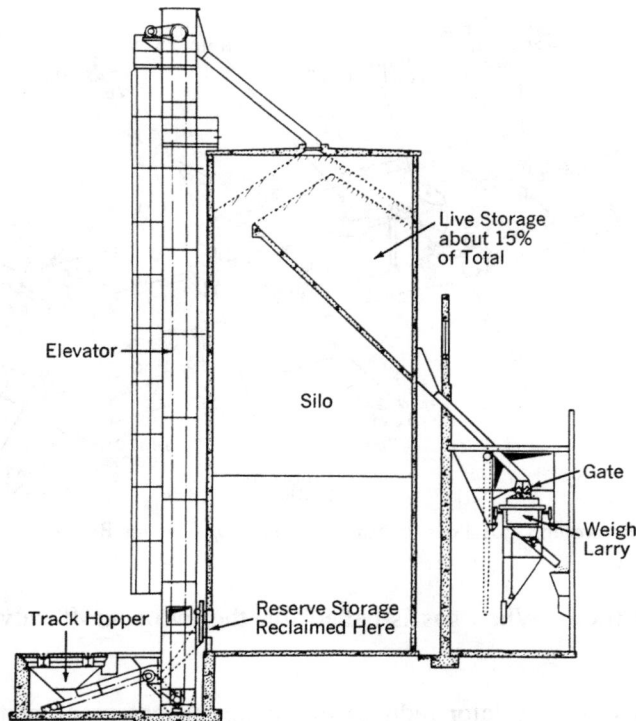

Figure 5-15. Silo that receives coal by truck or rail (Courtesy, Babcock & Wilcox Company)

the lower part of the silo. With this arrangement, the coal received first is used first. The approximate capacity of this silo is 600 tons.

Fluidized bed boiler systems often use pneumatic conveyors to unload the coal and limestone into their silos, to transfer the coal and limestone to the boiler hoppers, and to load the ash into its silo.

PACKAGED BOILER CONTROL SYSTEMS

Packaged boilers are equipped with combustion controls to:

- automatically start the burner;
- adjust the firing rate to maintain desired pressure;
- maintain proper air/fuel ratio over the load range;
- shut off the burner when the load is below the minimum firing rate;
- shut off fuel valves if flame failure or any other abnormal condition occurs.

Programmers

The brain behind the combustion control system is the programmer. Receiving information from pressure and temperature sensors and the flame scanner, it decides when to start and stop the burner, how much fuel to deliver to maintain pressure, and when to initiate shutdown for abnormal conditions.

The older electromechanical designs have cams driven by a clock motor. These cams open and close switches that control the programmer's functions. Older programmers also contain vacuum tube circuits, which amplify the flame scanner or flame detector signals to a usable level. In newer electronic models, switches and vacuum tubes have been replaced with solid state devices, but their functions essentially remain the same. Along with sensor advancements, these programmers can be networked to allow remote monitoring of all functions.

Flame Detection

The purpose of the flame detection system is to sense the presence or absence of a flame. This information is passed to the programmer, which decides whether operation should continue. If the programmer thinks there is a conflict, the boiler is shut down.

All flames have the following characteristics:

- Heat production
- Atmosphere ionization around the flame
- Light emission (infrared to visible to ultraviolet)

Flame detection systems designed for home heating systems use heat from the flame as the method of detection. The heat can be converted to a physical force by using bimetal or hydraulic sensors, or to an electrical signal by using a thermocouple. The main disadvantage of these methods is that between one and three minutes are required for the sensor to heat up or cool down. This amount of time is not excessive when controlling the modest amount of fuel used in small domestic systems, but it is too long in larger commercial and industrial systems.

Flame rod flame detection systems depend on the ability of the flame to conduct electricity, which is called flame ionization. When connected to a suitable electronic control circuit, a response time of two to four seconds is achieved. These systems can monitor both gas pilot and gas main flames but are used primarily in pilot flame service. Oil flames are too hot for flame rods.

Optical sensors are divided into three light groups: visible, infrared, and ultraviolet. Sometimes these devices are called flame scanners.

Photocells detect visible light and are used on domestic and small commercial oil burners. They can't be used on gas burners, because well-adjusted gas flames do not emit sufficient light. Photocells must be mounted so they have an unobstructed view of the flame, but must not be pointed toward any refractory. Light emitted from a hot refractory may be mistaken for a flame.

Infrared detectors, unlike photocells, can be used on both oil and gas flames and are commonly used on most packaged boilers. These detectors can respond to infrared rays emitted by a hot refractory, even when the refractory has ceased to glow visibly. However, infrared radiation from a hot refractory is steady, whereas radiation from a flame often exhibits a flickering characteristic. The infrared detection circuit only responds to flickering infrared radiation, rejecting the steady signal from a hot refractory. There are still times when the steady infrared signal of a refractory can fluctuate. For instance, an infrared signal can be reflected, bent, or blocked by smoke or fuel mist within the combustion chamber. Consequently, when installing an infrared system, care must be taken to ensure that it only responds to flame.

The material used to detect the invisible infrared radiation is lead sulfide. When exposed to infrared radiation, the electrical resistance of lead sulfide decreases. Since lead sulfide can only pass a few microamps without burning up, vacuum tubes or transistors have to amplify its output to a usable level.

Ultraviolet detectors also monitor oil and gas flames. This type only responds to a hot refractory over 2,500°F, but it can pick up a signal from spark ignitors. Some consider ultraviolet detectors to be the best systems.

Figure 5-16 shows the light output of a hot refractory and gas and oil flames. Figure 5-17 is a comparison chart of the various detection methods. Only abide by recommendations made by qualified flame detection authorities.

Automatic Programmer Sequence

Before detailing the start-up sequence for a typical boiler, it is necessary to understand the following terms:

- Pressuretrol and vaporstat — Pressure switches that measure different pressure ranges. Pressuretrols are calibrated in pounds per square inch. Vaporstats are calibrated in ounces per square inch or inches of water column.

- Operating limit control — A pressuretrol that signals the programmer to either turn the burner on or off. The high-pressure limit switch is set higher than the operating control and is the safety backup in case the operating limit control fails.

- Modulating or firing rate motor — Positions the burner air register and fuel valve so they produce the desired firing rate.

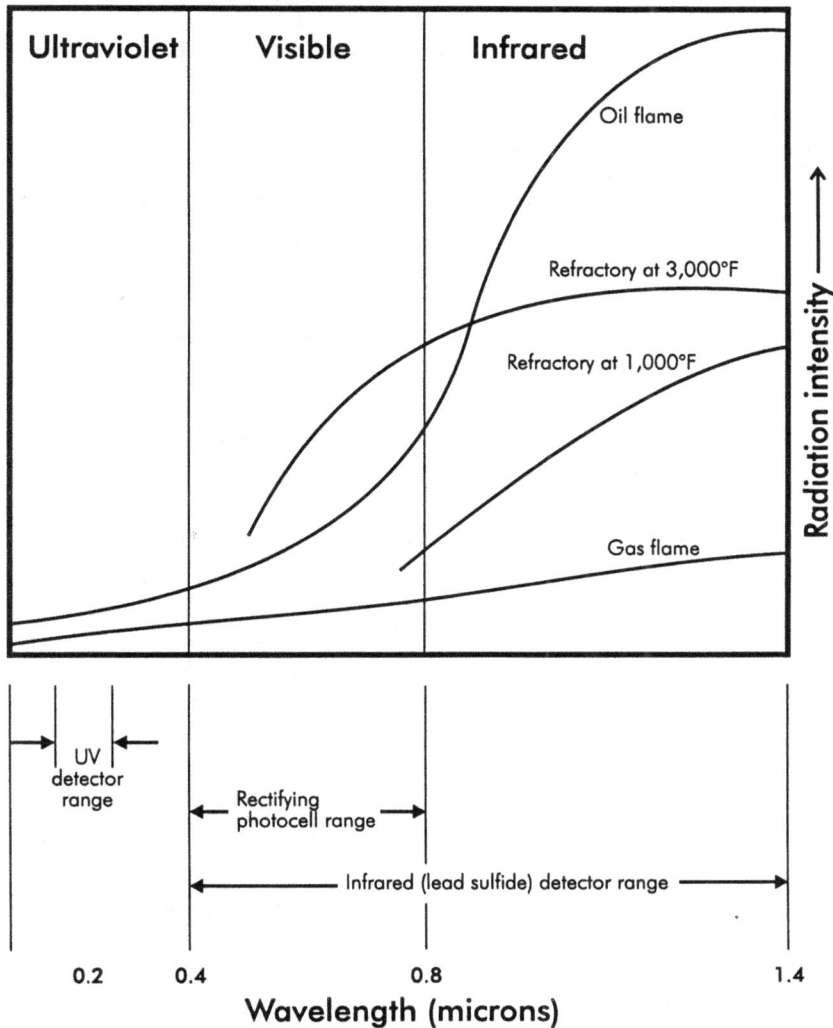

Figure 5-16. Light output of a hot refractory and gas and oil flames

The following start-up sequence is assumed to be for a typical boiler that has been on standby and on which the operating limit control has just signaled the programmer to start the burner:

1. Prepurge — The purpose of the prepurge is to prevent explosions by removing any unburned fuel in the boiler before ignition occurs. The programmer starts the forced draft fan, and the draft is verified by a vaporstat or an airflow switch. Proof that the forced draft motor is running is not sufficient because the fan's V-belt or shaft coupling could be broken. The modulating motor then drives the burner's register to the high-fire position, opening it for maximum airflow. A high-fire position switch in the modulating motor tells the programmer that it is in the high-fire position, and it stays there for 30 seconds, or

	Flame rod	Rectifying photocell	Infrared (lead sulfide) detector	Ultraviolet detector
Supervises oil	No	Yes	Yes	Yes
Supervises gas	Yes	No	Yes	Yes
Detects ignition spark	No	No	No	Yes
Signal type	DC	DC	AC	DC
Light responsive	None	Visible	Infrared	Ultraviolet
Max. ambient temp. of cell	500°F	165°F	125°F	125° to 250°F
False flame signals by:				
inductive pickup	No	No	Yes[1]	No
capacitive pickup	No	No	Yes[2]	No
refractory glow	No	Yes	Yes[3]	No[4]

1. Leads must be run in grounded conduit.

2. Leads must be shielded, twisted pair, or BX.

3. Doesn't sense hot refractory alone, but turbulent hot air, steam, smoke, or oil spray may cause the radiation to fluctuate, simulating flame.

4. Ultraviolet detectors respond to hot refractory over 2,500°F.

Figure 5-17. Flame detector comparison chart

long enough to allow four air changes. The purge time for gas and oil are the same.

2.   Ignition — After the prepurge, the modulating motor drives the register to the low-fire position. A low-fire position switch tells the programmer that the register is in the minimum airflow position. The ignition of the pilot and main burner must take place at low-fire. The ignition transformer and the pilot gas valve are then energized. If no flame is detected after 10 seconds, the programmer goes to postpurge, shuts off the burner, and sounds an alarm. An operator must correct the problem, then push the reset button on the programmer for another try. If the flame detector sees a flame, the programmer opens the main fuel valve. The pilot and the main fuel valve remain open together for 10 or 15 seconds, after which the pilot gas valve closes. If the main flame goes out at this time, the programmer then goes to the postpurge mode, shuts off the burner, and sounds an alarm. An operator must correct the problem, then push the programmer's reset button.

3.   Run — If the scanner continues to receive a flame signal, the programmer turns control of the firing rate motor over to either the automatic modulating pressure controller (modulating pressuretrol) or to the manual firing rate control. If the main flame goes out when the

boiler is in the run mode, the programmer shuts off the main fuel valve, goes into postpurge, and sounds an alarm. An operator must correct the problem, then push the reset button on the programmer for another try.

4. Standby — When steam pressure continues to increase even at low fire, the operating pressuretrol tells the programmer to shut the burner off. The programmer closes the main fuel valve, goes into a 15-second postpurge, then goes into standby, waiting for the operating limit control to start the cycle again.

5. Postpurge — After the fuel valve is closed, the fan continues to operate for an additional 15 seconds to purge the boiler of any remaining unburned fuel.

PRESSURE SETTINGS

The burner pressure controls should accomplish the following tasks:

- Properly set the firing rate to match load demands
- Have the burner at low fire at each start
- Have the burner at low fire before shutdown

The following describes the pressure relationships between the settings of the safety valves, high limit, operating limits, and modulating control *(Note: these values were picked for this example, and the settings on your equipment will probably be different.):*

- 150 psi — Safety valve(s) opens
- 125 psi — High-limit switch opens
- 115 psi — Operating limit switch turns off
- 110 psi — Operating limit switch turns on
- 105 psi — Low-fire modulating range
- 95 psi — High-fire modulating range

In normal operation, the burner shuts down at 115 psi. As the pressure drops to 110 psi, the operating limit switch closes and restarts the burner at low fire. If the load exceeds the low-fire input, the modulating control responds by increasing the firing rate to match the demand. As the load changes, the firing rate changes accordingly. If the pressure drops below 95 psi, the load is beyond the capacity of the boiler.

The high-limit control provides a safety factor that shuts the burner off if the operating limit switch fails. The setting of this control should be sufficiently above the operating control to avoid nuisance shutdowns. The setting, however, must be within the limits of the safety valve setting and must not exceed 90% of its setting. Having the operating pressure too close to the set safety valve pressure increases the possibility of leakage, which necessitates early valve replacement.

When starting a cold boiler, it is recommended that the controls be on manual and set at low fire until the boiler is warmed up. If the burner is on automatic control during a cold start, it moves to high-fire immediately. This rapid heat input subjects the boiler to undesirable stresses.

AIR/FUEL RATIO CONTROL SYSTEMS

There are two general types of combustion control systems for oil- and gas-fired packaged boilers: positioning and metering.

The positioning or jack shaft systems have proven to be a simple and reliable method to establish an air/fuel ratio over a wide load range for single-burner boilers up to 150,000 lb/hr. As pictured in Figure 5-18, a sensing device such as a modulating pressuretrol or pneumatic transmitter sends a signal to a positioner. Often called a modulating motor, it simultaneously positions both the forced draft damper and fuel valve through a series of levers and connecting rods.

One disadvantage of this method is that the air/fuel ratio is fixed and can't be adjusted to compensate for air density and fuel variations. Several manufacturers have developed oxygen trim systems that maintain the proper air/fuel ratio under all operating conditions.

An electrical positioning system consists of two parts: a pressure sensor and a modulating motor. A potentiometer, sometimes called a slide wire, is the pressure sensor. Its wiper arm is positioned by the boiler's steam pressure. There is another potentiometer in the modulating motor whose wiper arm is positioned by the motor's shaft. The electrical output from these two potentiometers is compared. If they are different, the modulating motor positions its wiper arm until their outputs are the same. In the process, the firing rate is also changed by the movement of the jack shaft. In Figure 5-19, the modulating motor is driven to the open or high-fire position (for example, when purging the boiler) by shorting out the "B" and "R" wires. It is driven to the closed or low-fire position (for example, when lighting off) by shorting out the "W" and "R" wires.

Reliability increases whenever an electromechanical component is replaced with electronics. The latest models replace the wiper arm with a solid state pressure sensor that has no moving parts.

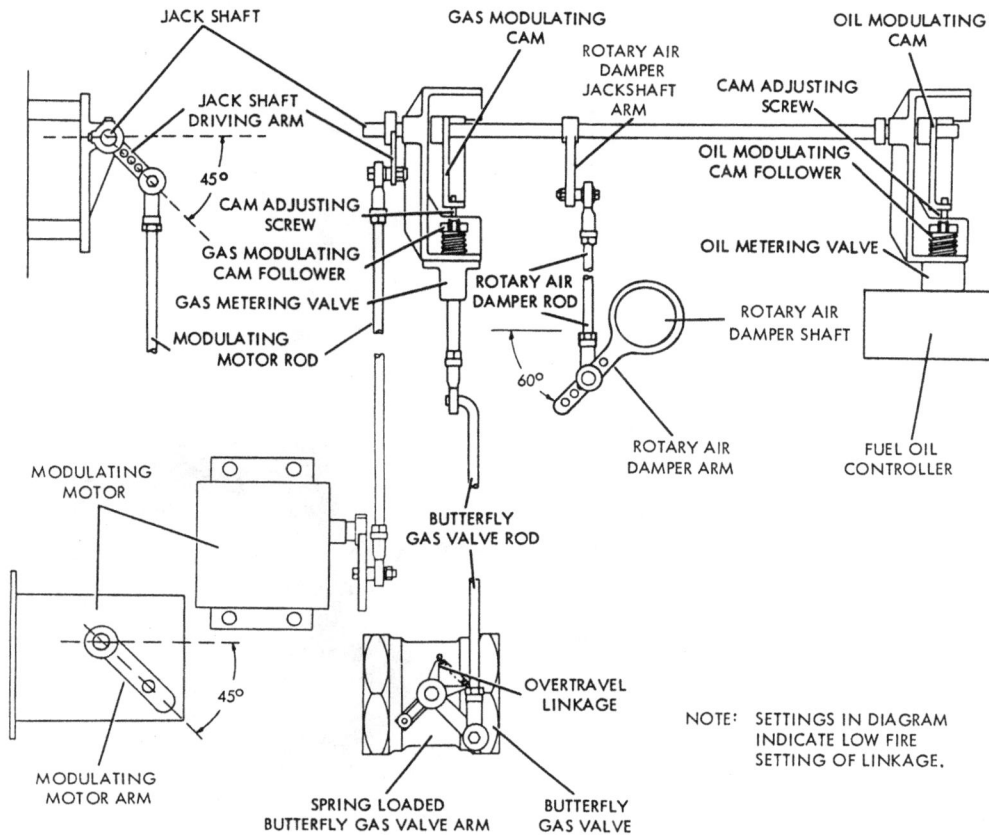

Figure 5-18. Jack shaft combustion control (Courtesy, Cleaver Brooks)

Metering control systems measure the amount of air and fuel being used, then proportion them for maximum combustion efficiency. This method compensates for changes in barometric conditions and fuel quality and is well suited for oxygen trim, carbon monoxide trim, and other sophisticated control schemes. Such systems are found on large multiburner boilers.

One feature often used on metering control systems is lead-lag air/fuel ratio control to prevent a fuel rich (smoky) condition when the boiler's load changes. During an increasing load, the air increases before the fuel increases. During a decreasing load, the fuel decreases before the air decreases. This way there is always excess air during the load swings.

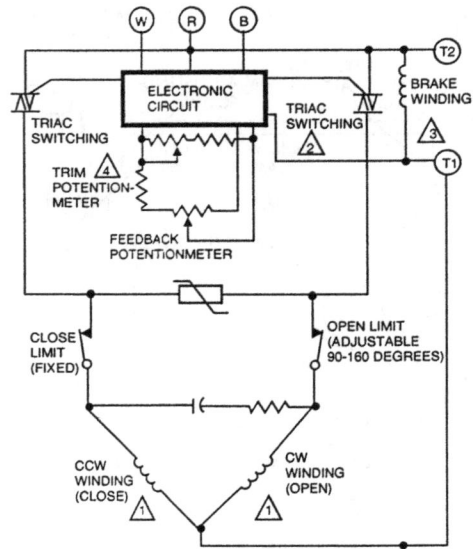

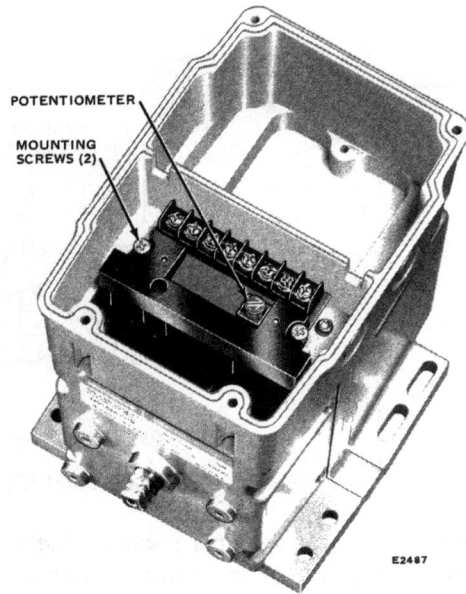

△1 DIRECTION OF MOTOR TRAVEL AS VIEWED FROM POWER END.

△2 CONNECT 24V POWER TO T1-T2 TERMINALS ONLY. DO NOT
CONNECT POWER SUPPLY TO CONTROLLER TERMINALS.
PROVIDE DISCONNECT MEANS AND OVERLOAD PROTECTION
AS REQUIRED.

△3 WHEN MOTOR IS POWERED, BRAKE HOLDS MOTOR POSITION
AGAINST SPRING RETURN ACTION.

△4 ON ADJUSTABLE STROKE MOTORS ONLY.

M794B

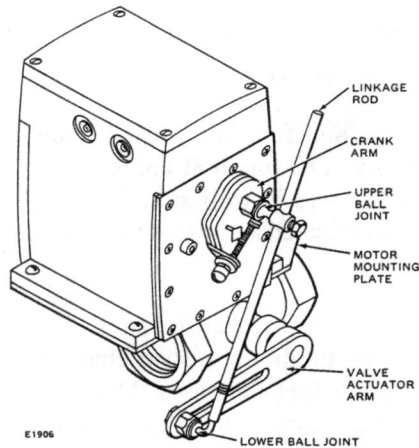

Figure 5-19. Modulating motor and schematic (Courtesy, Honeywell)

CHAPTER SIX

Fuel-Saving Equipment

High stack temperature is an indicator of a low-efficiency boiler. Soot blowers, economizers, air heaters, and turbulators are devices that reduce stack temperature and increase efficiency.

Stack temperature increases as boiler tubes become dirty from the insulation effect and as soot and fly ash accumulate on the heat-absorbing surfaces. Table 6-1 shows how efficiency is lost.

Soot on Boiler Tubes (in.)	Heat Loss (%)	Increased Fuel Consumption (%)
1/32	12	2.5
1/16	24	4.5
1/8	47	8.5

Table 6-1. Heat loss as related to soot build-up

SOOT BLOWERS

Soot blowers direct steam or high-pressure air to eliminate soot build-up. A fixed-position rotating soot blower, like the one in Figure 6-1, can be placed in the convection section of boilers, economizers, and air heaters. (Economizers and air heaters are described later in this chapter.) This type of rotating blower consists of a tube (sometimes called the soot blowing element), with a series of holes or nozzles in it, which is connected to a rotating valve. The soot-blowing medium, steam or high-pressure air, is applied, and the tube is rotated. The holes in the soot blowing element are aligned so the steam is discharged between the boiler tubes. A favorite test question is, "What if steam were allowed to strike a tube directly?" The answer is that it would eventually erode the metal and cause a leak.

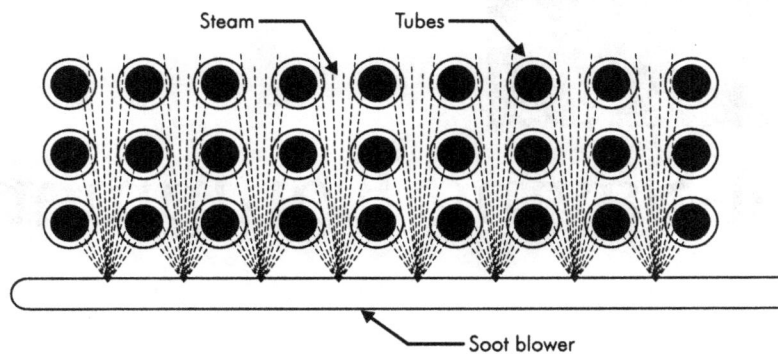

Figure 6-1. Fixed-position rotating soot blower

Since rotating blower elements are always exposed to flue gases, they are installed in the cooler sections of the boiler. Where it is too hot for a fixed blower, retractable soot blowers are used. The soot-blowing element, or lance, is outside the boiler when not in use. Air or steam is blown through one or two nozzles at the end of the lance as it rotates during insertion and retraction.

The frequency of soot blowing depends on the fuel burned and firing rate. The steam line has to be drained of condensation before starting so that slugs of water do not damage the equipment. Sufficient draft has to be established to carry the loosened soot through the boiler. With multiple blowers, start from the furnace end and work toward the stack. On larger installations, the entire soot-blowing process is automated. Steam or air work equally well for soot blowing. The decision to use one or the other is based on economic factors.

If the stack temperature suddenly increases, the probable cause is a broken baffle. This is common on dry-back fire-tube boilers but can occur in water-tube boilers also.

Instead of the flue gas being directed to every region of the convection section as intended, it takes the path of least resistance and short circuits part of the boiler on the way to the stack. Higher stack temperature results because some of the tubes were robbed of their opportunity to absorb heat from the flue gas. Boiler efficiency is increased when additional heat is added without burning additional fuel. Economizers and air heaters recycle the heat of the flue gas by returning it to the boiler.

ECONOMIZERS

The function of an economizer is to recover heat energy that would otherwise be lost in the flue gas by heating the incoming feedwater. Each 10°F increase in the boiler feedwater cuts fuel consumption by at least 1%.

Figure 6-2 depicts the operating characteristics of a typical boiler. The resulting 7.5% savings in fuel is based on 6,000 hours of operation per year, operating at 70% average load and a fuel cost of $0.60 per gallon.

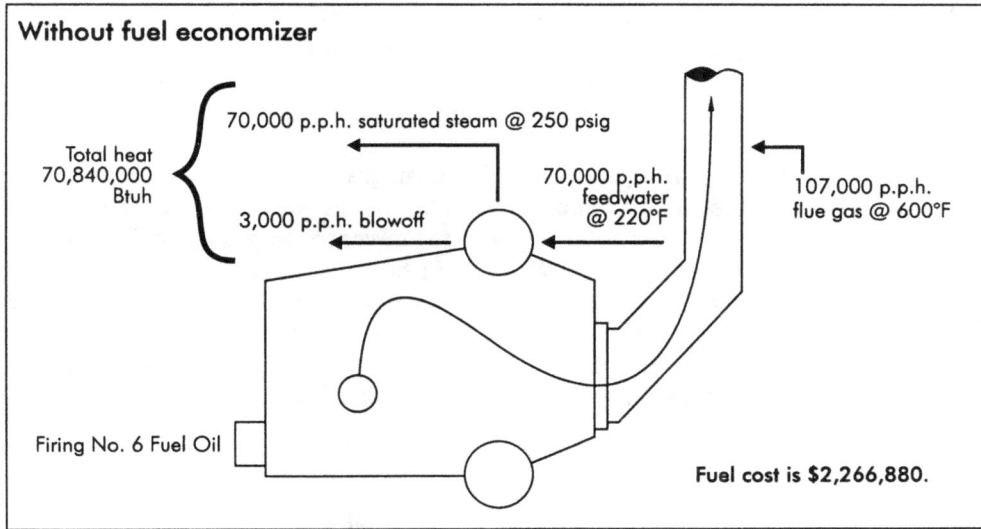

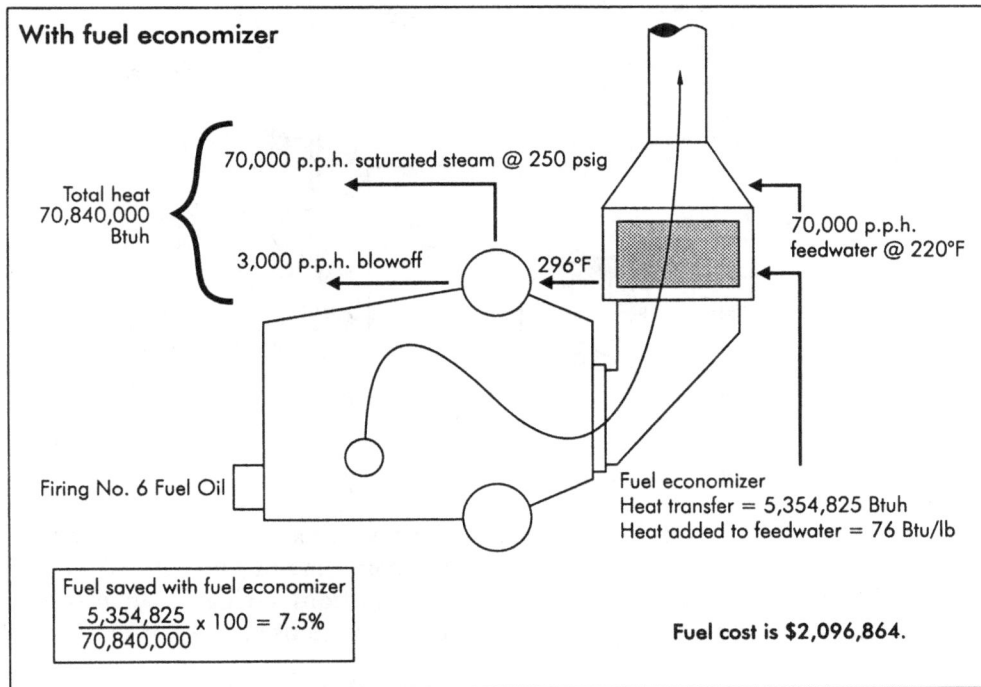

Figure 6-2. Fuel-cost savings resulting from use of a fuel economizer

A main consideration in economizer design is the prevention of corrosion. During combustion, sulfur in the fuel combines with oxygen to form sulfur dioxide (SO_2) and sulfur trioxide (SO_3). To prevent these compounds from further conversion into sulfurous (H_2SO_3) and sulfuric (H_2SO_4) acids, the surface temperatures of the stack and economizer must be kept high enough to prevent the water vapor in the flue gas from condensing. Stack and economizer corrosion can be prevented by regulating the economizer feedwater temperature to keep the flue gas well above its dew point.

Figure 6-3 shows a typical economizer piping diagram in which the feedwater is preheated to the inlet of the economizer. This is one way of keeping the metal surfaces in the stack area hot enough to prevent flue gas condensation and the corrosion that would result. Sensors in both the economizer and stack control a valve that supplies enough steam to a feedwater preheater to keep all metal surfaces above the flue gas dew point. There should be three valve bypasses installed around the steam control valve, preheater, and economizer so that maintenance and inspections can be carried out while the boiler is still in service.

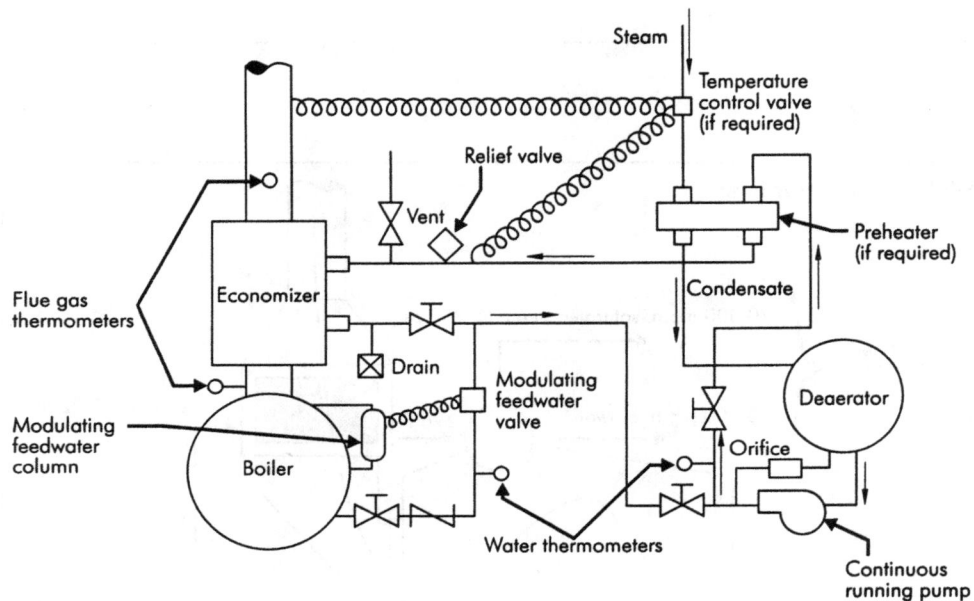

Figure 6-3. Typical economizer piping diagram

Finned tubes are often used in economizers to provide extended tube surface. With 8 to 12 times more heating surface than a bare tube of the same diameter, they provide better heat transfer. Finned tubes also afford savings in tube weight and space compared to ordinary tubes. Soot blowers are used to keep the tubes clean.

AIR HEATERS

Like economizers, air heaters reclaim some heat from the flue gas that would otherwise be lost. In this case, the combustion air is preheated. Practically all pulverized coal-fired units require preheated air at 300° to 600°F for drying the fuel. Preheated air is not essential for stoker units, but larger units work better with it.

There are two classifications of air heaters: recuperative and regenerative.

In recuperative air heaters, the heat is transferred directly from the hot gases on one side of a surface to air on the other side. These types of air heaters normally operate on the counterflow principle, in which air and flue gas travel in opposite directions. This type of flow results in greater heat transfer, because hot combustion gases pass by the warmer air that is leaving the heater. Recuperative air heaters can be either the tubular or plate type, Figure 6-4. The tubular type routes the flue gas through tubes. Air passes over the outside of the tubes, flowing several times across the tubes to increase heat transfer. Plate-type air heaters use plates that are placed in a structural steel frame. Air and flue gas flow past each other on opposite sides of the plate, causing heat transfer to occur.

In regenerative heaters, heat is transferred indirectly from the hot gases to the air through an intermediate heat storage medium. The regenerative heater consists of a rotating plate, in which heat storage plate elements are first heated by the hot flue gas, then rotated into the combustion air stream where the stored heat is released. The cooled plates continue to rotate back to the flue gas stream where they are reheated.

TURBULATORS

When flue gases travel through fire-tube and cast iron boilers, their turbulent flow makes good contact with the boiler metal and efficient heat transfer results. However, in boilers where there is not enough velocity to maintain turbulence, the hot gases flow away from the metal, in the center of a fire tube, for example. A layer of cooler insulating gas forms between the gas in the center of the tube and the metal.

Turbulators are angular metal strips inserted into the sections of a boiler where the flue gas is not turbulent, Figure 6-5. The installation of turbulators must be done professionally. Before this or any other fuel-saving modification is installed, a boiler must be cleaned and adjusted so it operates at maximum efficiency. A set of preinstallation readings should be taken (stack temperature and O_2 or CO_2) to establish combustion efficiency. Turbulators change the firing characteristic of the boiler, and postinstallation adjustments are often necessary.

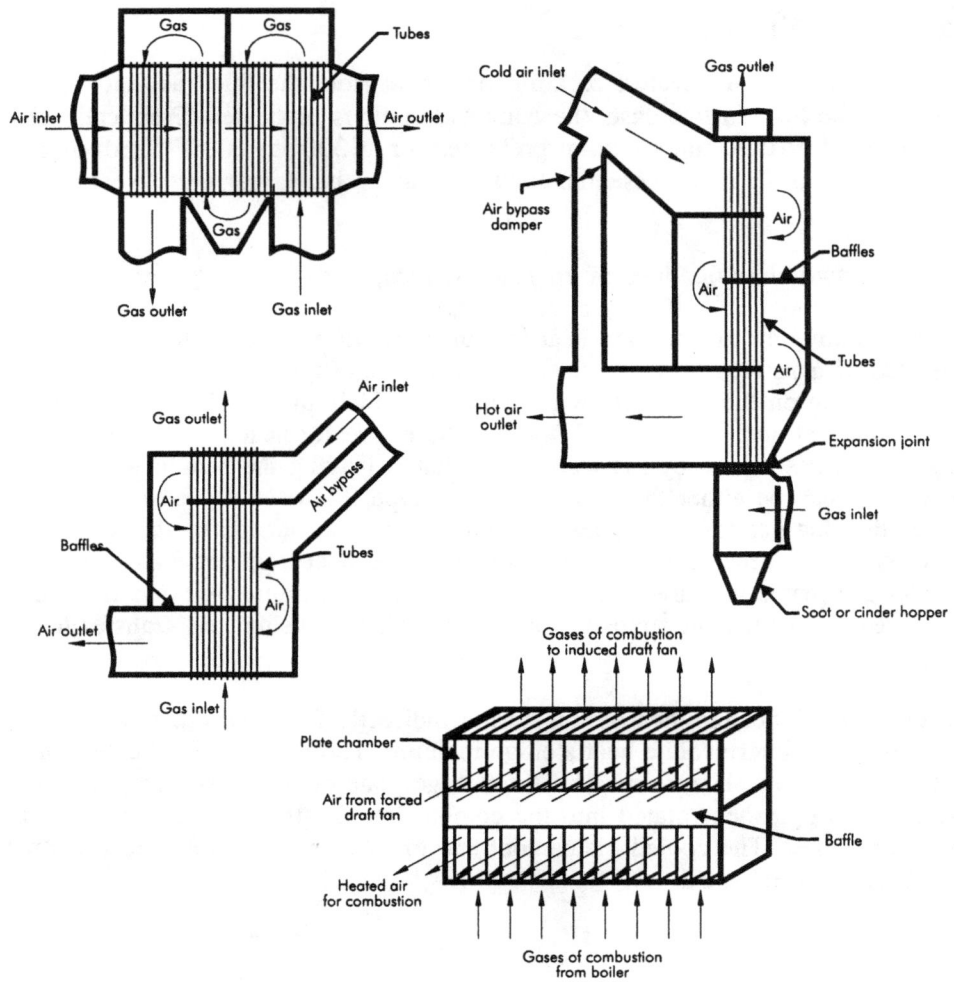

Figure 6-4. Tubular and plate air heaters

Again, test the combustion efficiency and compare it against the original readings. If the modifications have been installed properly, efficiency should increase.

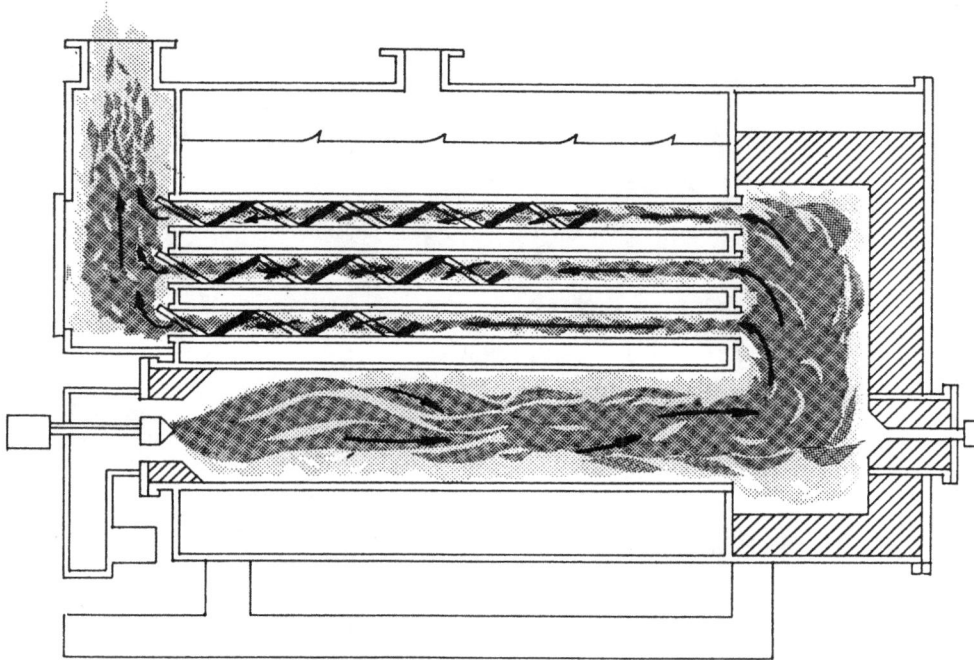

Figure 6-5. Turbulators (Courtesy, Fuel Efficiency, Inc.)

CHAPTER SEVEN

Steam Traps and Pipes

A steam trap is an automatic valve that senses the difference between steam and condensate. The trap discharges the condensate with little or no loss of steam, which contributes to high operating efficiency. Steam traps are divided into three main groups: thermostatic, mechanical, and thermodynamic.

THERMOSTATIC TRAPS

Thermostatic traps measure temperature. The balanced-pressure thermostatic trap shown in Figure 7-1 has a liquid-filled bellows that expands and contracts. When steam is in contact with the liquid-filled bellows, the steam causes the bellows to expand, which closes the valve. If condensate or air is in contact

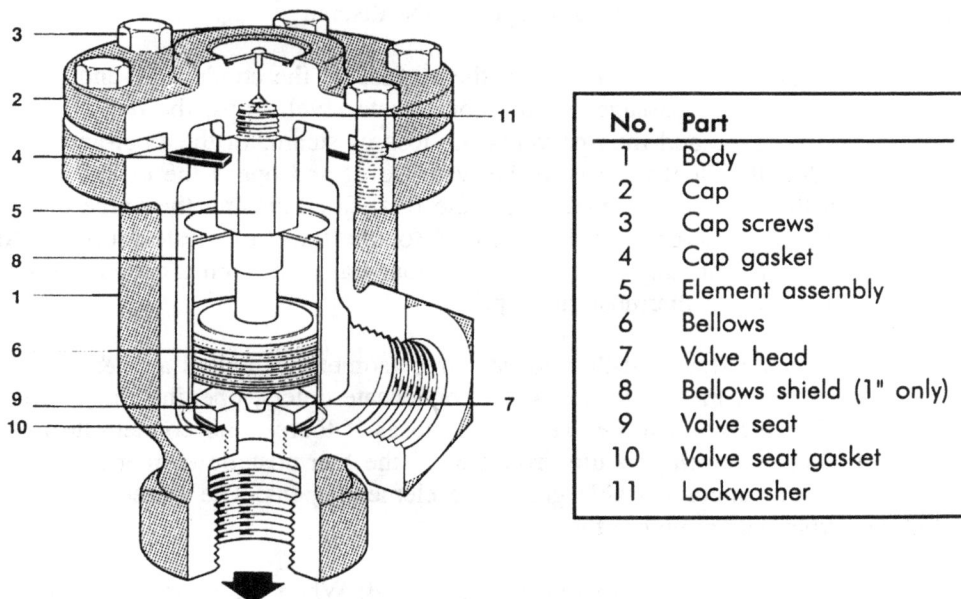

No.	Part
1	Body
2	Cap
3	Cap screws
4	Cap gasket
5	Element assembly
6	Bellows
7	Valve head
8	Bellows shield (1" only)
9	Valve seat
10	Valve seat gasket
11	Lockwasher

Figure 7-1. Balanced-pressure thermostatic trap (Courtesy, Spirax Sarco, Inc.)

with the bellows, the bellows contracts and condensate is discharged. The cycle then repeats itself. The working steam pressure does not affect the operation of this trap, because it is the difference in temperature between the steam and condensate which operates the bellows.

The bimetallic trap also works according to the thermostatic principle. In a bimetallic strip, two strips of suitably different metal are bonded together. The top strip expands more than the bottom one when heated, allowing cool air and condensate to pass through the trap. As steam enters the trap and heats up the bimetallic strip, the strip bends and closes off the valve.

The liquid expansion thermostatic trap is operated by the expansion and contraction of a liquid-filled thermostat, which responds to the temperature difference of steam and condensate. When the steam is turned on, air and condensate pass through the open trap. As the condensate temperature increases, the oil in the thermostatic element expands and closes off the valve. An adjusting nut positions the valve relative to its seat, which allows the trap to be set at a given temperature, usually 212°F or lower.

Some liquid expansion traps are used for freeze protection. When the temperature drops to 40°F, the trap opens, creating enough flow to prevent freeze-up.

MECHANICAL TRAPS

Mechanical traps distinguish between steam and condensate by their different densities. Various floats are used to operate the discharge valve.

In the loose float type, a ball floats on the surface of the condensate and uncovers the discharge passage. As the condensate level drops, the ball covers the discharge passage and thus prevents the loss of steam. In the float-and-lever trap, Figure 7-2, the float rises with the water level and opens the discharge valve. When the condensate level drops, the float falls, closing the valve and preventing steam loss. Air must be removed for the trap to operate properly. Air can be vented automatically from the loose float trap and from the float-and-lever trap by adding a thermostatic trap.

Figure 7-3 shows a float-and-thermostatic trap (sometimes called an F&T trap). In this type of trap, the float rises when condensate enters, opening the valve and discharging the condensate. The valve closes if there is no condensate in the trap. If there is a temperature drop around the thermostatic element that is caused by air or noncondensable gases, the element opens. The element expands and closes when steam enters the trap.

The inverted bucket trap is pictured in Figure 7-4. When the system is turned on, the bucket is at the bottom and the discharge valve is open. Air is vented

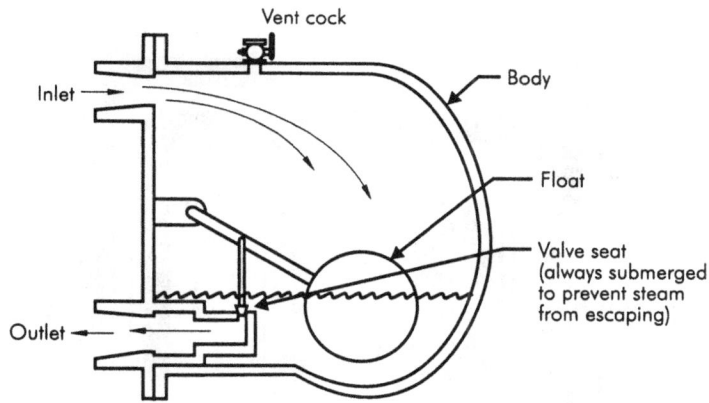

Figure 7-2. Float-and-lever trap

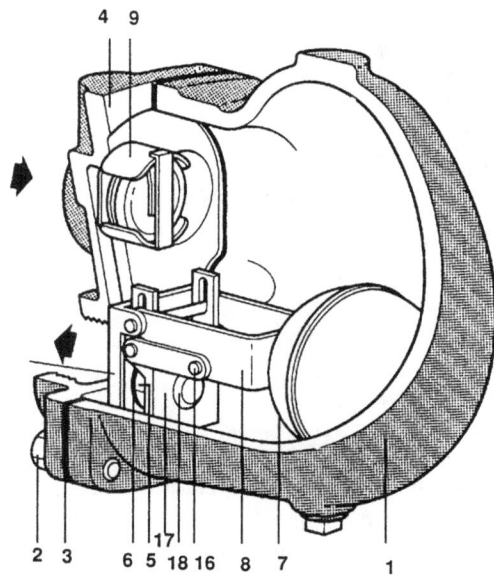

No.	Part
1	Body
2	Cover screws
3	Cover gasket
4	Cover
5	Valve seat
6	Valve seat gasket
7	Ball float
8	Float arm
9	Air vent assembly
	Air vent head
	Air vent seat
16	Pivot pins
17	Head bracket, stop, link
18	Valve head

Figure 7-3. Float-and-thermostatic trap (Courtesy, Spirax Sarco, Inc.)

through a small hole on top of the inverted bucket. As condensate enters the trap through the bottom, the water level rises on both the inside and outside of the bucket and then leaves through the discharge valve at the top. Entering steam fills the inverted bucket and makes it float, closing off the discharge valve. Steam slowly escapes out of the bucket through the vent hole. If the escaping steam is replaced by more steam, the trap remains closed. If the escaping steam is replaced by condensate, the bucket sinks, opens the valve, and discharges the water.

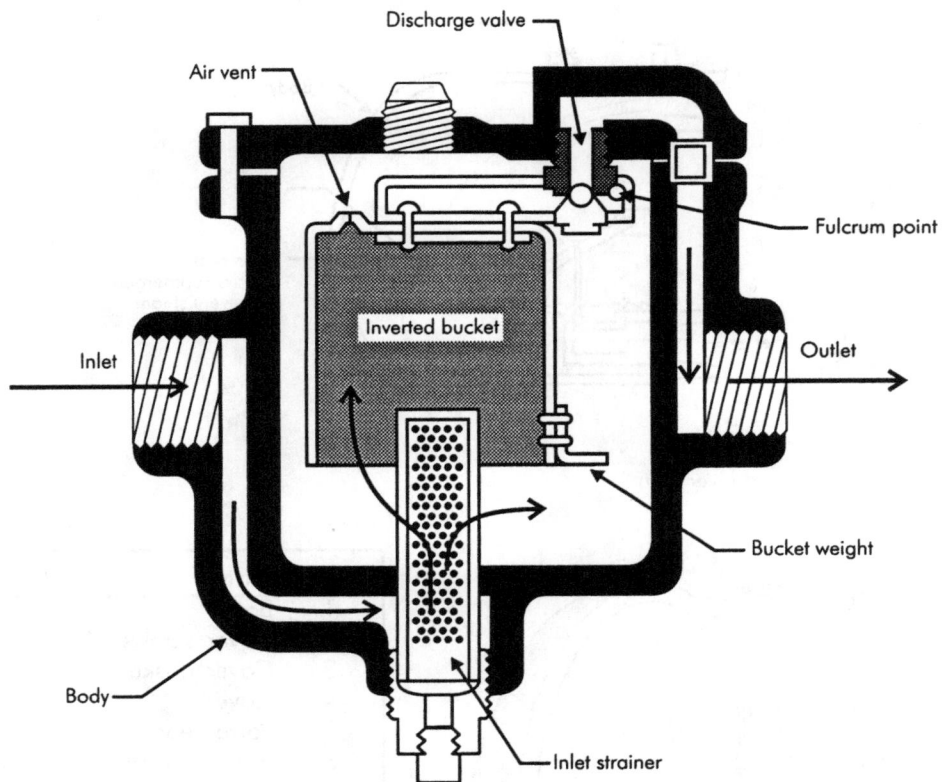

Figure 7-4. Inverted bucket trap

THERMODYNAMIC TRAPS

Thermodynamic traps, or disk traps, identify steam and condensate by the difference in their kinetic energy or velocity as they flow through the trap, Figure 7-5. Low pressure flash steam pushing down on the large surface on top of the disk overcomes the force of the live steam pushing up on the smaller, exposed disk area.

After startup, air and cool condensate lift the disk off its seat. The air and condensate are then discharged. As the temperature of the condensate increases, some of it flashes into steam. The resulting mixture of steam and condensate flows outward across the underside of the disk. Because flash steam has a larger volume than condensate of the same weight, the flow increases as more flash steam is formed. This high velocity causes a low-pressure area to be formed under the disk and the expanding flash steam exerts pressure on top of the disk, forcing the disk downward against its bearing surface and stopping all flow.

Although the flash steam pressure is much lower than the pressure pushing against it from the inlet side, the large exposed area overcomes the system

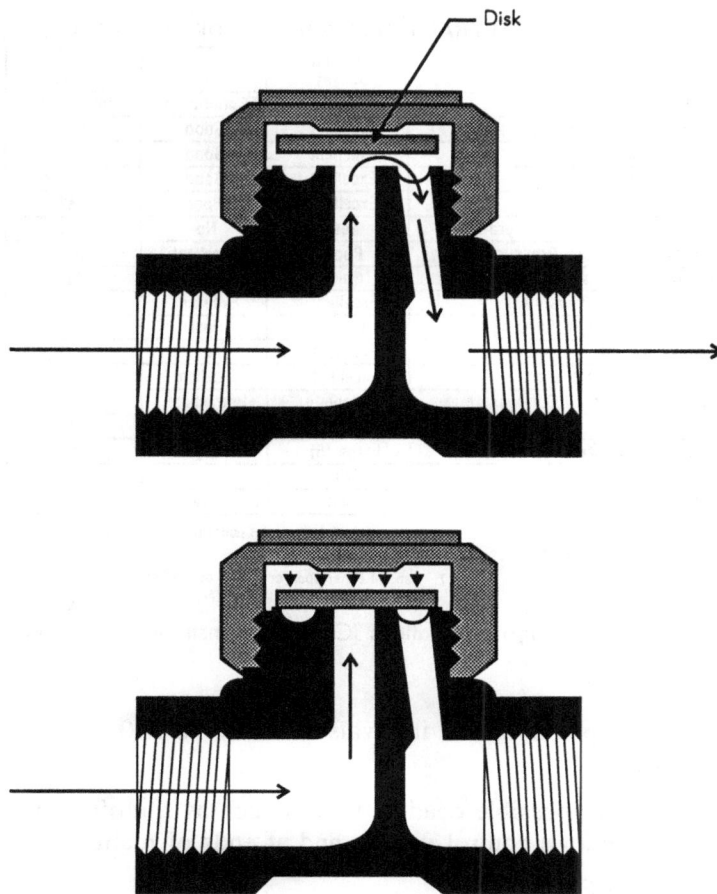

Figure 7-5. Thermodynamic trap

pressure. As the flash steam above loses heat, some of it condenses, reducing the pressure above the disk. The disk is again lifted off its seat, and the cycle repeats itself.

There is no one ideal trap for every application. Figure 7-6 outlines some of the characteristics of the different traps. Steam trap vendors are the best sources of information regarding which trap is best for each application.

TRAP LOCATION

Traps are required at all locations where condensate is formed and collected. These include the following locations:

- Steam Distribution Mains
  - At elevation changes such as risers and expansion loops.

HOW VARIOUS TYPES OF STEAM TRAPS MEET SPECIFIC OPERATING REQUIREMENTS

	Characteristic	Inverted Bucket	F&T	Disc	Bellows Thermostatic
A	Method of Operation	Intermittent	Continuous	Intermittent	(1) Intermittent
B	Energy Conservation (Time in Service)	Excellent	Good	Poor	Fair
C	Resistance to Wear	Excellent	Good	Poor	Fair
D	Corrosion Resistance	Excellent	Good	Excellent	Good
E	Resistance to Hydraulic Shock	Excellent	Poor	Excellent	Poor
F	Vents Air and CO_2 at Steam Temperature	Yes	No	No	No
G	Ability to Vent Air at Very Low Pressure (¼ psig)	Poor	Excellent	NR(3)	Good
H	Ability to Handle Start-up Air Loads	Fair	Excellent	Poor	Excellent
I	Operation Against Back Pressure	Excellent	Excellent	Poor	Excellent
J	Resistance to Damage from Freezing (4)	Good	Poor	Good	Good
K	Ability to Purge System	Excellent	Fair	Excellent	Good
L	Performance on Very Light Loads	Excellent	Excellent	Poor	Excellent
M	Responsiveness to Slugs of Condensate	Immediate	Immediate	Delayed	Delayed
N	Ability to Handle Dirt	Excellent	Poor	Poor	Fair
O	Comparative Physical Size	Large (5)	Large	Small	Small
P	Ability to Handle "Flash Steam"	Fair	Poor	Poor	Poor
Q	Mechanical Failure (Open - Closed)	Open	Closed	(6) Open	(7)

1. Can be continuous on low load.
3. Not recommended for low pressure operations
4. Cast iron traps not recommended.

5. In welded stainless steel construction—medium.
6. Can fail closed due to dirt.
7. Can fail either open or closed depending upon the design of the bellows.

Figure 7-6. A guide to steam trap applications (Courtesy, Armstrong International, Inc.)

- At all low points and at intervals of 200 ft to 500 ft on long horizontal runs.

- Ahead of all possible dead-end areas such as shutoff valves, pressure and temperature control valves, and at ends of mains.

• Steam-Operated Equipment

- Draining directly below heat exchangers, coils, unit heaters, cooking kettles, dryers, etc.

- Ahead of turbines, humidifiers, etc.

Water Hammer

Water hammer describes the condition that occurs when a slug of water is carried by flowing steam at a velocity higher than 60 mph. This slug then strikes an obstruction such as an elbow or pressure-reducing valve with enough force to cause physical damage.

Regardless of how well a steam line is insulated, some heat escapes. If condensate forms, it must be removed before water hammer occurs. Figure 7-7 shows how a slug of water is formed by the build-up of condensate. To collect condensate as it travels along a steam main, a large diameter drip leg is required. Condensate is collected in drip legs and then discharged via a steam trap.

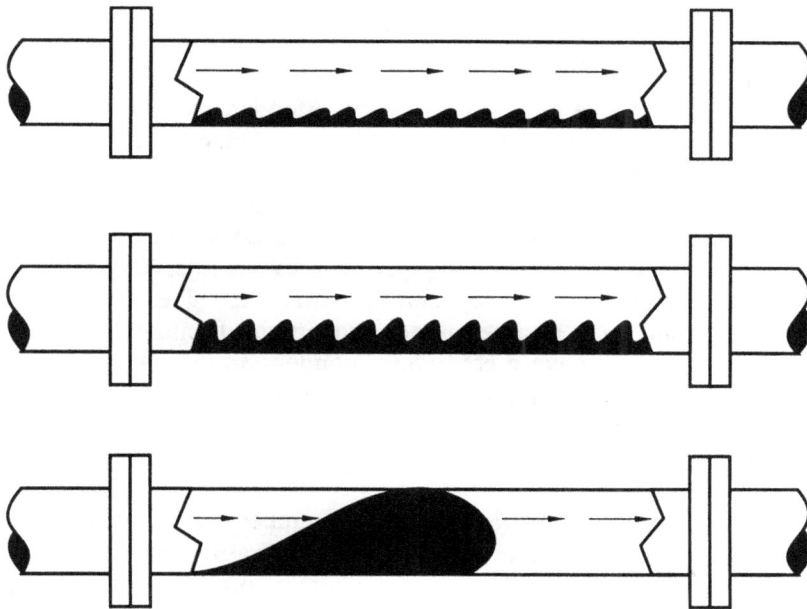

Figure 7-7. A slug of water is formed by the build-up of condensate

TRAP INSTALLATION RECOMMENDATIONS

The following recommendations should be considered when installing traps:

- Use strainers and/or dirt pockets to protect all traps and valves from pipe scale and dirt. These devices require periodic cleaning to remain effective.

- Install shutoff valves and test valves on both sides of the trap. Use gate valves for unrestricted flow.

- Install unions on each side of the trap. Uniform face-to-face dimensions permit fast trap replacement and maintenance at the shop rather than in the field.

- Take steam-supply piping from the top of the supply main to obtain dry steam.

- Bypasses around traps are discouraged. If they are left open or they leak, they will defeat the function of the trap. It is better to use two parallel traps.

- Use a trap on each piece of equipment. Differences in condensing rates cause uneven pressure drops, resulting in air-binding and partial flooding when several pieces of equipment are drained into a common trap.

Testing Traps

Faulty steam traps are a major source of waste in a steam distribution system. A trap that is blowing live steam is the worst offender, but traps that are plugged or stuck closed can also be costly. When condensate backs up into heat-transfer equipment, the decreased efficiency results in lost production.

Steam traps are best tested visually. This is easy to do with traps that discharge to the atmosphere, but most traps are piped to a return line. To test the latter, isolate the trap from the return line, then open the test valve. Inverted bucket and disk traps should have an intermittent condensate discharge; float-and-thermostatic traps should have a continuous condensate discharge; while thermostatic traps may be either continuous or intermittent, depending on the load. If nothing comes out of the test valve, open up the valve before the trap to make sure the trap is not plugged or stuck closed.

Do not mistake flash steam for live steam. Remember that condensate under pressure has more Btu per pound than condensate at atmospheric pressure. When discharged, these extra Btu will re-evaporate some condensate and produce flash steam. How can you tell the difference between flash and live steam? Flash steam will float out intermittently each time the trap discharges. Live steam will blow continuously in a blue stream.

Another way to test traps is to use a stethoscope or an ultrasonic listening device. On inverted bucket traps, a definite bursting sound is produced when the discharge valve is opened. On disk traps, the cycle rate can easily be heard. Listening devices don't work as well on thermostatic traps and float-and-thermostatic traps because their normal discharge is modulating and their cycles are gentle.

Measurement of temperature is not always a sure way to test steam traps. The temperature of condensate and flash steam from a properly working trap is around 212°F, very close to the temperature of condensate and live steam from a defective trap. Regardless of the type of measurement device used (pyrometers, temperature-sensitive crayons, etc.) surface temperature measurements are subject to error, usually on the low side.

Piping

When working out pipe size problems, make sure that all the units of measure are correct. For example, find pipe size in inches, given flow in pounds per hour, density in cubic feet per pound, and velocity in feet per minute.

The key to this problem is to find the area of the pipe first:

$$\text{Area} = \frac{\text{Flow } (ft^3/lb)}{\text{Velocity } (ft/min)}$$

Steam flow is usually given in pounds per hour, so it is necessary to convert the units of measure. The piece of information needed is steam density in cubic feet per hour:

$$\text{Area} = \frac{(lb/hr)(ft^3/lb)}{ft/min}$$

As the units are worked out, notice that the hours and minutes do not cancel out. Therefore, either the hours have to be changed to minutes, or the minutes have to be changed to hours. Let's change the minutes to hours:

$$\text{Area} = \frac{(lb/hr)(ft^3/lb)}{(ft/min)(60 \text{ min/hr})}$$

This is the cross-sectional area of the pipe in square feet. The next step is to convert square feet into square inches:

$$(\text{Area, } ft^2) \ (144 \ in^2/ft^2) = \text{Area, } in^2$$

Area equals $0.7854d^2$ (where d = diameter), so the diameter can be calculated as follows:

$$d = \frac{\sqrt{\text{Area}}}{0.7854}$$

Notice that all units of measure were kept consistent throughout.

Pipe Expansion

Metal expands when heated. Different steels have different coefficients of expansion. For the purposes of our example here, we will use the coefficient of 0.0000065 inch per inch-°F. As usual, watch your units.

Consider a 30-ft line that will carry steam at 300°F. The pipe's temperature is now 70°F. How much will it expand?

$$(30 \text{ ft}) \ (12 \text{ in./ft}) \ (0.0000065 \text{ in./in.-°F}) \ [(300° - 70°F)] = 0.5382 \text{ in.}$$

This expansion and contraction has to be considered to prevent damage. The following three methods will take care of this problem:

- Expansion loops can be installed at critical points to compensate for the change in length, Figure 7-8. Their spring action absorbs the pipe's movement.

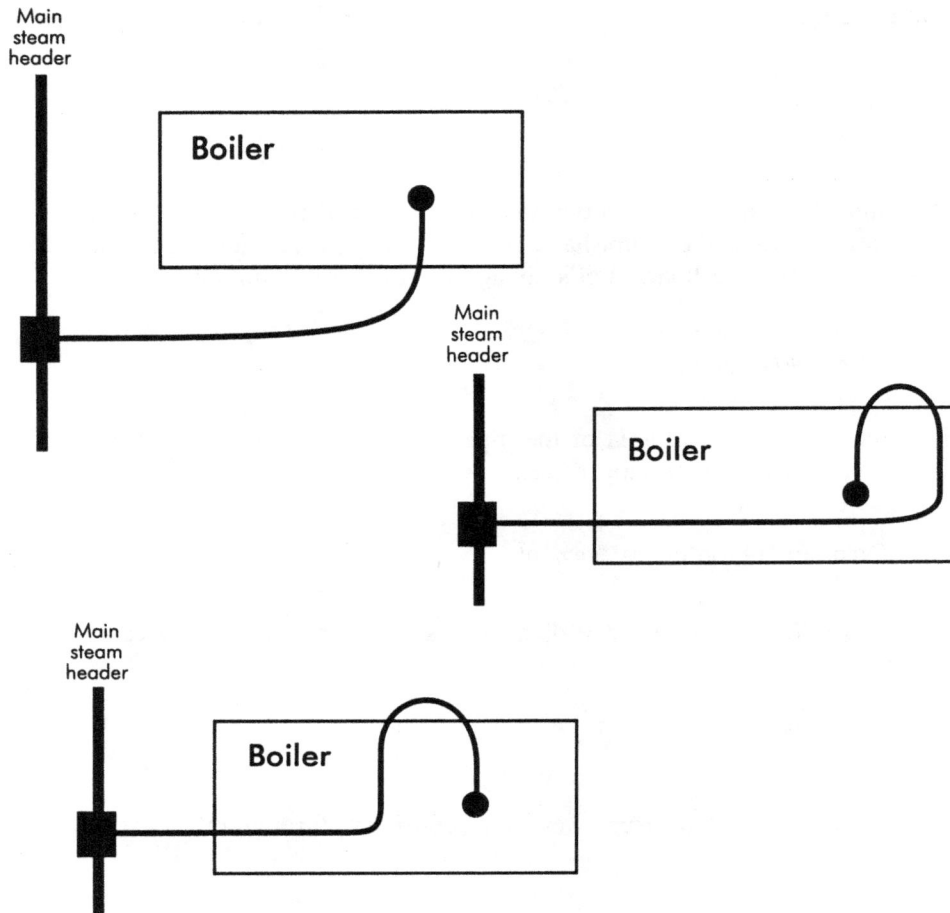

Figure 7-8. Expansion loops

- A packed slip joint may be installed, which provides a telescoping action. The packing is similar to that used for valve stems.

- Similar to the packed joint is the corrugated expansion joint. The accordion-like action takes care of the pipe movement and won't leak like a packed joint could.

Water Treatment

Water has been called the universal solvent, because it dissolves many substances. It never occurs in nature in a pure state, because groundwater picks up impurities as it seeps through rock strata. Surface water contains organic matter and insoluble suspended matter such as sand and silt. Even rain picks up oxygen and carbon dioxide as it falls to Earth.

When water is turned into steam, the minerals previously dissolved in the water are left behind and deposit a scale on the hot boiler surfaces. This scale, mostly calcium and magnesium, is an excellent insulator and slows the transfer of heat to the water. When boiler surfaces are covered with scale, heat normally absorbed by the water goes up the stack instead, and the tube metal temperature rises to the point of failure. Water must be treated to prevent scale deposits and to control corrosion.

Table 8-1 shows how the thickness of boiler scale corresponds to fuel waste.

Thickness of Boiler Deposits (in.)	Fuel Wasted (%)
1/32 (0.031)	7
1/25 (0.040)	9
1/20 (0.050)	11
1/16 (0.063)	13
1/11 (0.091)	15
1/9 (0.111)	16

Table 8-1. Fuel wasted due to boiler deposits (Reprinted from the U.S. Bureau of Mines)

The amount of scale dissolved in the feedwater is measured in parts per million (ppm). While parts per million may seem like an insignificant measurement, consider the accumulation of scale for a 100-hp boiler producing 3,450 lb of steam per hour with the feedwater hardness at 50 ppm:

Time frame	Scale accumulation (lb)
1 hour	0.1725
24 hours	4
1 week	29
1 month	126
1 year	1,507

Boiler failure occurs when scale is allowed to accumulate. Steel retains its strength up to 700°F, and it starts to weaken above that point. At 1,000°F it has a hard time supporting its own weight. How does steel survive in a boiler when the flame temperature is 3,000°F? It survives because water absorbs the heat from the metal fast enough so the temperature of the metal does not reach the danger point. Start adding insulation such as scale between the metal and water, and the metal temperature increases. With enough scale, the metal will overheat causing blisters, bags, and eventually a ruptured pressure part.

EXTERNAL TREATMENT

Water conditioning may either be external or internal. External treatment is the treatment of water before it enters the boiler. The raw water is filtered to remove suspended matter. Pressure filters often use sand to act as a filter.

Softeners

Water hardness is determined by calcium and magnesium levels in the water. Water softening is the process that removes hardness from the water by changing the calcium and magnesium carbonates into sodium carbonates. The sodium cycle ion exchange process is the term used to describe the simple water softening process used in thousands of homes and industries.

From the outside, softeners look like pressure filters. The main difference is that a softener can introduce brine into the vessel for regeneration purposes.

The water softening process takes place as follows:

1. In the service stage of the water softening process, water enters the top of the softener, runs through the zeolite resin bed, exits through the bottom, then goes on to the deaerator or feedwater pump, Figure 8-1. The zeolite resin trades its sodium for the calcium and magnesium in the raw water. In other words, the raw water enters the softener with its harmful hardness and leaves with harmless sodium. This process continues until the zeolite resin runs out of sodium. This is called the endpoint, and it is when the softener must be regenerated. If the process is allowed to continue, the hardness will break through. The endpoint is determined by a simple timing cycle; by measuring how much water has been softened; or by testing the output for hardness.

2. When the softener reaches its endpoint, it is taken off line and backwashed for about 10 minutes to loosen and regrade the zeolite to prevent packing and channeling. Backwashing also removes accumulated dirt. The backwash flow is from the bottom to the top, and then out to a waste drain, Figure 8-2.

3. Plain salt (NaCl) is used for regeneration, because it is an inexpensive source of sodium. The zeolite resin is regenerated by passing a saturated salt solution (brine) downward through the exchanger bed. This strong brine solution removes the calcium and magnesium from the zeolite and replaces them with sodium, Figure 8-3. Many installations use a brine pump.

4. The softener is then placed on rinse, and water is admitted from the top to flush out the calcium, magnesium, and remaining brine. The rinse water leaves through the bottom and goes to the waste drain, Figure 8-4. Conductivity or hardness measurements determine when the softener has been rinsed enough. The softener is then ready to go back into service.

Opening and closing separate valves to direct the flow of water into the softener can be a confusing and tedious process. This is why pressure filter and softener manufacturers either automate the valves, or put all the valves in the same body in a multiport design.

When there is a high demand for soft water, two or more softeners are usually installed in parallel to ensure an adequate supply. When one softener has reached its endpoint it is taken off line and regenerated, and a previously regenerated softener is placed on line.

Water scale kills boilers. The best way to prevent scale is to remove it before it gets into the boiler. Keep in mind that hardness will be removed from the feedwater; either the softener can remove it, or the boiler can remove it by converting it into scale.

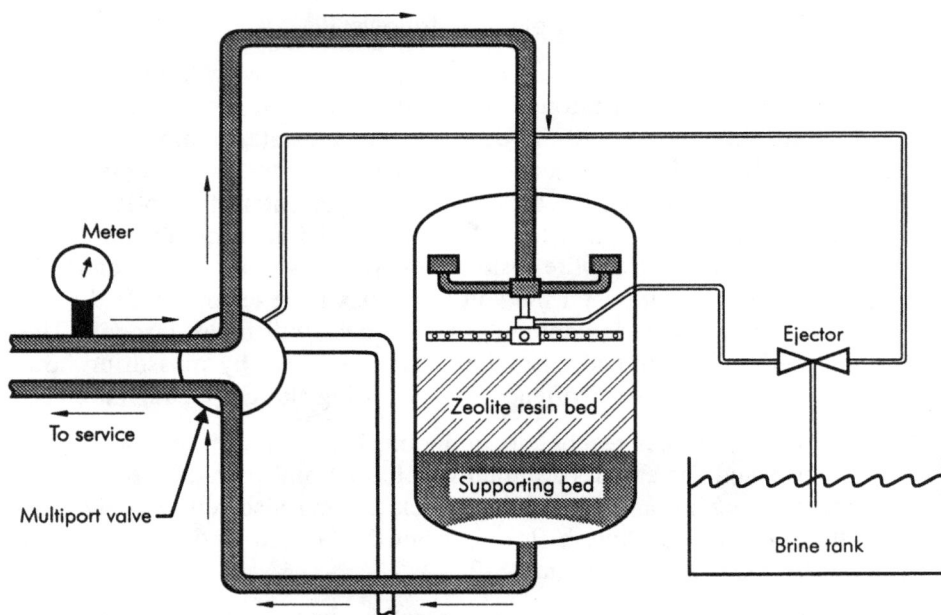

Figure 8-1. Service stage in the water softening process

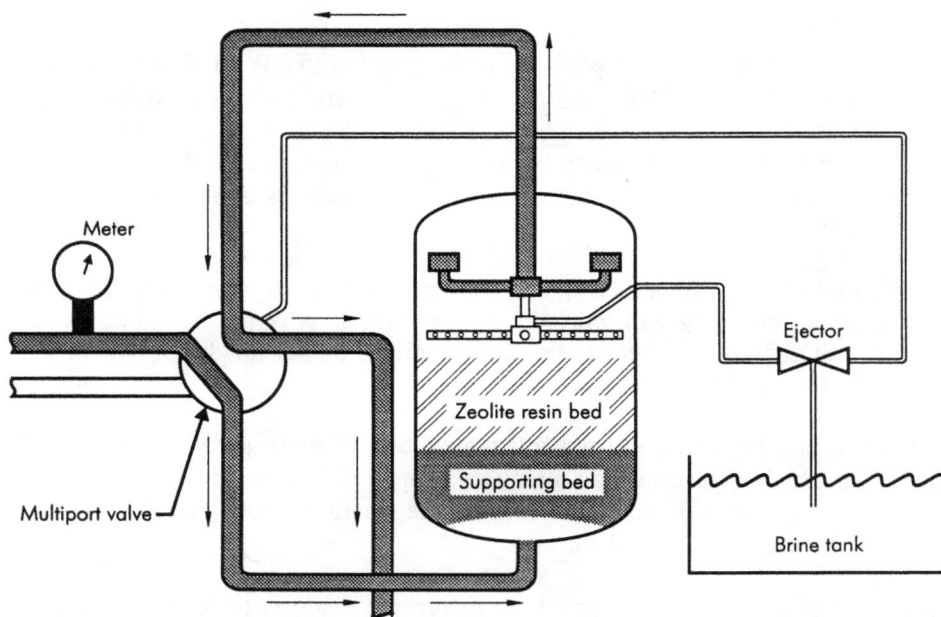

Figure 8-2. Backwashing

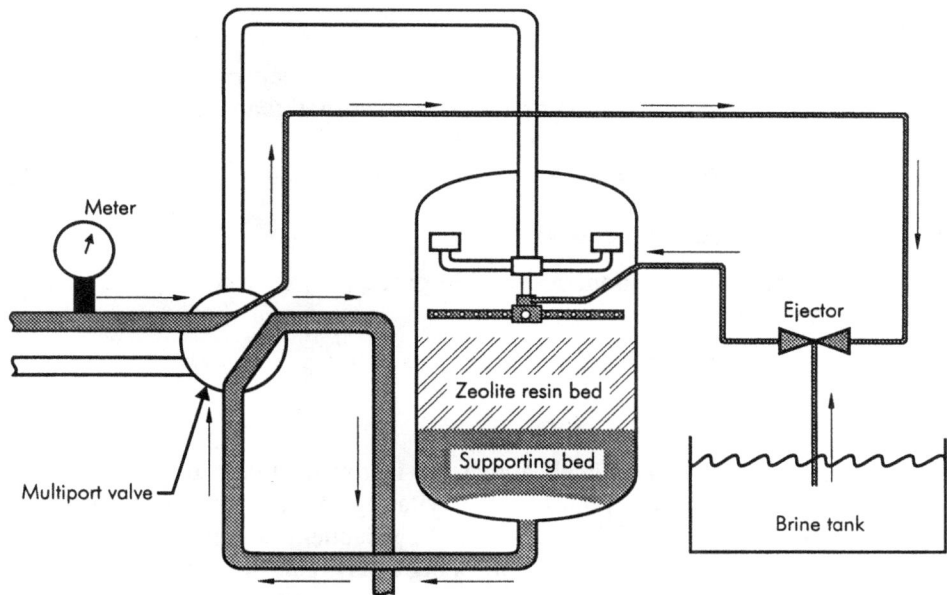

Figure 8-3. Regeneration

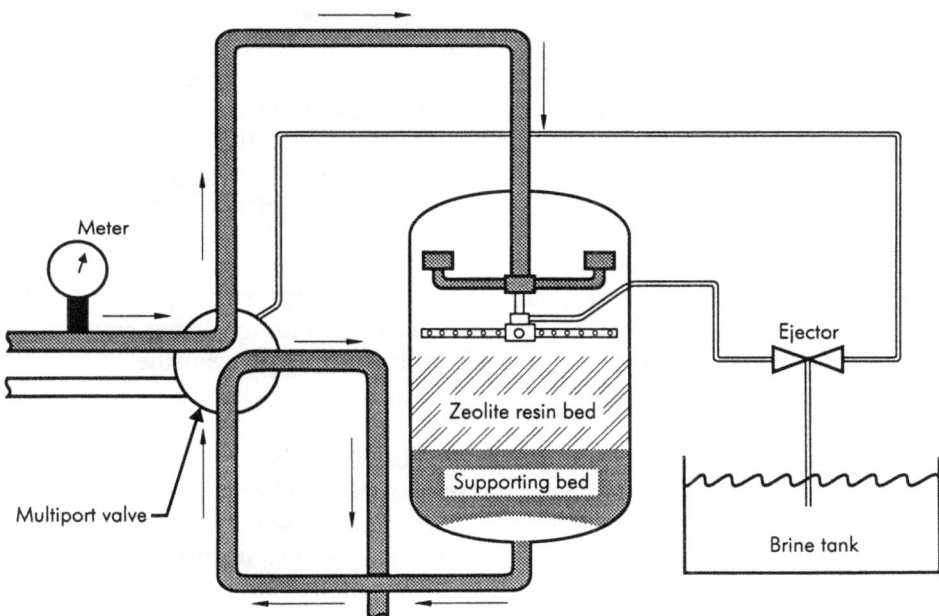

Figure 8-4. Rinse

Deaeration

Oxygen in the boiler causes corrosion in the form of deep pits. Deaerators work on the principle that hot water holds less dissolved gas than cold water. Heated water in a pot has bubbles that form under the surface long before the water boils. These are dissolved gases coming out of solution as the water warms up.

Deaerators perform three functions:

- Removal of dissolved noncondensable gases (air) such as oxygen, nitrogen, and carbon dioxide

- Heating of feedwater

- Storage of feedwater

Deaerators are always located well above the feedwater pump to provide enough head pressure to prevent the pump from cavitating and/or from becoming vaporbound. Cavitation is the formation and subsequent collapse of steam bubbles in the water, which, apart from being noisy, can destroy an impeller in a few weeks of continuous operation. Centrifugal pumps begin to cavitate when the suction pressure is too low to keep the hot water above its vapor pressure. When a pump is vaporbound, its impeller is filled with steam instead of water. Centrifugal pumps cannot operate when that happens.

Under certain circumstances, a deaerator may be considered to be an open feedwater heater; that is, a closed vessel where steam and water come into contact. An open feedwater heater is a pressure vessel but is not open to the atmosphere. Steam is usually supplied at 5 to 15 psi, and the temperature of the feedwater is very close to the saturated temperature of the supply steam. Add the proper internals and a vent, and the open feedwater heater becomes a deaerator.

In closed feedwater heaters, the steam and water are separated by tubes so they do not come in contact. They are used in electric generating plants between the feedwater pump and the boiler. The tube section must withstand the full boiler feed pump pressure, which can be up to 5,000 psi.

When supply water enters the deaerator, it is broken down into very small drops, either by being forced through a spray nozzle and/or by flowing over a series of trays. The water then comes in direct contact with the steam and is heated to within a few degrees of the steam. Whether the incoming water is distributed by sprays or by a combination of sprays and trays, the released gas works its way to the top of the vessel. The gas then passes through an internal vent condenser cooled by the incoming water. The condensable vapor (steam) falls into the storage section. The noncondensable gases (oxygen, nitrogen, and carbon dioxide) are vented to the atmosphere.

Some deaerators have an internal vent condenser, which is located in the deaerator vent line, either inside or outside the deaerator itself. It saves energy by condensing the vent steam and returning the condensate back to the system, while allowing the noncondensable gases to escape. Cold make-up water is used to condense the steam, and this heated water is also returned to the system.

It is very important to keep the deaerator vent open, otherwise there can be no deaeration. The deaerator vent may be kept open by:

- maintaining a healthy visible vapor plume out of the vent;
- locking the vent valve open;
- drilling a hole in the vent valve's gate;
- removing the vent valve's gate;
- not installing a vent valve in the first place.

Deaerators remove oxygen down to 0.005 cc per liter, but to make sure none of it reaches the boiler, an oxygen scavenger is added to the storage section of the deaerator. Sodium sulfite is used because it works well and is inexpensive. It does add solids to the boiler water, which is not good when the pressures exceed 1,000 psi. Hydrazine, which is expensive and a little tricky to handle, is used in high-pressure boilers. It doesn't add solids because it turns into water and nitrogen when it absorbs oxygen. The following chemical equations show how these oxygen scavengers work:

$$2Na_2SO_3 \text{ (sodium sulfite)} + O_2 \text{ (oxygen)} \longrightarrow 2Na_2SO_4 \text{ (sodium sulfate)}$$

$$N_2H_4 \text{ (hydrazine)} + O_2 \text{ (oxygen)} \longrightarrow N_2 \text{ (nitrogen)} + 2H_2O \text{ (water)}$$

In a typical boiler room installation, condensate is returned to a surge tank or condensate receiver. The condensate loss is made up automatically by a float-controlled valve using city water, which first goes through a softener. This water is then pumped to the deaerator through another float-controlled valve, which maintains the deaerator level. The feedwater pumps located under the deaerator then transfer the water to the boilers. Some localities require a backflow preventer in the city water line feeding the softener. This is to prevent contaminated water from entering the potable (drinking) water supply.

INTERNAL TREATMENT

Internal water treatment is applied inside a boiler and is used to prevent corrosion and the formation of scale. The most common method of internal control is the use of phosphate to remove scale formation. Phosphate changes the calcium carbonate and magnesium carbonate into a precipitate or sludge. Compounds are added to keep this precipitate or sludge from sticking to metal surfaces. The sludge is then removed by bottom blowdown. The chemical reaction works best when the pH level is between 10.5 and 11.2. Caustic soda is added to maintain this pH level.

The pH level is a number between 0 and 14 that indicates the degree of acidity or alkalinity. Numbers 0 to 7 denote acidity with 0 being the strongest acid. Numbers 7 to 14 denote alkalinity with 14 being the most alkaline. The mid-point of the scale is 7 and means the solution is neutral. Figure 8-5 shows approximate pH values of acids and bases.

The phosphate concentration in boiler water should be between 30 and 60 ppm. One ppm means one million pounds of water contain one pound of dissolved solids. Grains per gallon is another unit of measurement for concentrations. One pound equals 7,000 grains, and one grain per gallon equals 17.1 ppm.

Chelants and polymer chemicals are also used for internal treatment and are better than phosphates at maintaining scale-free metal surfaces. They form soluble complexes with the calcium carbonate and magnesium carbonate and are removed through the continuous blowdown. Caustic soda, phosphates, chelants, and polymers are usually added directly to the boiler.

Chemical Feed

Shot feeding can be accomplished by either bypassing a pump or bypassing the pump's discharge line. Figure 8-6 shows a feeder connected to the suction and discharge lines of a pump. Chemical is added to the system by bypassing part of the pump's discharge through the feeder and then back to the pump's suc-

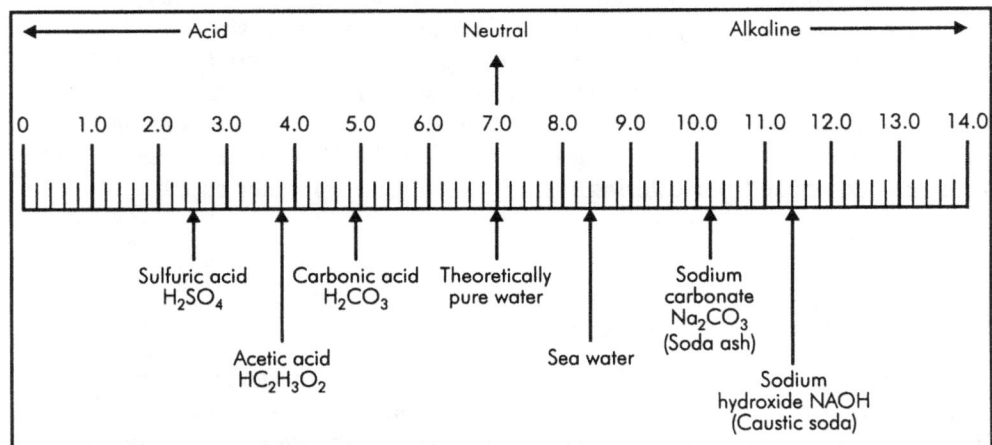

Figure 8-5. Approximate pH values of common acids and bases

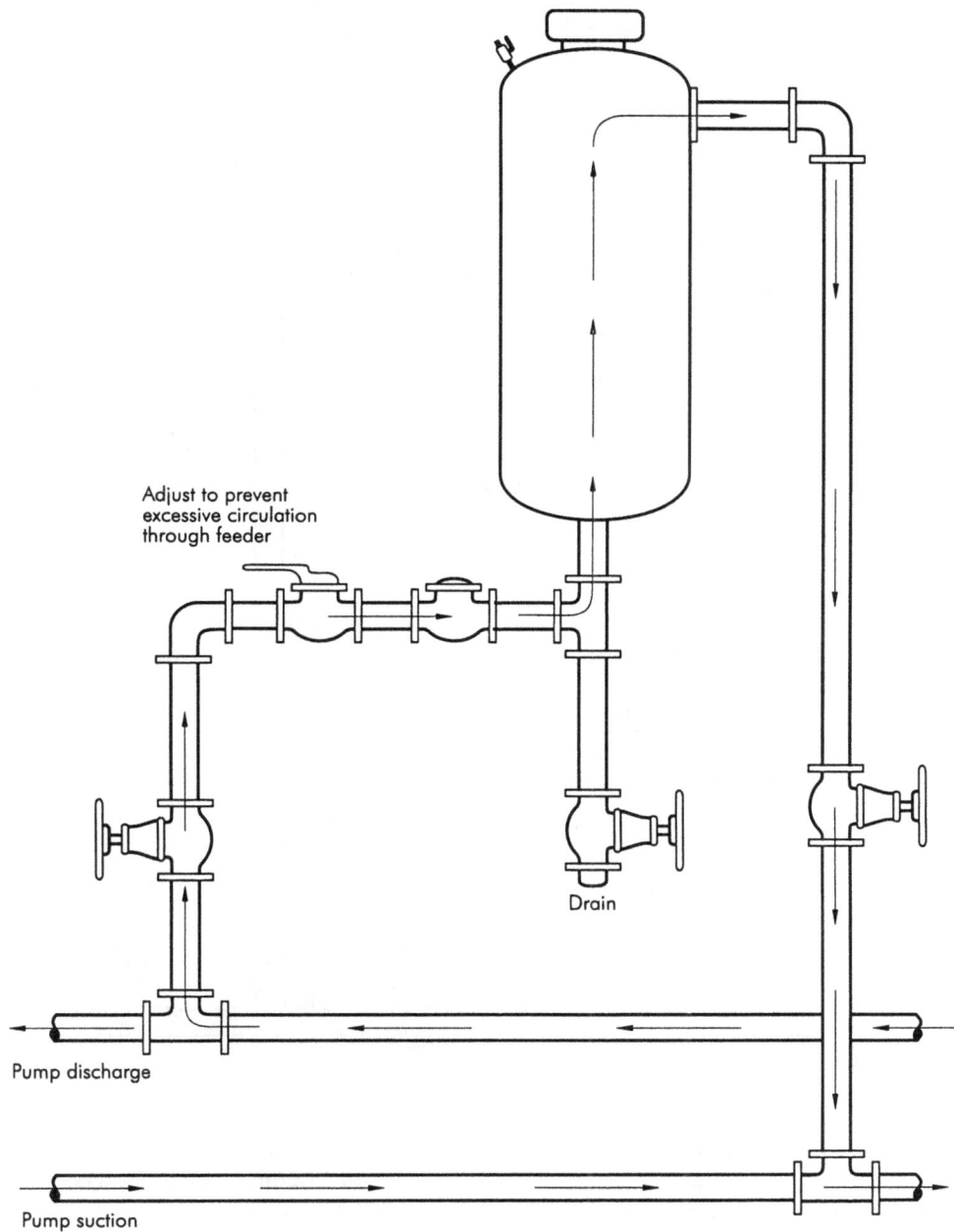

Figure 8-6. Pump bypass shot feeder

tion. A word of caution is that most concentrated boiler chemicals can eat the bronze pump and valve parts.

Figure 8-7 shows a bypass feeder piped into the pump's discharge line. By opening the two feeder valves and closing the bypass valve, the pump's discharge flows through the feeder, thus adding chemical into the system.

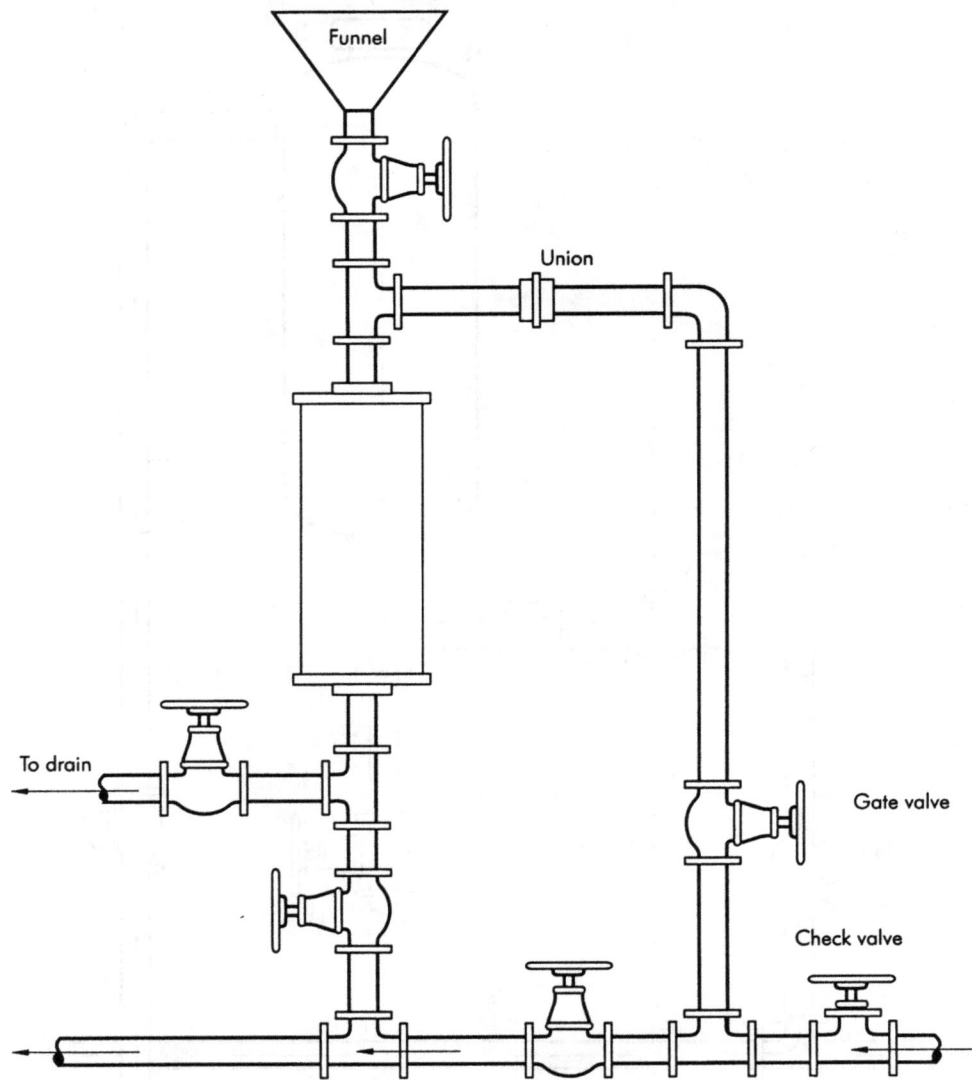

Figure 8-7. Bypass feeder

The best way to add chemicals to the boiler water is to use a chemical feed pump. These pumps can feed the precise amount of chemical into the system continuously. They are usually placed under a small tank equipped with a mixer to blend the chemicals.

Sampling

To add the right amount of chemical, it is necessary to know how much chemical is in the boiler water initially. The first step in water testing is to obtain a good sample.

If hot boiler water is used, errors are incurred by flashing. For example, if a water sample is taken from a 150 psi boiler, the dissolved solids will appear higher because about 16% of the water evaporates due to flashing. Before testing, boiler water should first go through a sample cooler. A sample cooler is a heat exchanger that uses cooling water (usually city water) to remove enough heat from the boiler water so it doesn't flash and is not too hot to handle. The sample should be collected in a plastic or glass container. Do not use aluminum containers, because the caustic soda in the boiler water reacts with aluminum.

The continuous blowdown line is the best point from which to draw sample water from a boiler. This line gives a representative sample from the area of highest concentration of dissolved solids that is undiluted by incoming feedwater.

The common water sample tests are for alkalinity, sulfite, total dissolved solids, and phosphate or chelant, depending on the treatment program. The chemical supply vendor will explain what level of chemical residue should be in the boiler. Adjust the feed to maintain the recommended level of chemicals.

A large part of the money paid for chemicals is for the knowledge and experience of the supply vendor and for the laboratory facilities that test their chemicals. The vendor should be ready, willing, and able to help correct existing problems and to offer advice on how to avoid potential problems. However, it is not all the vendor's responsibility; after proper training, plant personnel have to do their part.

CAUSTIC EMBRITTLEMENT

Caustic embrittlement is a form of boiler metal failure that is normally undetectable until the metal fails suddenly. Three conditions must be present for caustic embrittlement to occur:

1. Leakage of boiler water to permit the escape of steam and subsequent concentration of boiler water at the point of leakage

2. Corrosion of the boiler metal by concentrated caustic soda originating from the concentrated boiler water

3. High metal stress in the area

Leakage and concentration of boiler water normally occur in steam-drum riveted joints and where boiler tubes are rolled into the drum or tube sheet. Expansion and contraction can create small crevices through which boiler water can escape. Modern boilers are now welded instead of riveted, so caustic embrittlement has become less of a problem.

TUBE CLEANING

Excessive waterside deposits must be removed or tube failures can be expected. These deposits can be removed mechanically or chemically.

Mechanical cleaners, like the one shown in Figure 8-8, consist of a rotating cutter head driven by an air-powered motor that is small enough to fit through the tube. The operation is sometimes called *turbining* tubes. The operator feeds the rotating cutter through the tube, along with a stream of water to keep equipment cool and to flush away the removed scale. The top cutter is used on fire-tube boilers; it knocks scale off the outside by hitting the inside of the tube. The other two cutters are for water-tube boilers.

The most commonly used chemical cleaning solution is a 5% hydrochloric acid solution with an inhibitor added to lessen the acid attack on the boiler metal. There are two basic acid-cleaning methods: circulation and static (sometimes called fill and soak). With both methods, pH measurements are taken periodically. When the readings remain constant, the cleaning is complete.

Which method is best? Acid cleaning does a very good job of cleaning every part of the boiler, including sections that mechanical cleaners can't reach. This method should be used by specialists who have the experience and equipment to do a proper job. One serious problem can be the spent cleaning solution. If there are no on-site disposal facilities, it must be trucked away. Mechanical cleaning may be done by plant personnel. One advantage of mechanical cleaning is that by running a cutter head through each tube, you can be assured that no tubes are blocked.

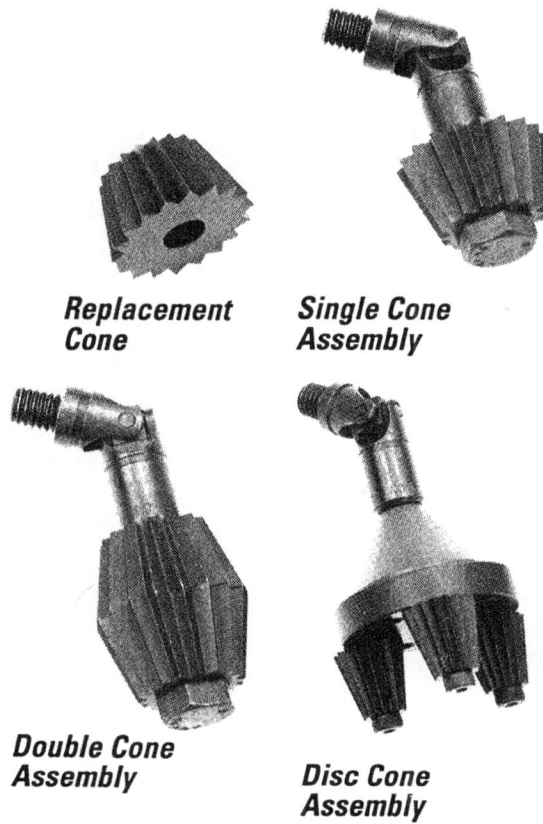

Figure 8-8. Tube cleaners (Courtesy, Goodway Technologies Corporation)

Turbines

A turbine is a rotational device that converts the heat energy contained in steam into mechanical energy. Turbines are classified by how the exhaust steam is handled and the way it is converted into mechanical energy. If the steam is exhausted into a condenser under a vacuum, it is called a *condensing* type. If it is exhausted into a steam line that is under pressure, it is called a *noncondensing* or *back-pressure* type. Utilities use condensing turbines to generate electricity, while industry uses back-pressure turbines to drive mechanical devices such as pumps and fans. There are also combinations of the above, such as the extraction turbine, from which some steam is extracted before it reaches the condenser.

Turbines can also be classified by the way steam is converted into mechanical energy. The two types are impulse turbines and reaction turbines. Imagine trying to push a car stuck on ice. You walk up to the car and give it a good shove. The *impulse force* you applied is enough to send the car on its way. Unfortunately, because you did not get a good foothold on the slippery surface, the *reaction force* sent you flying in the opposite direction.

The first known steam-driven device was designed by Hero around 175 B.C. It consisted of a hollow sphere supported on its axis. Steam exiting through two nozzles resulted in a reaction that caused the sphere to turn. It was really a toy that developed no usable power. The second turbine of historical significance was the Branca impulse wheel of 1629 A.D. Steam left a fixed nozzle and hit the blades of a wheel. The change of the steam's momentum when it hit the blades resulted in an impulse force, which caused the wheel to turn.

As shown in Figure 9-1, an impulse turbine has nozzles that direct steam against crescent-shaped blades (sometimes called buckets) of a rotating wheel.

In a velocity-compounded impulse turbine (Curtis stage turbine), Figure 9-2, the steam passes through a set of stationary blades. The stationary blades guide the steam to another row of rotating blades to absorb the kinetic energy of the high-velocity steam. If the steam pressure is high enough, additional stages can

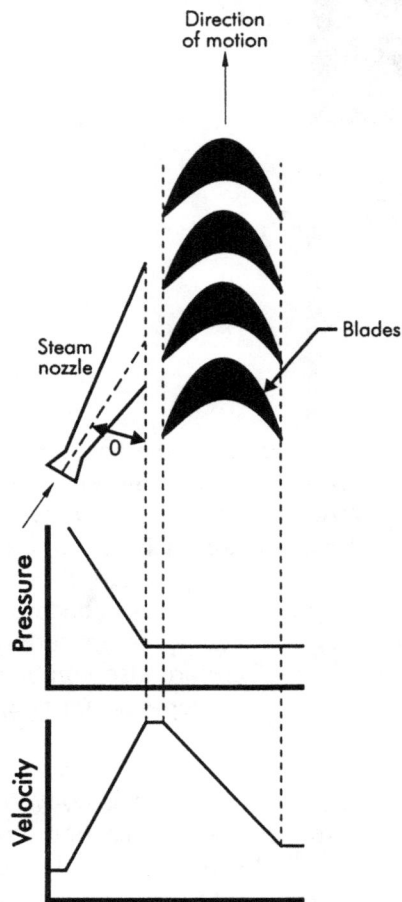

Figure 9-1. Impulse turbine

be used. Efficiency is increased by allowing only a limited pressure drop in one set of nozzles. Note that with impulse turbines, the pressure drop is at the nozzle; no pressure drop occurs while the steam passes through the blades. Figure 9-3 shows a typical impulse mechanical-drive turbine with two rows of rotating blades and one row of stationary blades.

Figure 9-4 shows a reaction turbine, which also uses alternating rows of stationary and rotating blades shaped like airfoils. The stationary blades direct the steam against the next row of rotating blades. Note that unlike the impulse design, there is a pressure drop in the moving blades. Because of the pressure difference across the moving blades, the rotor is subjected to an axial force, which may be very large.

An axial force is a force in the direction of a shaft. It tries to push the shaft out of the casing. Thrust bearings are used to counteract this force and hold the shaft in place. Bearings will be discussed in greater detail later.

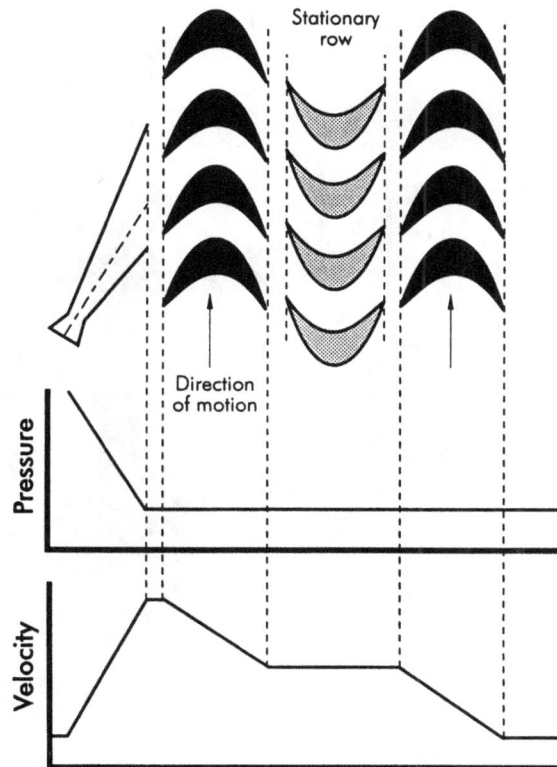

Figure 9-2. Curtis stage turbine

On large machines where the axial force could be so large that the thrust bearings would be too large, a special design is used. By having the steam enter the casing in the middle, then exit at the ends, the two axial forces are canceled out. Figure 9-5 shows the configuration of a typical large turbine. The axial thrust in the high-pressure turbine stage is absorbed by a thrust bearing. The axial thrust in the low-pressure section is canceled by having the steam enter the middle and exit the ends.

Large turbines can take advantage of the inherent efficiencies in each design. High-pressure steam first goes through a few impulse stages and then many reaction stages. The impulse stages are more efficient at higher pressures; the reaction stages are more efficient at lower pressures.

The efficiency of mechanical-drive turbines is given in steam rate or water rate (pounds of steam or water per horsepower-hour). The lower the rate, the higher the efficiency. The term *water rate* stems from the method of measurement. It is much easier to weigh the condensed water leaving an engine or turbine than it is to measure the entering steam. Steam rates are generally reduced with

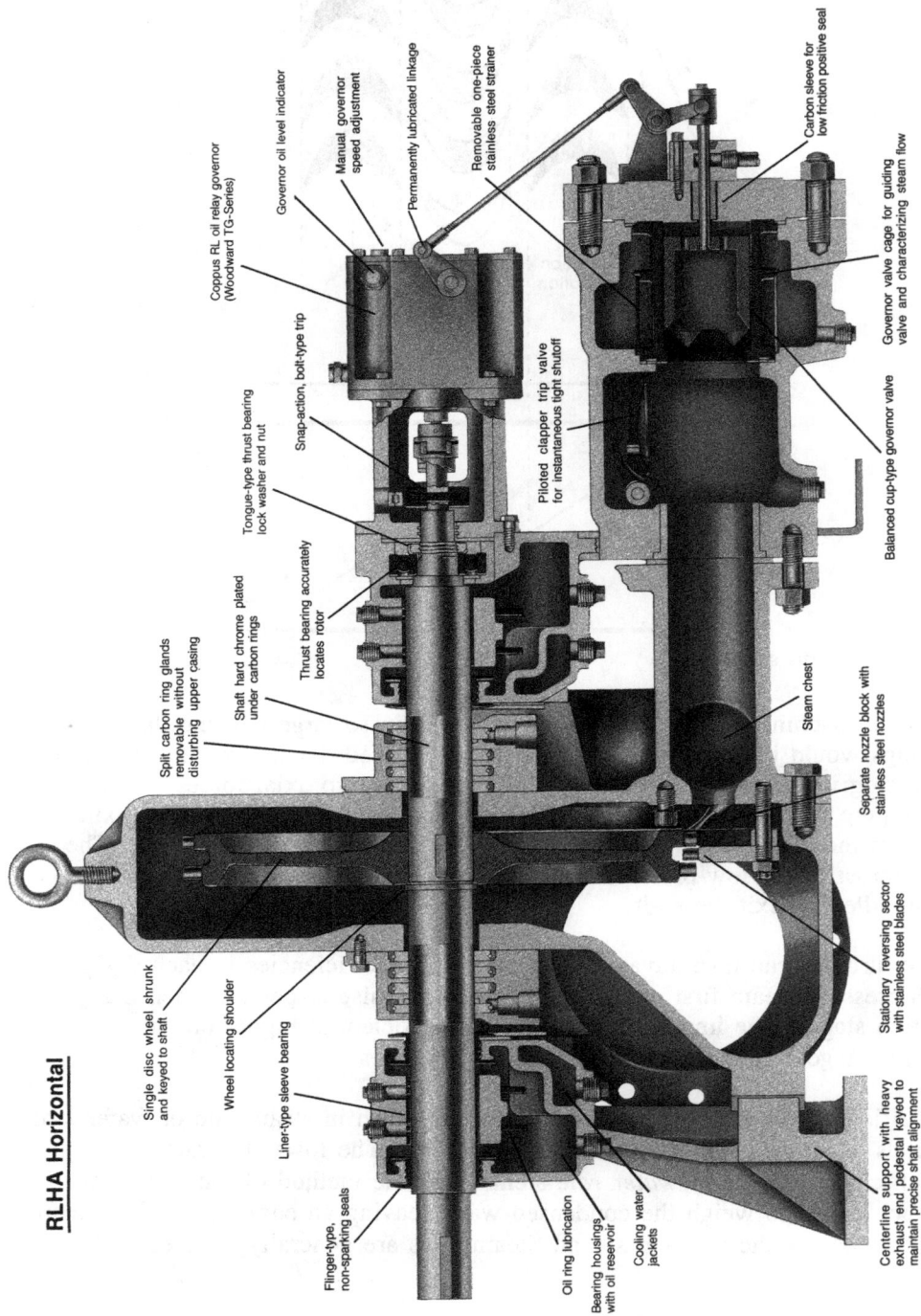

RLHA Horizontal

Single disc wheel shrunk and keyed to shaft

Wheel locating shoulder

Liner-type sleeve bearing

Flinger-type, non-sparking seals

Oil ring lubrication

Bearing housings with oil reservoir

Cooling water jackets

Centerline support with heavy exhaust end pedestal keyed to maintain precise shaft alignment

Stationary reversing sector with stainless steel blades

Separate nozzle block with stainless steel nozzles

Steam chest

Balanced cup-type governor valve

Governor valve cage for guiding valve and characterizing steam flow

Carbon sleeve for low friction positive seal

Piloted clapper trip valve for instantaneous tight shutoff

Coppus RL oil relay governor (Woodward TG-Series)

Governor oil level indicator

Manual governor speed adjustment

Permanently lubricated linkage

Removable one-piece stainless steel strainer

Snap-action, bolt-type trip

Tongue-type thrust bearing lock washer and nut

Thrust bearing accurately locates rotor

Shaft hard chrome plated under carbon rings

Split carbon ring glands removable without disturbing upper casing

Figure 9-3. Typical impulse mechanical-drive turbine (Courtesy, Tuthill Corporation, Coppus Turbine Division)

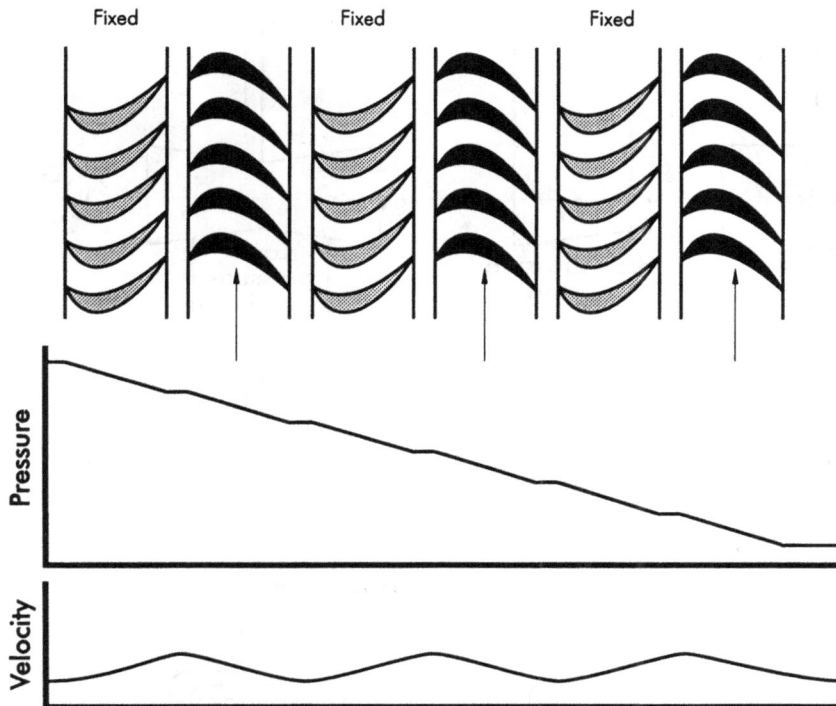

Figure 9-4. Reaction turbine blades

higher turbine speeds, greater number of stages, larger turbine size, and always when there is a higher difference in heat content between entering and leaving steam conditions.

Most turbines can easily operate at 6,000 to 7,000 rpm, and some go up to 10,000 rpm. Speed-reduction gears are used to match the turbine's most efficient operating speed with the requirements of what is being driven.

TURBINE GOVERNORS

Governors maintain constant turbine speed in spite of load fluctuations by regulating the steam flow. All governors have a speed sensing device and a means to position the throttle valve.

On small machines, the direct-acting flyball type is used as both the speed sensing element and throttle positioner. As shown in Figure 9-6, centrifugal force acts upon spring-loaded weights, which are attached to the end of the shaft. Increasing the speed forces the weights outward, which in turn positions the throttle valve via a lever and link.

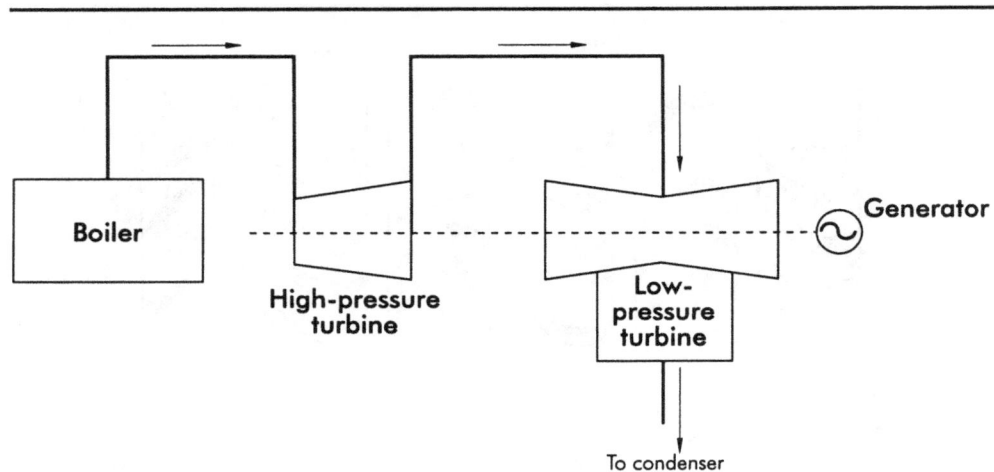

Figures 9-5. Typical large turbine

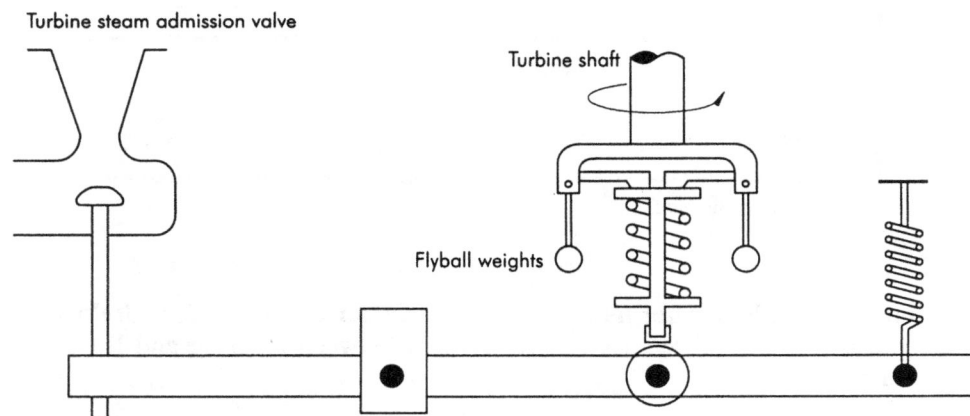

Figures 9-6. Direct-acting flyball governor

For more precise speed control, the direct-acting hydraulic governor is used, Figure 9-7. Here speed sensing is accomplished with a hydraulic oil pump directly coupled to the turbine's shaft. The faster the speed, the higher the output pressure. This pressure directly positions the throttle valve.

For larger machines where considerable force is required to position the throttle valve, an oil-relay governor is used, Figure 9-8. All the speed sensing device has to do is to position the relay valve, which in turn positions the throttle valve's hydraulic power cylinder. The hydraulic motive power is supplied by a separate oil pump which often doubles as the bearing lubrication pump.

On smaller machines, a single governor-controlled steam-admission valve is sufficient for satisfactory operation. On larger machines, multiple nozzle valves are used for more precise control. Some turbines have manual valves to close

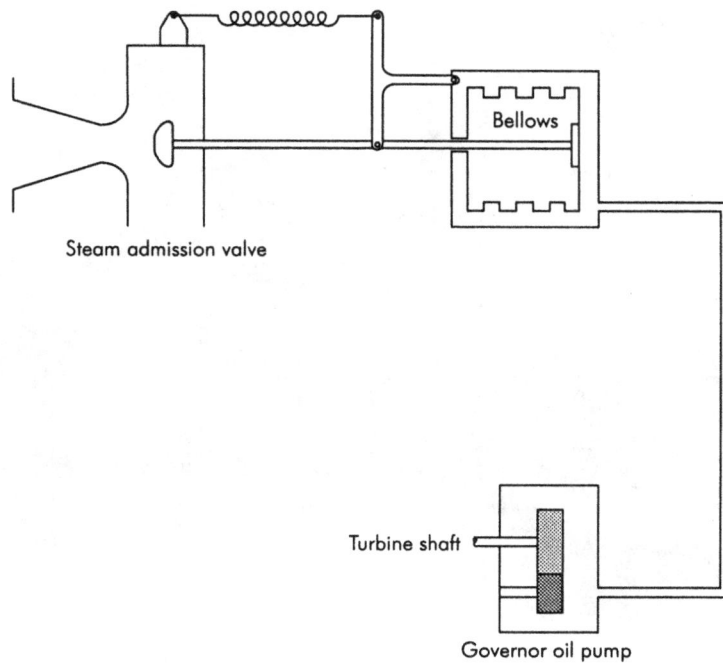

Figures 9-7. Direct-acting hydraulic governor

off nozzles for better efficiency at low loads. A strainer should be located ahead of the throttle to prevent solid particles from entering.

Safety Devices

All turbines are equipped with overspeed trips to protect against possible damage caused by the sudden loss of load, Figure 9-9. If, for instance, a coupling or shaft were to break and the speed control governor was not able to close the throttle valve fast enough, the turbine would accelerate so quickly that it would destroy itself and possibly injure personnel in the area.

Overspeed trip mechanisms are entirely independent of the speed control governor. Even if the governor becomes disabled, the overspeed trip will still function. Unless otherwise stated, for test purposes, the overspeed trip is set at 10% above the operating speed. For example, a turbine operates at 3,600 rpm. At what speed does the overspeed trip kick in? The answer is:

10% x 3,600 rpm = 360 rpm

3,600 rpm + 360 rpm = 3,960 rpm (or 3,600 rpm x 110% = 3,960 rpm)

Figure 9-8. Turbine governor (Courtesy, Tuthill Corporation, Coppus Turbine Division)

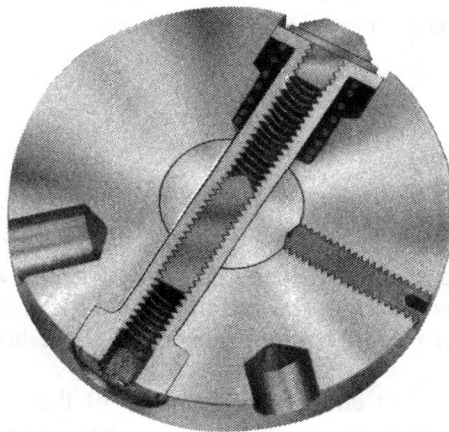

Figure 9-9. Overspeed trip mechanism (Courtesy, Tuthill Corporation, Coppus Turbine Division)

In smaller turbines, the overspeed trip actuating mechanism is a spring-restrained bolt located in the shaft. When centrifugal force overcomes the spring force, the bolt will move outward and trip the overspeed valve closed.

Once a turbine is placed in service, it is usually on line for months at a time. This is why the overspeed trip is tested whenever a turbine is started or stopped. The stored energy released when rotating turbine blades fail due to overspeed is enough to shatter the casing and propel metal pieces around like a bomb.

On startup, one operator observes a tachometer, while another operates the throttle. Once brought up to operating speed, the throttle operator deliberately causes the machine to overspeed, while the other operator notes when the overspeed trip operates. If the operation is not satisfactory, the problem must be corrected before the turbine is placed on line.

The overspeed trip is manually operated when the turbine is shut down. Depending on the machine, it is tripped either with the steam valve wide open or when the machine is coasting to a stop. This verifies that the mechanism still operates after extended service.

Besides the overspeed trip, turbines also have pressure-relieving devices attached to their casing. Back-pressure turbines have sentinel valves to alert the operator of an overpressure condition. These valves are required because the casing is not designed to withstand the steam main pressure. Overpressure occurs when the steam supply valve is opened before the exhaust valve is opened.

Condensing turbines have rupture disks attached to their low-pressure portion. Again, this portion of the casing is not designed to withstand high pressures. Unlike safety valves, rupture disks are good just one time. They are thin sheets of metal designed to fail at a specific pressure. One ruptured, must be replaced.

CONDENSERS

A prime mover converts energy into shaft horsepower. Motors, engines, and turbines are examples of prime movers. The purpose of the condenser is to lower the back pressure of the prime mover. This allows the steam to drop to a lower pressure and temperature, thereby increasing both efficiency and capacity. Equally important, it also recycles the pure hot condensate, which goes back to the boiler.

Condensing water, circulating water, and cooling water are used interchangeably to refer to the water that carries the rejected heat away. Do not confuse these terms with the term condensate, which is the pure hot water resulting from the condensed steam.

There are two types of condensers: contact and surface condensers. In contact (jet or barometric) condensers, steam and condensing water come in contact with each other. Unless there is a need for a lot of hot process water, this condenser is no longer used. Its greatest disadvantage is that the condensate is unsuitable for boiler feedwater, because it is contaminated with the condensing cooling water.

In surface condensers, steam and condensing water never make contact with each other. Therefore, the condensate remains uncontaminated and is suitable for reuse as boiler feedwater. The heat in the condensate is also recycled and directed back to the boiler. After the steam is exhausted from the turbine, it passes around the tubes (called the shell side). Cooling water circulates inside tubes (called the tube side).

Cooling water is often not very clean. Condensers are designed so access to the tube sheet is relatively easy, and the inside of the tubes can be cleaned periodically. If the cooling water is very dirty, the tube side of the condenser is divided in half. The output of the turbine is reduced, half of the condenser is taken out of service and cleaned, then the procedure is repeated for the other half.

The condenser shell is designed to operate under a vacuum, not under pressure. An atmospheric relief valve, located on the condenser shell, automatically opens to relieve pressure when the condenser is no longer in a vacuum.

Air invariably finds its way into the condenser and must be removed to maintain condenser performance. As air accumulates, it increases the back pressure on the turbine, reduces the rate of heat transfer, and therefore, the capacity of the condenser. The most common way to remove the noncondensable gases is with a multistage steam-jet air pump. Air is drawn into the path of steam as it leaves a nozzle, then to a condenser where the air is separated. The process is repeated in additional stages until the air can be discharged above atmospheric pressure.

A hot well is at the bottom of the condenser where the condensate accumulates. A specially designed hot well pump removes the condensate for its ultimate return back to the boiler. This pump operates under very difficult conditions since it has to handle hot water under a vacuum.

To find the amount of cooling water required under specified conditions, the following equation may be used:

$$Q = \frac{H - (t_o - 32°F)}{T_2 - T_1}$$

where:

Q = weight of cooling water to condense a pound of steam
H = heat content of exhaust steam
t_o = temperature of condensate
T_1 = temperature of cooling water entering condenser
T_2 = temperature of cooling water leaving condenser

After the amount of cooling water is found to condense one pound of exhaust steam, multiply by the amount of pounds of steam per hour to complete the problem.

Example 9-1. A 50,000 kW plant has a steam rate of 14 lb/kWh. Its condenser operates at 2 in. Hg. The 75°F circulating water will pick up 10°F as it passes through the condenser. How much cooling water is required in gpm?

Solution 9-1. To solve this problem, we must assume the following: 2 in. Hg steam contains 1,105 Btu/lb and is 101°F. Use the above stated equation:

$$\frac{50,000 \text{ kW} \times 14 \text{ lb}_{steam}/\text{kWh} \times 1,105 \text{ Btu/lb}_{steam} - (101° - 32°F)}{10 \text{ Btu/lb}_{water} \times 500 \text{ lb/hr/gpm}} = 145,040 \text{ gpm}$$

Example 9-2. Heat rate measures a plant's performance in Btu per kilowatt-hour. Find the efficiency for a plant with a 50-megawatt turbine-generator that is supplied with 950 psi, 850°F steam and exhausts to the condenser at 1.5 in. Hg absolute. The heat rate is 12,000 Btu/kWh.

Solution 9-2. This problem only requires one equation, which is as follows:

$$\text{Efficiency} = \left(\frac{3,413 \text{ Btu/kWh}}{12,000 \text{ Btu/kWh}} \right)(100) = 28.44\%$$

The 28.44% value is considered fair. Typical efficiencies are between 25% and 35%. *Note: It is common for examiners to include information in a question that is not required for the solution.*

Example 9-3. A 100-megawatt turbine-generator unit is supplied with steam at 1,200 psia and 1,000°F and exhausts into a condenser at 2 in. Hg absolute. At the rated load, the turbine uses steam at the rate of 1 million lb/hr. What is the efficiency of this unit?

Solution 9-3. It is good practice to first sketch the system and label its various components, Figure 9-10.

The efficiency is output divided by input. We know the output is 3,413 Btu/kWh, so all we have to do is calculate the input (one kWh equals 3,413 Btu). The first step is to calculate the water rate (w_R).

$$\text{Water rate} = \frac{1,000,000 \text{ lb/hr}}{100,000 \text{ kW}} = 10 \text{ lb/kWh}$$

Notice that megawatts (millions of watts) had to be converted to kilowatts (thousands of watts) in order to keep the units consistent.

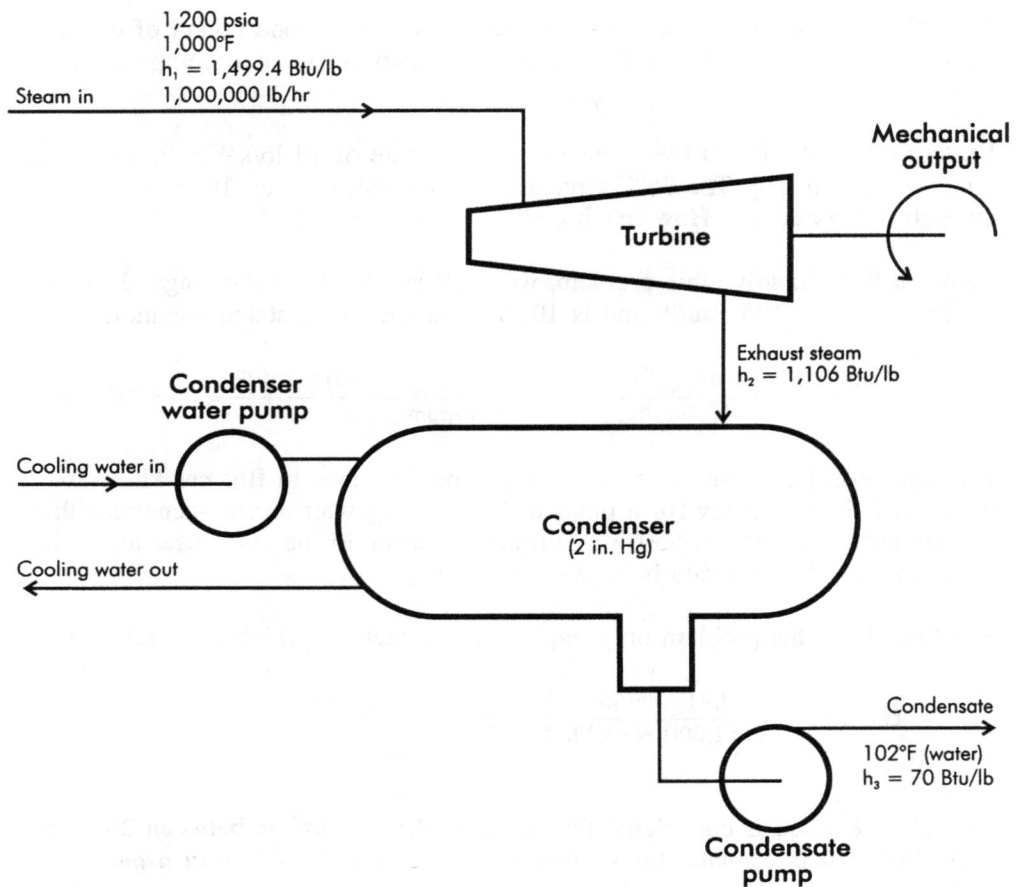

Figure 9-10. Sketch system components

Next calculate the amount of heat that the turbine consumes. The superheated steam table states that steam at 1,200 psia and 1,000°F contains 1,499.4 Btu/lb. Label this "h_1." Remember the symbol "h" is the abbreviation for enthalpy (Btu/lb).

The amount of heat returned to the boiler from the condenser must be credited. Since the pressure in the condenser is 2 in. Hg absolute, assume that the water returning to the boiler from the condenser is equal to the heat of saturated water at that pressure. Label this point "h_3." Most steam tables only give pressures in psia, not in. Hg; however, 1 psi equals approximately 2 in. Hg.

1 in. Hg = 0.5 psia = 80°F = 48 Btu/lb (water) = 1,096 Btu/lb (steam)

2 in. Hg = 1.0 psia = 101°F = 69 Btu/lb (water) = 1,106 Btu/lb (steam)

The water rate times the net heat used by the turbine-condenser is the heat input in Btu/kWh [$(w_R)(h_1 - h_3)$]:

$$\text{Efficiency} = \frac{\text{Btu/kWh}_{out}}{\text{Btu/kWh}_{in}} = \frac{3,413 \text{ Btu/kWh}}{(w_R)(h_1 - h_3)}$$

where:

w_R = steam or water rate (lb_{steam}/kWh)
h_1 = supply steam's heat content (Btu/lb)
h_3 = condensate's heat content (Btu/lb)

therefore:

$$\frac{3,413 \text{ Btu/kWh}}{(10 \text{ lb/kWh})(1,499.4 \text{ Btu/lb} - 69 \text{ Btu/lb})} = \frac{3,413 \text{ Btu/kWh}}{14,304 \text{ Btu/kWh}} = 23.9\%$$

Most examiners do not permit steam tables in the test; therefore, an educated guess is required to obtain the enthalpy. You are not expected to memorize the steam tables, but there is a pattern to them. It is very important to let the examiner know the basis of your estimates and to show all work.

The pattern for enthalpy is that starting at 80 psia, enthalpy starts at 1,183.1 Btu/lb, peaks at 450 psia at 1,204.6 Btu/lb then declines back to 1,183.4 Btu/lb at 1,200 psia. You will be very close if you state the approximate estimated enthalpy of saturated steam at 1,190 or 1,200 Btu/lb. The same holds true for the vacuum. Most problems will give 1 or 2 in. Hg for the condenser condition. The approximate enthalpy values for these pressures are 48 and 69 Btu/lb respectively.

Superheated steam also has a pattern. Starting from 80 psia to 1,200 psia, steam at 1,000°F has enthalpy values from 1,530 Btu/lb to 1,500 Btu/lb. Steam at 700°F has enthalpy values from 1,377 Btu/lb to 1,311 Btu/lb.

Example 9-4. Using the information in Example 9-3, how much heat is rejected by the condenser?

Solution 9-4. We know the water rate and the heat contained in the water leaving the condenser, so just find out how much heat is entering the condenser.

Assume that the exhaust steam entering the condenser is saturated at 2 in. Hg. (Be sure to state your assumption.) For problems like this, it is good to know the following figures from the steam table: 101°F for temperature, 1,100 Btu/lb (h_g) for the heat entering the condenser, and 70 Btu/lb (h_f) for the heat leaving the condenser.

$$(w_R)(h_2 - h_3) = (1 \text{ million lb/hr})(1,100 \text{ Btu/lb} - 69 \text{ Btu/lb}) = 1,031 \text{ MBtuh}$$

All heat engines convert only about one-third of the input heat into shaft horsepower; the rest of the heat is rejected, resulting in inefficiency. For example, in an internal combustion engine, one-third of the input heat goes to shaft horsepower, one-third goes out the exhaust, and one-third is rejected by the radiator.

Example 9-5. How many gallons per minute of condenser cooling water are required for a 2,500-kW generator with a steam rate of 12 lb/kWh, 400 psi, 500°F steam inlet and 27 in. Hg outlet steam conditions? The condenser is rated for a 10°F cooling-water temperature rise.

Solution 9-5. The steam inlet condition can be ignored, because we are only concerned with the condenser. It can be assumed that the 27 in. Hg is vacuum (pressure below atmospheric), but be sure to state this assumption. To convert to absolute pressure, subtract 27 from 30 to obtain 3 in. Hg (29.92 in. Hg is standard atmospheric pressure, but 30 is close enough).

Remember that 1 in. Hg equals 0.5 psia and 3 in. Hg equals 1.5 psia. The latent heat of evaporation (h_{fg}) of 1.5 psia steam is about 1,112 Btu/lb. Assume that the condensate leaving the condenser is 1.5 psia water at its saturation point. Again, state all assumptions. Now we can set up the problem.

Calculate how much heat must be absorbed by the condenser cooling water:

$$(1,112 \text{ Btu/lb}) (12 \text{ lb/kWh}) (2,500 \text{ kW}) = 33,360,000 \text{ Btuh}$$

Each pound of cooling water is expected to increase 10°F, and since the definition of a Btu is the amount of heat required to raise one pound of water one degree Fahrenheit, then each pound of water will absorb 10 Btu:

$$(10°F) (1 \text{ Btu/°F-lb}) = 10 \text{ Btu/lb}$$

$$\frac{33,360,000 \text{ Btuh}}{10 \text{ Btu/lb}} = 3,336,000 \text{ lb/hr}$$

The problem with this answer is that it must be in gpm, so hours must be converted to minutes, and pounds must be converted to gallons. Dividing by 60 will convert hours to minutes, and dividing by 8.33 will convert pounds to gallons. The shortcut is to multiply 60 by 8.33, which equals 500 lb/hr/gpm. Therefore, to convert pounds per hour to gallons per minute, divide by 500:

$$\frac{3,336,000 \text{ lb/hr}}{500 \text{ lb/hr/gpm}} = 6,672 \text{ gpm}$$

Pumps and Motors

Imagine slinging a bucket of water around in a circle. Experience tells you that if this circular motion is fast enough, no water falls out due to centrifugal force. This action occurs in a centrifugal pump. Liquid is directed into the center of a rotating impeller and slung out, away from the impeller by centrifugal force. The task of the pump is to convert the readily moving liquid (velocity energy) into discharge pressure (pressure energy). In a volute pump, this is accomplished in the stationary casing with the use of a volute; in a diffuser or turbine pump it is done with a set of diffusion vanes that surrounds the impeller, Figure 10-1.

VOLUTE PUMPS

The volute pump is the most common centrifugal pump and comes in several configurations. The two designs most encountered are the end-suction type and the single-stage, double-suction type.

Regardless of the volute pump's configuration, the impeller is always offset from the center of the casing. The point where the wall of the casing is closest to the impeller is the tongue of the casing. Between the tongue of the casing and a point to its left, Figure 10-1, liquid is discharged from the rotating impeller. This liquid rotates with the impeller and is then discharged from the pump.

As additional liquid is slung out from the rest of the rotating impeller, the volume of liquid in the casing increases. To maintain a constant velocity inside the pump, the area from the casing tongue to the discharge nozzle gradually increases. At the pump discharge, the diffuser converts the accumulated velocity energy into discharge pressure.

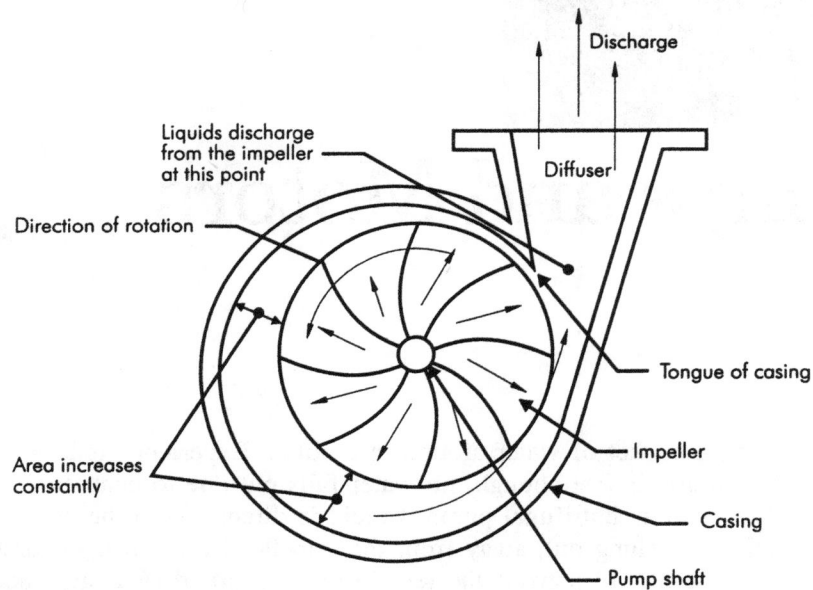

Figure 10-1. Volute pump

DIFFUSER PUMPS

In diffuser pumps, vaned diffusers positioned around the impeller convert the velocity energy of the liquid being slung from the impeller into pressure energy. The flow from the vaned diffusers is collected in a circular casing and then discharged through the outlet pipe, Figure 10-2. Instead of one throat section (as is common in volute pumps), a diffuser pump has several.

Diffuser pump design is primarily used in vertical turbine pumps, Figure 10-3. Other common names for this design are deep well or turbine well pumps. By using the diffuser design, several stages can be bolted together to obtain higher pressure. For example, if one stage adds 80 psi, three stages bolted together in series add 240 psi. Stages in series increase pressure but not flow. To add flow, parallel pumps have to be connected. Parallel pumps increase flow but not pressure. For example, if water must be pumped from a well several hundred feet deep, the pump would be built with enough stages to produce the pressure required to push the weight of water to the surface.

The hollow shaft motor, is the most commonly used driver for vertical turbine pumps. This design supports the weight of the drive shaft and pump; absorbs the hydraulic thrust developed when operating; and allows the shaft to be raised or lowered for the proper position of the impeller in the stage bowl. When engine driven, a right angle hollow shaft gear is used.

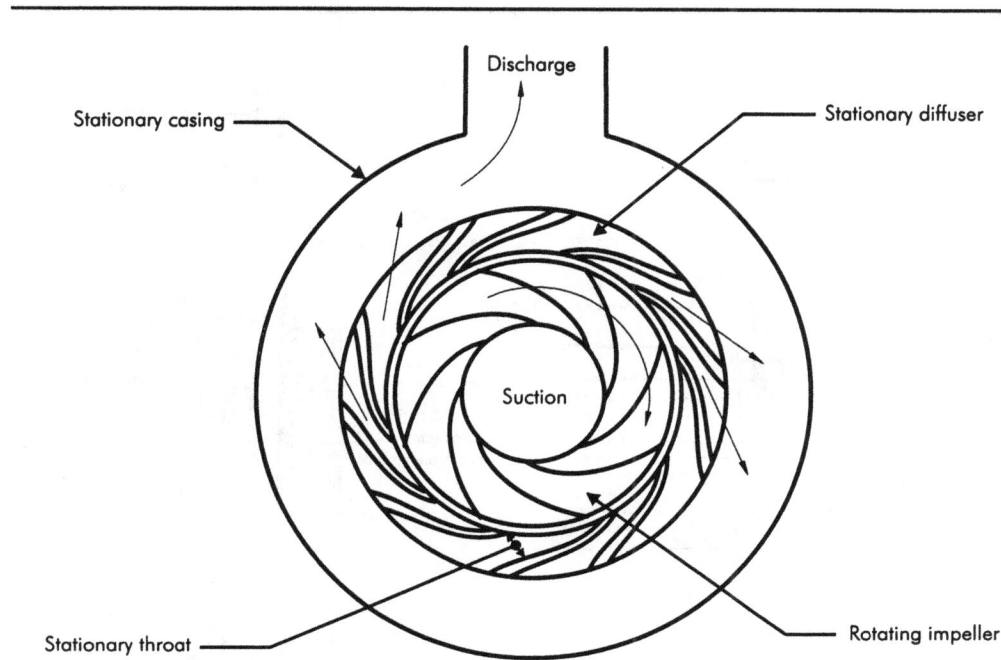

Figure 10-2. Diffuser pump

PROPELLER (AXIAL-FLOW) PUMPS

Axial-flow pumps are high-volume, low-head machines used for irrigation and storm water installations. Though from the outside they look like vertical turbine pumps, they are much different. Their impellers look like propellers and operate in a straight pipe without vanes, Figure 10-4.

When more than two stages are required to deliver the required pressure, other designs such as mixed-flow pumps are used. As the name implies, mixed-flow pumps are a cross between turbine and propeller pumps. They are used when moderate head and high flow rates are required, such as in condenser water circulating service.

MULTISTAGE PUMPS

The discharge pressure of a pump can be raised by increasing the revolutions per minute (rpm) and/or increasing the impeller diameter. When the required head cannot be developed by a single impeller, single-stage pumps piped in a series or multistage pumps must be used. Each impeller adds its net rated head to its suction pressure. An example is a pump rated at 300 ft of head taking its suction at atmospheric pressure. If liquid coming from the previous pump were discharged into an identical pump, an additional 300 ft of head would be added to the pumped liquid, for a gross total of 600 ft.

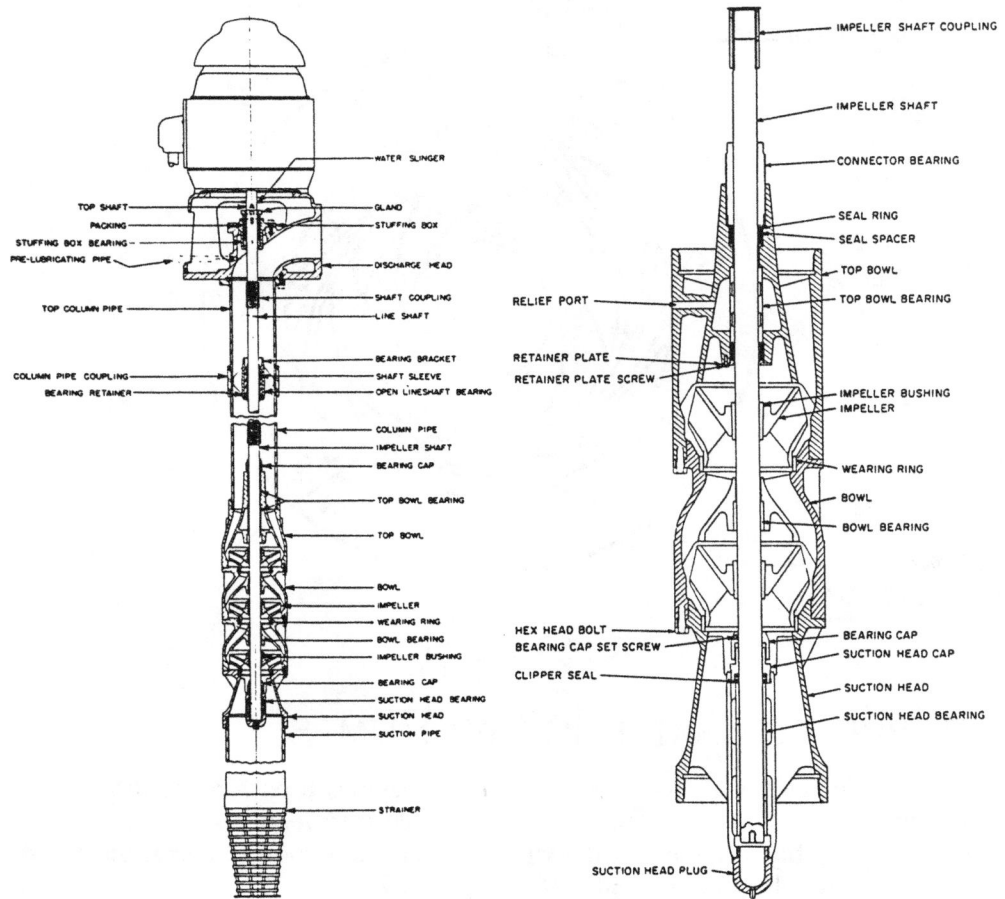

Figure 10-3. Vertical turbine pump and detail (Courtesy, Ingersoll-Dresser Pump Co.)

Figure 10-4. Axial-flow impeller

It is cheaper and more efficient to manufacture multistage pumps in one pump body. One problem with this arrangement is that a lot of axial thrust is developed due to unequal forces on the faces of the impellers. Three methods of curtailing this problem are the use of 1) an opposing force on an equal number of impellers; 2) a hydraulic balancing device; or 3) a large thrust bearing.

Figure 10-5 shows a four-stage pump with opposed impellers, which are impellers facing in different directions. The axial thrust from half of the impellers is countered by thrust in the opposite direction by the other impellers. Usually the first two stages are at the ends to cut down on stuffing box pressure (the stuffing box stops leakage where the shaft passes through the pump's casing).

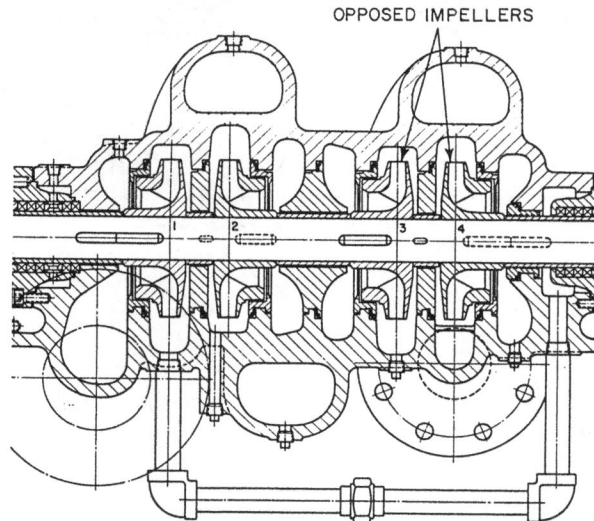

OPPOSED IMPELLERS

Figure 10-5. Four-stage pump with opposed impellers (Courtesy, Ingersoll-Dresser Pump Co.)

If all the impellers of a multistage pump face in the same direction, the total hydraulic axial thrust acting toward the suction end would be the sum of the individual impeller thrusts. A hydraulic balancing arrangement can be used to balance this axial thrust and to reduce the pressure on the stuffing box next to the last stage. This arrangement consists of a chamber at the back of the last stage impeller that is separated from the rest of the pump's interior by a drum or disk, which rotates with the shaft.

The balancing chamber is connected either to the pump suction or to the vessel from which the pump takes its suction. As a result, the backpressure in the balancing chamber is very close to the suction pressure. The leakage between the last pump stage and balancing chamber depends on the pressure across the balancing device and the internal clearances.

Figure 10-6 shows how a balancing drum is used to cancel out the axial thrust. The force on Area C is opposite and nearly equal to the axial force at the pump's suction port. A small thrust bearing is commonly used to handle the remaining load.

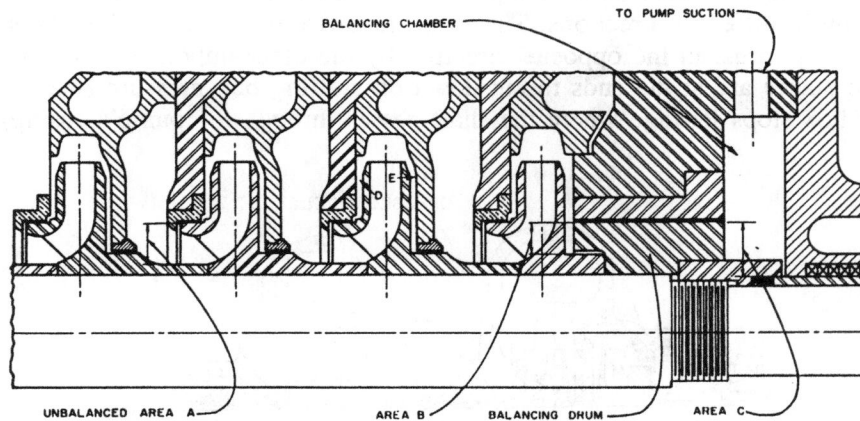

Figure 10-6. A balancing drum is used to cancel out axial thrust (Courtesy, Ingersoll-Dresser Pump Co.)

A vertical pump motor is an example of the third method of compensating for the axial thrust imbalance. In this type of pump, a large thrust bearing is used to absorb all the axial thrust.

Notice that up until now, the term head (in feet) has been used instead of pressure (in psi). This is because centrifugal pumps, with a given impeller and rpm, develop the same head regardless of the liquid being pumped. For example, if a heavy ball and a light ball are thrown straight up with the same velocity, both reach the same height. Naturally the heavier ball would require more power to achieve the same velocity. The same is true for centrifugal pumps. More horsepower is required for heavier liquids, but all liquids would be pumped to the same height.

To illustrate this, consider pumping gasoline, water, and brine using identical pumps, Figure 10-7. Assume a flow of 200 gpm; head of 100 ft; and pump efficiency at 70%. For a given pressure (in psi), the head varies according to the specific gravity and the liquid, and the required horsepower remains the same, Figure 10-8. Figure 10-8 assumes 200 gpm at a pressure of 43.3 psi and a pump efficiency of 70%.

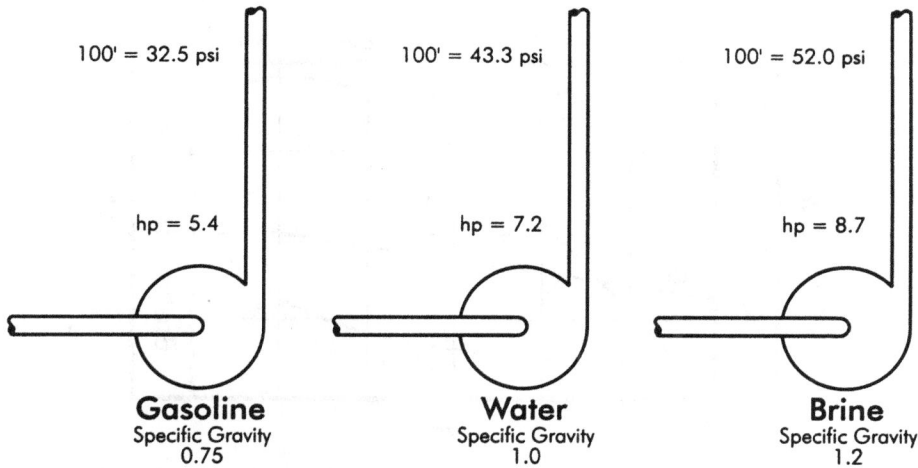

Figure 10-7. Gasoline, water, and brine pumped in identical pump

100' = 32.5 psi   100' = 43.3 psi   100' = 52.0 psi

hp = 5.4   hp = 7.2   hp = 8.7

Gasoline
Specific Gravity
0.75

Water
Specific Gravity
1.0

Brine
Specific Gravity
1.2

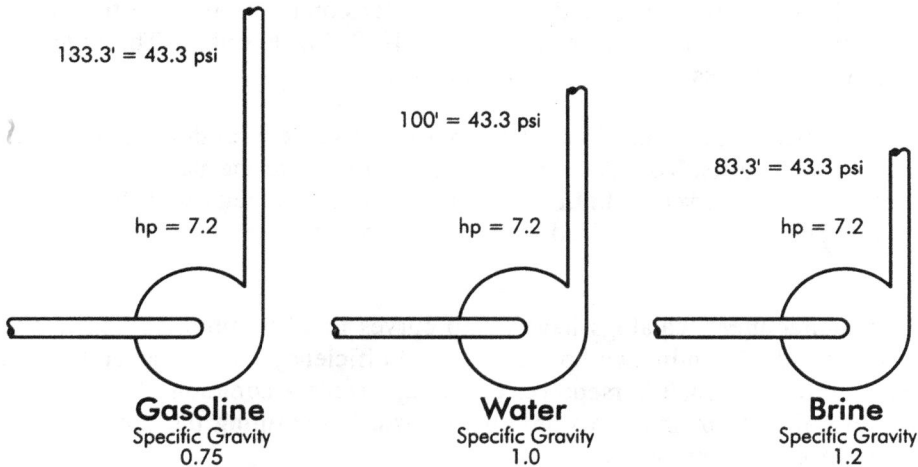

Figure 10-8. Head varies among gasoline, water, and brine

133.3' = 43.3 psi   100' = 43.3 psi   83.3' = 43.3 psi

hp = 7.2   hp = 7.2   hp = 7.2

Gasoline
Specific Gravity
0.75

Water
Specific Gravity
1.0

Brine
Specific Gravity
1.2

PUMP CURVES

All centrifugal pumps have performance curves that show the relationship between flow and factors such as head, efficiency, horsepower, and net positive suction head, Figure 10-9. Net positive suction head (NPSH) is the amount of head (in absolute pressure) of the liquid at the pump's suction.

For a particular rotating speed and impeller diameter, all centrifugal pumps have head-capacity rating curves that show the relationship between the discharge head (or pressure) developed and the flow. As capacity increases, the head pressure decreases. The highest developed head occurs at zero flow.

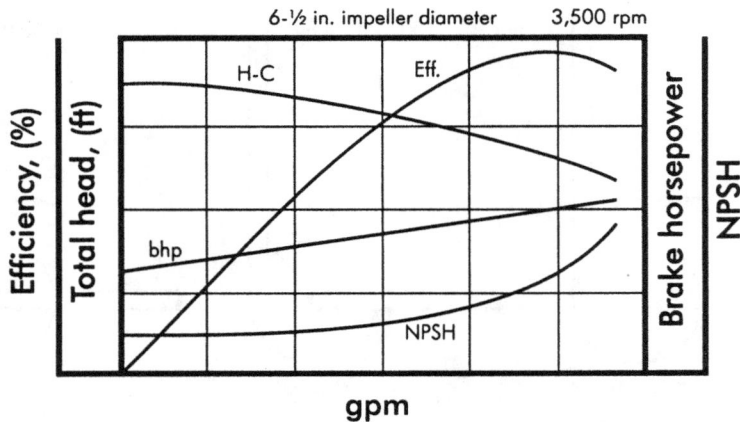

Figure 10-9. Performance curves for centrifugal pumps

To find the horsepower required for a particular condition, refer to the brake horsepower capacity curve shown in Figure 10-9 (labeled bhp). The required horsepower increases with an increase in flow.

Both the head-capacity and brake horsepower curves are plotted from the data of actual pump tests. This data can be used to calculate the pump efficiency curve. Efficiency increases until about 90% of rated capacity and then starts to drop off. Pumps should be sized to operate at or near the peak of the efficiency curve.

Most manufacturers' catalogs have pump curves that resemble the one in Figure 10-10. Instead of continuous horsepower and efficiency curves, several separate curves are drawn. Each horsepower curve represents a common electric motor size. Several sizes of impellers are also shown, representing the range that can be used in a given pump size.

To select a pump once the head and flow have been established, refer to the manufacturer's pump curves. Make sure the motor is large enough to do the job, but not so large that it wastes energy. Also be sure to select the proper impeller size. If it is too large, the motor can be overloaded and/or the discharge will have to be throttled. If the impeller is too small, it can't do the job.

If it is necessary to use the smallest impeller available, consider the next smaller pump for increased efficiency. The head and flow should be as close to maximum efficiency as possible. The maximum efficiency is sometimes called the *eye of the curve*. If the operating point is outside the efficiency curves (either too far left or right), select another pump. If the point is too far left of the efficiency curve, the pump will eventually be damaged due to unbalanced hydraulic loading, Figure 10-11. Also verify that the system's net positive suction head (NPSH) is greater than the required net positive suction head. (Net positive suction head is explained in more detail later in this chapter.)

1,750 rpm

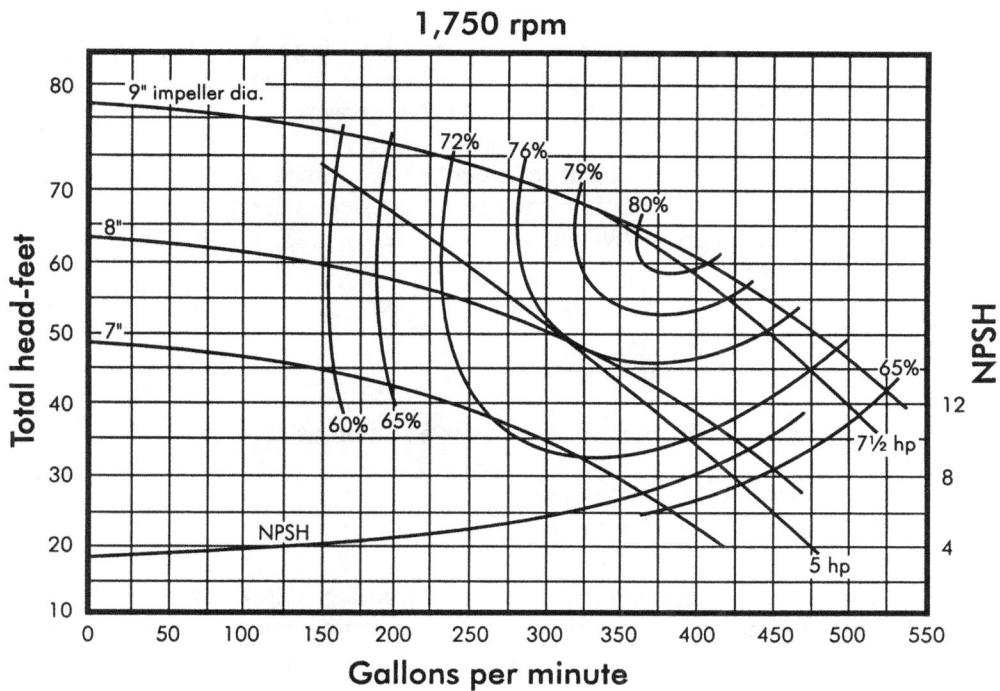

Figure 10-10. Common manufacturer pump curves

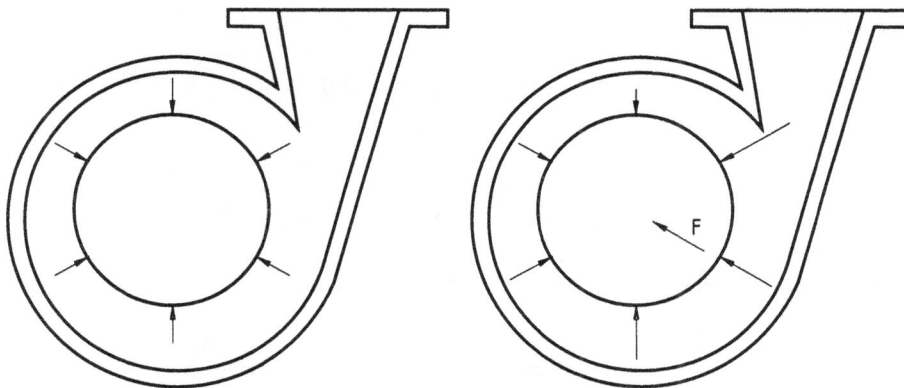

Figure 10-11. Unbalanced hydraulic loading

Horsepower pump problems are inevitable on advanced exams. Pump curves will not be furnished, so the answer must be calculated.

It is first necessary to know that one horsepower is equal to 33,000 ft-lb/min. Therefore, if a pump lifts 33,000 lb of liquid in one foot per minute, it just did one horsepower worth of work.

If you are asked to find water or hydraulic horsepower, determine how much liquid (in pounds) was pumped in a minute, and how high the liquid was pumped in a minute. This should give you an answer in units of ft-lb/min. All that remains is to divide by 33,000 to obtain the answer.

Because no pump is 100% efficient, to calculate the power supplied to the pump, just divide the hydraulic horsepower by the decimal equivalent of efficiency. That is, if a pump is 73% efficient, divide by 0.73. Pump horsepower is expressed as follows:

$$hp = \frac{(Head,\ ft)(lb\ of\ liquid\ pumped\ per\ min)}{(33,000\ ft\text{-}lb/min)(Pump\ efficiency)}$$

Perhaps you have seen the pump horsepower equation as follows:

$$hp = \frac{(gpm)(Head,\ ft)(Specific\ gravity)}{(3,960)(Pump\ efficiency)}$$

In the second equation, the 3,960 comes from dividing 33,000 by 8.333 (the weight of one gallon of water in pounds). The specific gravity is there in case a liquid other than water is being pumped. (The specific gravity of water is 1.0.)

The first equation is easier to use, because the units can be checked. This is important because the pump's output could be given in pounds per square inch instead of feet of head, and sometimes flow is given in gph (gallons per hour) instead of gpm. However, both equations are correct.

To convert pounds per square inch (psi) to feet of head, use the following equation:

$$Head,\ ft = \frac{(psi)(2.31\ feet\ of\ water\ per\ psi)}{Specific\ gravity}$$

If water is the medium used, the equation is simplified:

$$Head,\ ft = (psi)\ (2.31\ ft\ of\ water\ per\ psi)$$

Example 10-1. What horsepower is required to produce 200 psi at 250 gpm with an 82% efficient pump?

Solution 10-1. Plug the numbers into the equations given earlier:

$$Head,\ ft = (200\ lb/in^2)\ (2.31\ ft/psi) = 462\ ft$$

Use the first horsepower equation given earlier:

$$hp = \frac{(462 \text{ ft})(250 \text{ gpm})(8.33 \text{ lb/gal})}{(33,000 \text{ ft-lb/min})(0.82)} = 35.6 \text{ hp}$$

Now use the second horsepower equation to make sure the answer is correct:

$$hp = \frac{(250 \text{ gpm})(462 \text{ ft})(1.0)}{(3,960)(0.82)} = 35.6 \text{ hp}$$

NET POSITIVE SUCTION HEAD AND CAVITATION

Cavitation is the formation and subsequent collapse of vapor-filled cavities or bubbles in the pumped liquid. These bubbles are formed when the pressure in the suction line falls below the pumped liquid's vapor pressure — the force exerted by a gas on its container. (Just as flash steam is produced when the pressure of saturated water is reduced.) When these bubbles violently collapse with velocities exceeding the speed of sound as they reach the high-pressure side of the pump, a tremendous amount of energy is quickly released in a small area. These pressures exceed the tensile strength of the metal and cause pitting by blasting out metal particles.

The most obvious effects of cavitation are noise and vibration. Cavitation often produces noise that sounds like stones passing through the pump. Over time, impellers are damaged beyond repair. Perhaps you have seen a bronze impeller that looks like it has been severely pitted by corrosion. It's likely that the damage was due to cavitation. The noise and vibration can also cause bearing failure, shaft breakage, and other fatigue failures. In addition, cavitation also reduces the pump's efficiency and capacity.

Required net positive suction head (NPSH) is a characteristic of the pump and is data supplied by the manufacturer. It is the amount of head needed to overcome friction and flow losses in the pump's suction, which keeps the pressure at the eye of the impeller above the vapor pressure of the liquid. If a pump is operated below its required net positive suction head, cavitation occurs. Available NPSH is a characteristic of the system and is calculated by the pump installer. It must be greater than the required net positive suction head at the suction connection of the pump.

Vapor pressure is the pressure at which liquid and vapor can coexist. The vapor pressure for water is on a steam table. For example, water at 212°F has a vapor pressure of 14.7 psia, while water at 50°F has a vapor pressure of 0.1781 psia.

It is easier to pump cold water than hot water. To cavitate water at 50°F, the pump's suction pressure would have to be below 0.1781 psia. However, a slight drop below atmospheric pressure would cavitate water at 212°F.

Cavitation is avoided with the proper installation design that provides sufficient NPSH. The equation for available net positive suction head is as follows:

$$\text{Available NPSH} = Z + \left[\left(\frac{P - P_v}{\text{Specific gravity}} \right) (2.31) \right] - h$$

where:
- Z = static suction head in feet (a positive number) or static suction lift in feet (a negative number)
- P = pressure at the liquid's surface in pounds per square inch absolute (psia)
- P_v = vapor pressure of liquid in pounds per square inch absolute (psia)
- h = pipe pressure loss in feet
- 2.31 = feet of water per psi

Example 10-2. Determine the available NPSH for an open tank at atmospheric pressure that is 10 ft above the pump suction, and the friction loss in the suction line is 8 ft. The water is 60°F, vapor pressure is 0.256 psia, and specific gravity of water is 1.0.

Solution 10-2. Because the tank is at atmospheric pressure, 14.7 psia is used in the equation:

$$\text{NPSH} = 10 \text{ ft} + \left[\left(\frac{14.7 \text{ psia} - 0.256 \text{ psia}}{1.0} \right) (2.31 \text{ ft/psi}) \right] - 8 \text{ ft} = 35.4 \text{ ft}$$

Example 10-3. Find the available NPSH for the system in Example 10-2, only this time the water is 200°F and vapor pressure is 11.53 psia.

Solution 10-3. Use the same equation to determine the available NPSH:

$$\text{NPSH} = 10 \text{ ft} + \left[\left(\frac{14.7 \text{ psia} - 11.53 \text{ psia}}{1.0} \right) (2.31 \text{ ft/psi}) \right] - 8 \text{ ft} = 9.32 \text{ ft}$$

As shown, hot water reduces the available net positive suction head. This is why deaerators are always elevated above feedwater pumps. The weight of the column of water in the line exerts enough pressure at the pump's suction to prevent the hot water from cavitating.

Pump Components

The important components in a pump include impellers, wearing rings, shaft sleeves, and stuffing boxes. It is important to know the function of each of the components.

Impellers

One way impellers are classified is by whether their sides are enclosed with shrouds. Enclosed impellers, Figure 10-12, maintain their efficiency even if slightly worn and are used for clear liquids.

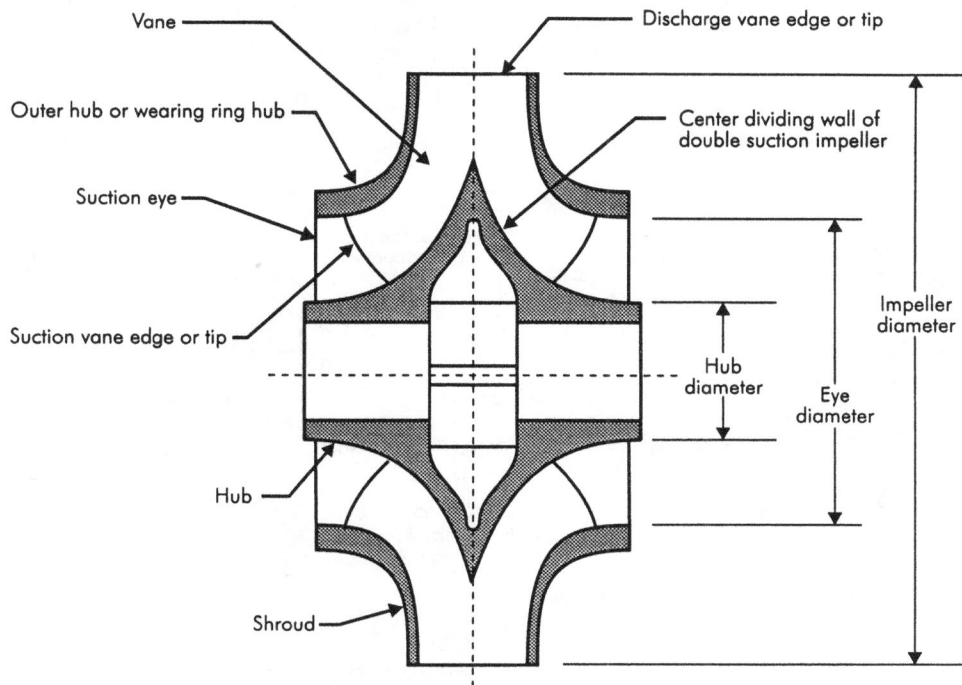

Figure 10-12. Enclosed impeller

The open impeller, Figure 10-13, is used when suspended matter in liquid has the potential to clog a closed impeller. A disadvantage of the open impeller is that its vanes are weak because they are not reinforced with a shroud. The semi-enclosed impeller, also shown in Figure 10-13, is a compromise between the two.

When installing a double-suction impeller, please note that it can be installed backwards. If an impeller is installed backwards, the pump still operates but at a much lower flow rate and pressure, Figure 10-14.

Wearing Rings

Due to wear, clearances between the rotating and stationary parts of a pump increase, which results in decreased pump efficiency. With small pumps, either the entire pump or the impeller and suction cover is replaced. With larger

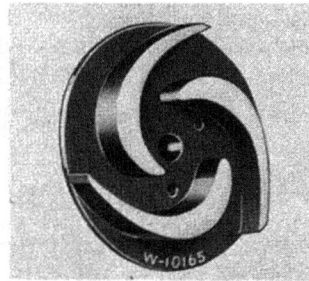

Figure 10-13. Open and semi-enclosed impellers (Courtesy, Ingersoll-Dresser Pump Co.)

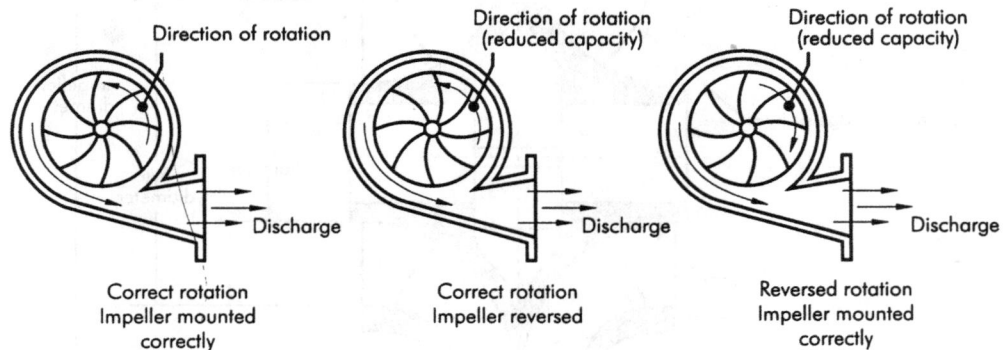

Figure 10-14. Impeller assembly & rotation

pumps, or pumps in which the stationary element of the leakage joint is part of an expensive casing, surfaces can be renewed. Figure 10-15a shows an impeller and renewable impeller ring. After the old, worn surface is machined true, the new wearing surface is fit on. Figure 10-15b shows how the renewable casing and impeller rings are installed to bring the pump back to its original specifications.

Shaft Sleeves

Even the best packing eventually wears down the rotating shaft surface. Instead of ruining an expensive shaft, a renewable protective sleeve is placed over the shaft in the stuffing box area. These sleeves, usually made of hard bronze or stainless steel, must have a fine finish. When they are no longer serviceable, they are replaced.

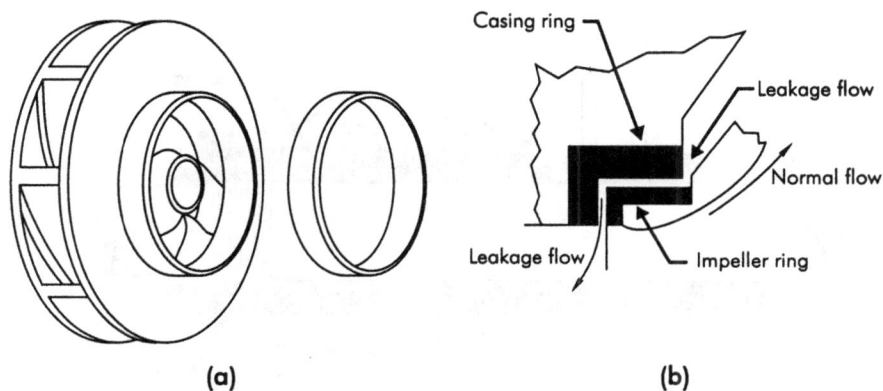

Figure 10-15. Impeller and renewable impeller and casing rings

Stuffing Boxes

The stuffing box, Figure 10-16, stops leakage where the shaft passes through the pump's casing. It is a cylindrical recess that accommodates many rings of packing around the shaft or shaft sleeve. The packing is compressed and held in place by an adjustable gland. Some leakage is necessary on packed pumps to cool and lubricate the packing.

Packing a Pump

When packing a pump, ensure that the shaft sleeves, bearings, and stuffing box are in good condition. A new shaft sleeve can make all the difference between success and failure. A shaft runout of more than 0.003" is too much and indicates new bearings are required. If the stuffing box is rough, it should be reworked, as it is very hard to seal the outside diameter of the packings where there is rust or pitting. The following procedure should be used when packing a pump:

1. Always use the correct size and proper packing (contact a packing vendor if there is any question).

2. Cut the packing rings so the ends are side by side and not overlapping. The preferred way is to wind the required number of rings on a mandrel that is the same diameter as the shaft. Cut the rings by making a straight cut along the mandrel, Figure 10-17. When removing the rings from the mandrel, slip them off without opening the rings. This is especially important when using a metallic packing.

3. Remove all the old packing.

4. Coat the new packing rings with a lubricant to make installation easier and to help establish a proper break-in.

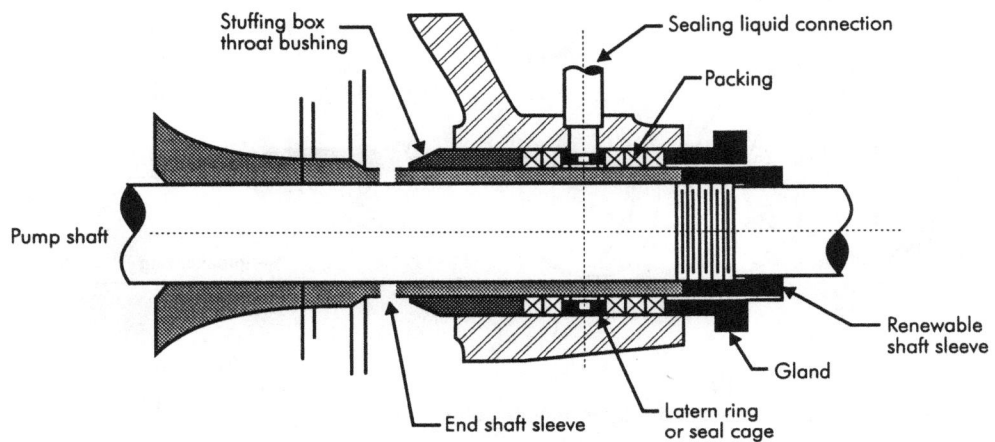

Figure 10-16. Conventional stuffing box

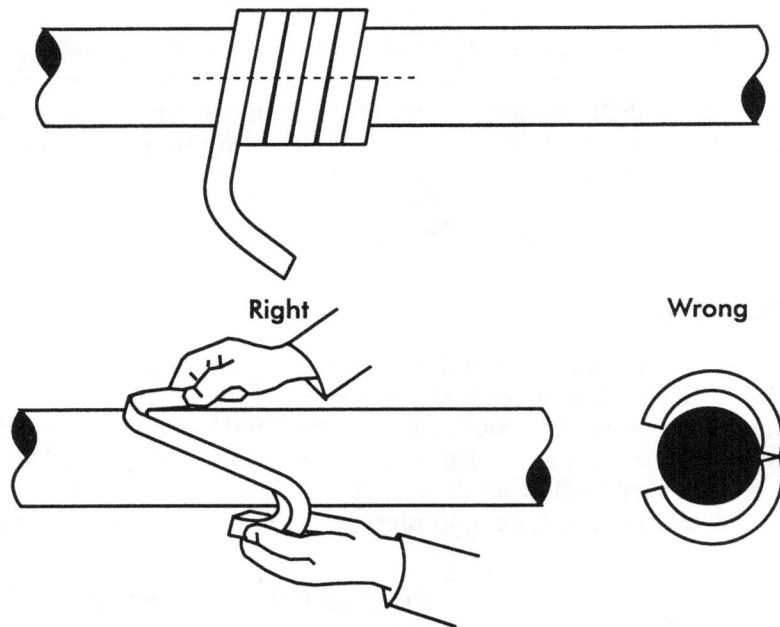

Figure 10-17. Cutting packing material and removing rings from mandrel after cutting

5. If a lantern ring is required, make sure it lines up with the flush connection.

6. Install the new rings over the shaft by twisting them open. This is especially important for metallic packings. The butts of succeeding packing rings should be staggered 90 degrees.

7. Seat each ring individually, compressing it in place with a tamping tool such as a split hollow cylinder.

8. Turn the shaft occasionally to help the seating. If each ring is not properly seated, the gland follower can't tighten the packing set and it leaves the gland too tight in the box. Keep in mind that, except for the action of abrasives, 70% of the wear takes place on the two rings nearest the gland.

9. Adjust the gland follower by tightening it with your hand. Permit generous initial leakage. Remember that the purpose of packing is to control leakage, not to prevent it. Packings must leak to perform properly, otherwise they burn up.

10. Gradually take up on the gland nuts one flat at a time (1/6 of a turn of a hex nut). Carefully monitor the temperature. Never permit heat to develop. If it does, back off the gland. As leakage levels off, tighten the follower at 15 minute intervals to get the leakage under control. (On a 1-inch shaft, allowable leakage is 5 drops to 20 drops per minute; a 2-inch shaft should be allowed to leak twice as much.)

Seal Cages (Lantern Rings)

When a pump operates under a suction, air can leak into the pump and cause loss of prime. In order for a pump to operate, its casing must be filled with water. If the casing is filled with air or steam, the pump will not function. Priming a pump is the act of filling the casing with water (or other liquid). In the case of old-fashioned hand crank pumps, all that was required was to pour a little water into the casing. One method of priming a centrifugal pump is to induce water into the casing by applying a vacuum to it. If air is sucked into the casing through the packing, then the priming process will take longer or even be unsuccessful. By using seal water to the lantern ring, air is prevented from entering the casing. Air also tends to enter the casing during normal operation. The continuous application to seal water via the lantern prevents loss of prime.

One solution is to separate the packing into two sections using a seal cage or lantern ring. Water or some other sealing fluid is introduced into the space created by the seal cage and causes sealing fluid to flow along the shaft in both directions. This way, sealing fluid, and not air, is sucked into the pump to prevent loss of prime. If the liquid being pumped is cool, clean, and nonhazardous, the sealing fluid can come from the pump itself. Figure 10-16 showed a stuffing box with a seal cage installed.

Lantern rings and sealing fluid are also used when the pumped liquid is abrasive and will damage the pump and packing if allowed to enter the stuffing box; and when the liquid being pumped is too dangerous to allow it to leak out. In both cases, the sealing fluid entering the pump prevents the pumped liquid from entering the stuffing box.

Warning: Never connect city water directly to a pump. City water supply must be isolated from any substance that is potentially hazardous to human health. It is against the law to make direct connections between potable water (water suitable for drinking) and contaminated liquid. There is always the chance that contaminated liquid can back up into the city supply when a direct connection exists. Many plants have process water stored up just for this use, or they use a seal water pump, Figure 10-18. Notice the air gap between the potable water inlet and the surface of the water in the tank. An air gap is the recommended way to isolate potable water from contaminated water. Another method is to use a backflow preventer.

Backflow Preventers

Any backflow preventer should conform to American Water Works Association (AWWA) standards and be certified by the American Society of Sanitary Engineers. Off-the-shelf check valves are not acceptable for use as backflow preventers.

Figure 10-19 shows a backflow preventer that uses the reduced pressure principle. When check valves are closed due to the contaminated water supply pressure being higher than the potable water pressure, a relief valve opens between the two check valves. If there is any leakage past the first check valve, the contaminated water exits out the open vent rather than building up pressure against the second check valve. Many variations of reduced pressure backflow preventers exist.

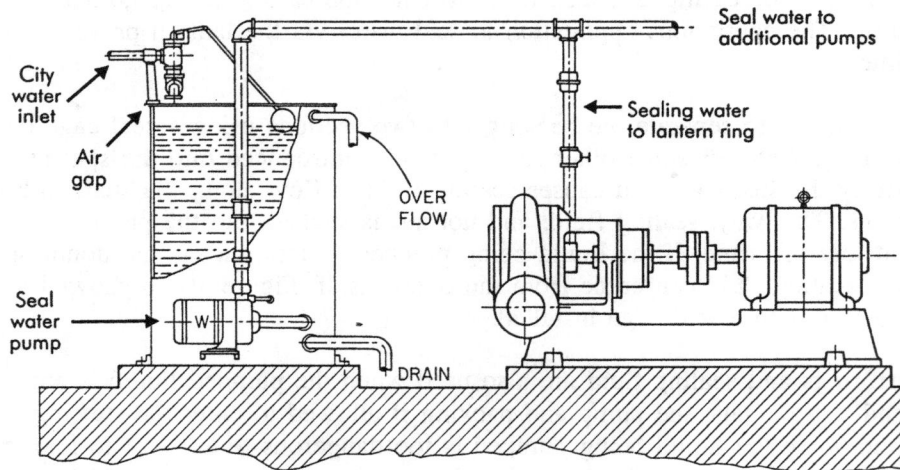

Figure 10-18. Seal water pump (Courtesy, Ingersoll-Dresser Pump Co.)

Negative Supply Pressure - Checks Closed - Vent Open

Static Pressure - No Flow

Normal Flow - Checks Open - Vent Closed

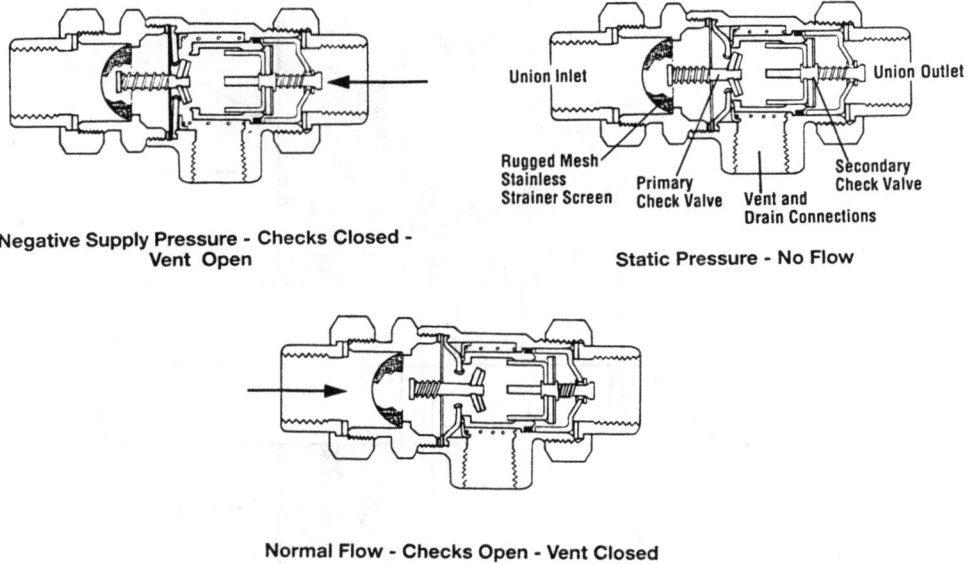

Figure 10-19. Backflow preventer (Courtesy, Watts Regulator Company)

Where backflow preventers are required by law, they are tested by the local health department and water company quarterly and are internally inspected each year.

Mechanical Seals

Under some circumstances, such as when pumping dangerous or toxic liquids, leakage must be as close to zero as possible. Since packing requires some leakage for lubrication and cooling, mechanical seals such as the one in Figure 10-20 were developed.

A mechanical seal consists of two highly polished flat surfaces that run against one another. A rotating surface is connected to the shaft, and a stationary surface is connected to the pump casing. These two surfaces, made of dissimilar materials, are held together by springs and form a tight seal.

Like packing, mechanical seals must be cooled and lubricated, usually by the liquid being pumped. When pumping liquids that contain solids, flushing water must be used to prevent early failure.

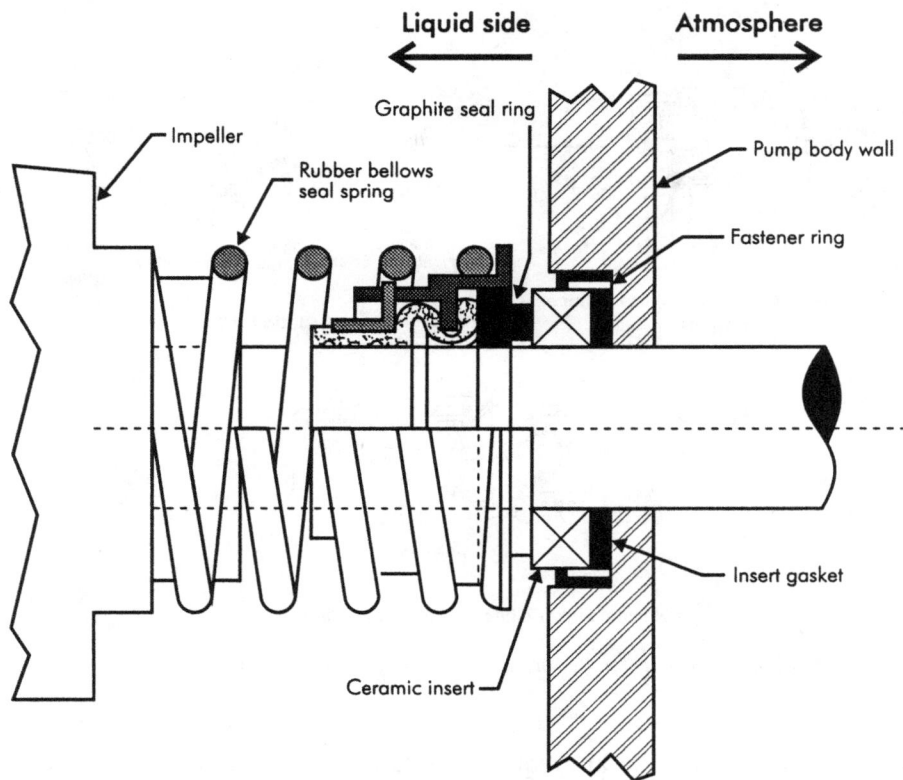

Figure 10-20. Mechanical seal

Extreme care must be taken when installing mechanical seals. One of the seal surfaces is usually made of a ceramic material that can be chipped or broken if it is put under unnecessary stress.

Centrifugal Machine Laws

All centrifugal pumps and centrifugal fans follow basic relationships that depend on flow, pressure, and the amount of power used. These are the basic relationships:

1. Flow is directly proportional to the speed.

2. Head is proportional to the square of the speed.

3. Horsepower is proportional to the cube of the speed.

The equation for these relationships is as follows:

$$\frac{rpm_1}{rpm_2} = \frac{gpm_1}{gpm_2} = \left(\frac{Head_1}{Head_2}\right)^{1/2} = \left(\frac{bhp_1}{bhp_2}\right)^{1/3}$$

where:
rpm = revolutions per minute
gpm = gallons per minute
Head = in feet
bhp = brake horsepower

Or to put it another way:

1.

$$\frac{rpm_1}{rpm_2} = \frac{Flow_1}{Flow_2}$$

2.

$$\left(\frac{rpm_1}{rpm_2}\right)^2 = \left(\frac{Pressure_1}{Pressure_2}\right)$$

3.

$$\left(\frac{rpm_1}{rpm_2}\right)^3 = \frac{hp_1}{hp_2}$$

Example 10-4. A pump is turning at 1,750 rpm, developing 50 ft of head at 100 gpm, and using 1.5 hp. If the speed is doubled to 3,500 rpm, what would be the gpm, head, and required horsepower?

Solution 10-4.

From equation 1, find the gpm at 3,500 rpm:

$$\left(\frac{3,500\ rpm_2}{1,750\ rpm_1}\right)(100\ gpm_2) = 200\ gpm_2$$

From equation 2, find the head at 3,500 rpm:

$$\left(\frac{3,500\ rpm_2}{1,750\ rpm_1}\right)^2(50\ feet_1) = (2)^2(50\ feet_1) = 200\ feet_2$$

From equation 3, find the horsepower at 3,500 rpm:

$$\left(\frac{3,500\ rpm_2}{1,750\ rpm_1}\right)^3(1.5\ hp_1) = (2)^3(1.5\ hp_1) = 12\ hp_2$$

VARIABLE-SPEED DRIVES

Systems are designed to meet peak demand. The problem is that peak demand occurs for only a few hours a year. The rest of the time the system is oversized.

If the fan or pump speed were reduced 20%, the theoretical power required would be reduced 48%. A dramatic savings! Studies have shown that many fan and pump systems operate at 70% to 80% of capacity. At 70% capacity, the theoretical power required drops to 35%. Notice the term *theoretical power* is used, because the variable-speed drive needs some power to operate. Check with the vendor to find out how much.

If your evaluation shows that a system runs at or near capacity, then a variable-speed drive is not justified for cost savings. Variable speed is also not good for feedwater pumps, because the rpm must be maintained to provide sufficient head to overcome the boiler's steam pressure.

To be technically correct, the term should be variable-*frequency* drives so they are not confused with other devices such as variable-diameter pulleys and hydraulic drives. Used in industry for many years for precise control of machinery such as printing presses, they are now used widely to save energy in pumps and fans.

As shown in Figure 10-21, fan dampers and pump throttle valves regulate flow and reduce power, but variable-speed drives do it more efficiently. Most off-the-shelf induction motors have no problem operating with a variable-speed drive; however, always check with the motor manufacturer.

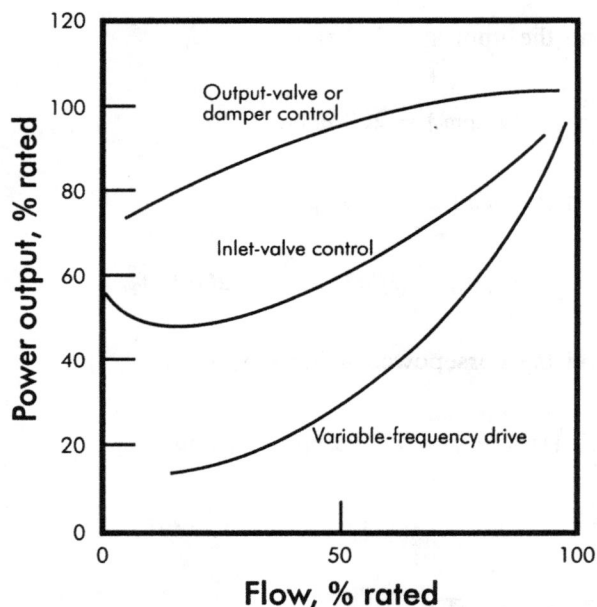

Figure 10-21. Pump control methods

Sophisticated motor protection comes standard with the new variable-speed drives, which is an advantage. When the microprocessor senses any abnormal situation, it shuts down the drive and a display states the problem.

POSITIVE-DISPLACEMENT PUMPS

Positive-displacement pumps come in three designs: rotary, screw, and reciprocating. Each revolution or stroke of a positive-displacement pump delivers the same quantity of liquid, regardless of changes in head. In rotary and screw pumps, the liquid is trapped between rotating elements (one or more) and the liquid is discharged in a relatively smooth flow. With reciprocating pumps, liquid is drawn into a cylinder through an inlet valve by the movement of a piston, When the piston reverses direction, the liquid is forced out of a discharge valve. The output flow is pulsating.

All pumps have slip, which is the leakage that occurs between the outlet and inlet due to internal clearances and leaks from valves, pistons, and packing. It is the difference between the theoretical and actual displacement expressed in percentage of the theoretical displacement.

ROTARY PUMPS

There are many types of rotary pumps. The most common type used in power plants are external and internal gear pumps. The external gear pump, Figure 10-22, works by trapping the fluid between the gear teeth and the pump casing. As the gears turn, the liquid is transferred from the inlet to the outlet of the pump. An electric motor drives one gear, which drives the other gear.

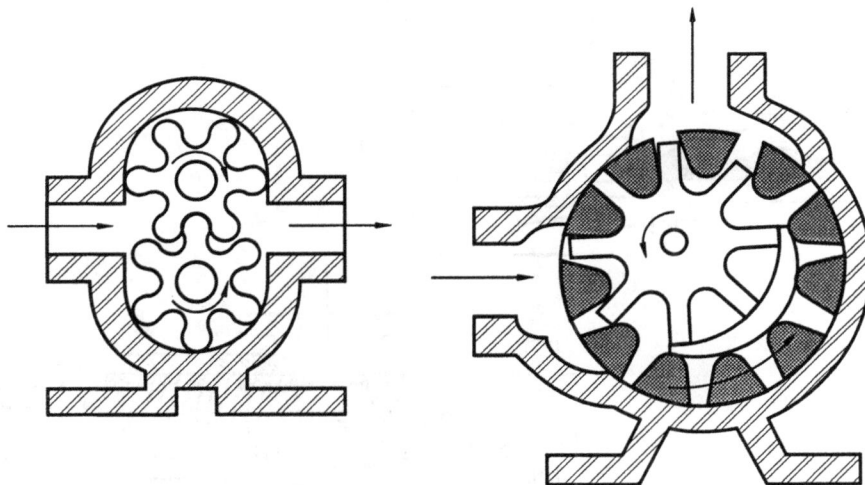

Figure 10-22. External and internal gear pump

The internal gear pump also works by trapping the fluid between gear teeth and casing, then transfers it to the pump's discharge as the internal parts rotate. The rotor (shown with 11 teeth) is driving the idler gear (shown with 8 teeth).

SCREW PUMPS

In screw pumps, liquid is trapped between the intermeshing threads on the rotors and the close-fitting housing. This forms one or more sets of moving seals in series that continuously move from inlet to outlet. Figure 10-23 shows a single-screw pump and a double-screw pump. Notice the double-screw pump requires driving gears, because the screws can't drive each other.

PUMP INSTALLATION

Caution must be exercised when installing rotary, screw, and reciprocating positive-displacement pumps driven by electric motors. These pumps must be equipped with relief valves to prevent excess pressure and damage to the system. These pumps deliver a constant volume and develop whatever pressure is required to maintain delivery. With a spring-loaded relief valve (either built into the pump or piped between the pump's discharge and return line), a safe pressure is maintained.

Relief valves built into the pump make for an easy installation but do not protect the pump from damage over a long period. The temperature of the pumped liquid increases slightly every time it passes through the pump. As this liquid is circulated again and again through the relief valve, it eventually gets

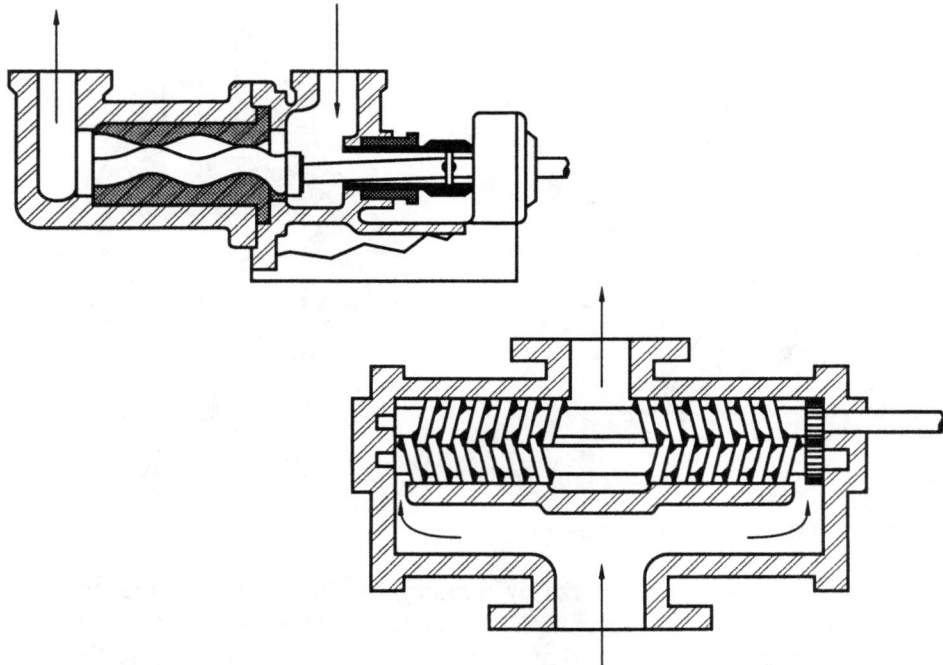

Figure 10-23. Single-screw pump, left; double-screw pump, right

hot enough to damage the pump. An external relief valve does not have this problem, since it draws liquid at a constant temperature from a reservoir and keeps the pump cool. The external relief valve must be connected directly to the pump's discharge and must go directly to the return line with no shutoff valves in between.

If possible, the pump's relief valve should not be used to regulate the supply-line pressure; a separate regulating or relief valve should be used instead. The relief valve is usually set 10% above the system's operating pressure.

RECIPROCATING PUMPS

Power pumps are reciprocating pumps driven by an electric motor. The most common power pump found in a boiler room is the type used for chemical feeding. In the pump shown in Figure 10-24, the concentrated chemicals are isolated from the pump mechanism by a diaphragm. The diaphragm is driven by a plunger acting on hydraulic fluid. Output pressure is determined by setting a bypass valve.

Air-driven diaphragm pumps are versatile and used for many services. These pumps are portable, easy to set up, self-priming, and can pump liquid of just about any viscosity. They can run dry without being damaged, or even run submerged. The capacity and discharge pressure are regulated by adjusting the inlet air supply to the pump. Though compressed air is expensive, it eliminates the danger of electrical shock and explosive hazards.

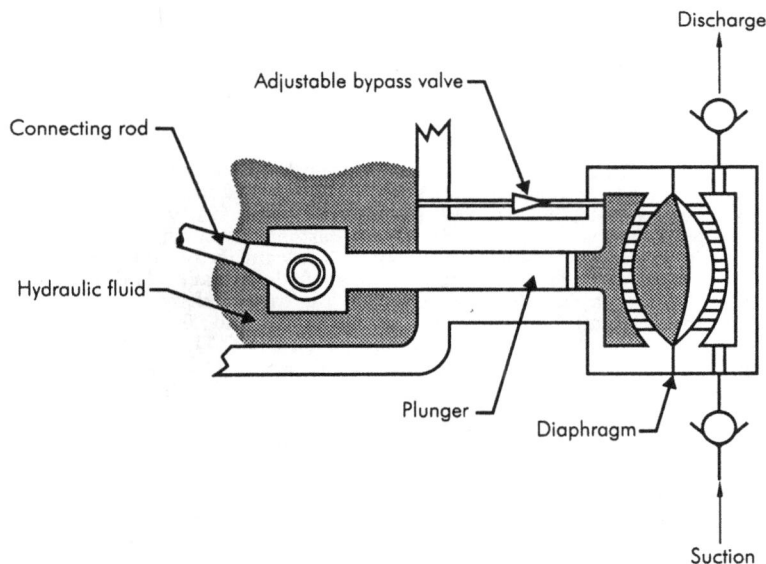

Figure 10-24. Power pump used for chemical feeding

RECIPROCATING STEAM-DRIVEN PUMPS

Although reciprocating steam-driven pumps are no longer found in most boiler rooms, they were universally used for boiler feed, condensate return, fuel oil supply, and fire protection. They are still used to supply the pressure for hydrostatic tests. Normally powered by steam, they can also be driven by compressed air. Figure 10-25 shows a duplex steam pump. Each of the two pumps in a duplex pump has a steam cylinder at one end and a liquid cylinder at the other. The rocker arm of one pump operates the other's steam valve. The action is positive, because a definite amount of liquid is displaced per stroke. D valves are used on the power end to control the steam or compressed air. One of the valves is always open, so no dead centers are encountered. The valves are

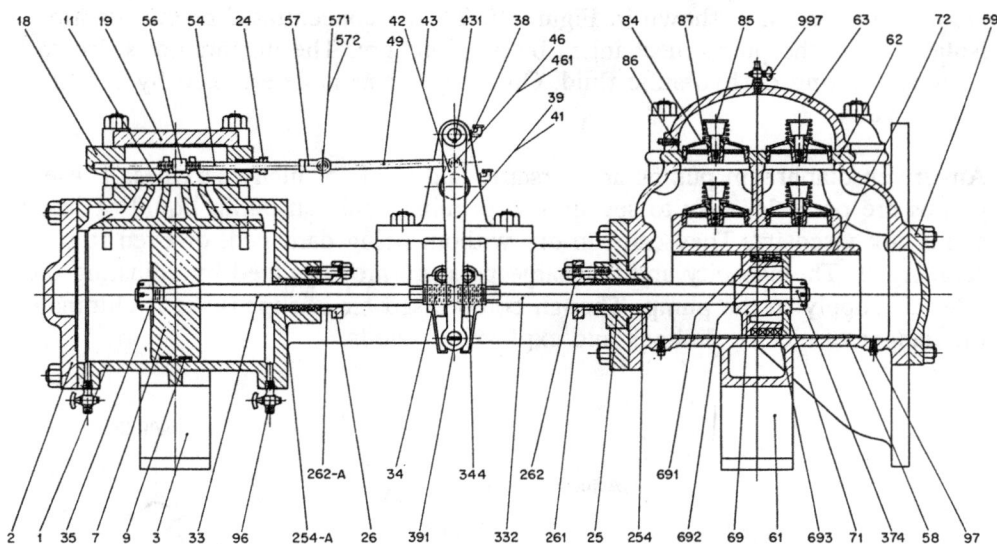

Typical section of duplex steam pump. 1, steam cylinder with cradle; 2, steam-cylinder head; 3, steam-cylinder foot; 7, steam piston; 9, steam piston rings; 11, slide valve; 18, steam chest; 19, steam-chest cover; 24, valve-rod stuffing box gland; 25, piston-rod stuffing box, liquid; 26, piston-rod stuffing box gland, steam; 33, steam piston rod; 34, steam piston spool; 35, steam piston nut; 38, cross stand; 39, long lever; 41, short lever; 42, upper rock shaft, long crank; 43, lower rock shaft, short crank; 46, crankpin; 49, valve-rod link; 54, valve rod; 56, valve-rod nut; 57, valve-rod head; 58, liquid cylinder; 59, liquid-cylinder head; 61, liquid-cylinder foot; 62, valve plate; 63, force chamber; 69, liquid piston body; 71, liquid piston follower; 72, liquid cylinder lining; 84, metal valve; 85, valve guard; 86, valve seat; 87, valve spring; 96, drain valve for steam end; 97, drain plug for liquid end; 254, liquid piston-rod stuffing box bushing; 332, liquid piston rod; 344, piston-rod spool bolt; 374, liquid piston-rod nut; 391, lever pin; 431, lever key; 461, crankpin nut; 571, valve-rod head pin; 572, valve-rod head-pin nut; 691, liquid piston snap rings; 692, liquid piston bull rings; 693, liquid piston fibrous packing rings; 997, air cock; 251, liquid piston-rod stuffing box; 254A, steam piston-rod stuffing box bushing; 261, piston-rod stuffing box gland, liquid; 262, piston-rod stuffing box gland lining, liquid; 262A, piston-rod stuffing box gland lining, steam. (Worthington Pump Corporation)

Figure 10-25. Duplex steam pump (Courtesy, Ingersoll-Dresser Pump Co.)

adjusted so that as one piston nears the end of its travel, the other piston starts. This minimizes discharge pressure fluctuation and ensures even flow.

The size of steam pumps is given in three numbers such as 3 x 2 x 3. Expressed in inches, the first number is the diameter of the steam end, the second is the diameter of the liquid end, and the last is the length of the stroke.

A pump with only one set of cylinders is called a simplex pump. A stroke for a simplex pump (one cylinder) is the movement of the piston from one end of the cylinder to the other. The stroke for a duplex pump (two cylinders) is the movement of both pistons moving once over their path.

An advantage of using steam- or air-driven pumps is they can build up pressure until they stall without damage. When the pressure drops below the stall point, they resume pumping. This is why duplex pumps were popular to keep plant fire mains pressurized, and why they are still used for hydrostatic tests of boilers and other pressure vessels.

While steam-driven pumps are obsolete, questions concerning their function still appear on licensing examination tests. The math used to solve problems concerning steam-driven pumps gives an examiner a good idea of your math skills.

Piston Pump Problems

Once two basic concepts are learned, then solving piston pump problems is straightforward. The first concept is how to calculate the volume of a cylinder. The equation below simply states that volume is equal to the circular area of the cylinder times its length:

$$\text{Volume} = (0.7854)\,(d^2)\,(\text{Length})$$

For example, what is the volume of a cylinder that is 5 inches in diameter with a 6-inch stroke? The answer is:

$$\text{Volume} = (0.7854)\,(5 \text{ in.})^2\,(6 \text{ in.}) = 117.81 \text{ in}^3$$

To convert cubic inches to gallons, divide by 231 in³/gal:

$$\frac{117.81 \text{ in}^3}{231 \text{ in}^3/\text{gal}} = 0.51 \text{ gal}$$

If the problem includes the piston rod diameter, then it would have to be included in the calculation. Since the volume of the rod is taking up space in the cylinder on half the strokes, its volume would have to be subtracted from the volume of the cylinder.

The second concept is the force produced by the power end piston is equal to the force put forth by the liquid end piston. To calculate the force on a cylinder, multiply the area of the cylinder in square inches times the applied pressure in pounds per square inch (psi):

$$Force_{steam\ end} = Force_{liquid\ end}$$

or

$$(Steam\ end\ area)\ (Steam\ pressure) = (Water\ end\ area)\ (Water\ pressure)$$

If the water-end piston is larger than the steam-end piston, the result is more capacity but at less output pressure. If the water-end piston is smaller than the steam-end piston, the result is higher output pressure but less capacity. For example, what is the output pressure of a 6 x 4 x 8 pump with 100 psi steam?

First, remember the equation:

$$(Steam\ end\ area)\ (Steam\ pressure) = (Water\ end\ area)\ (Water\ pressure)$$

Then, plug in the numbers:

$$(6\ in.)^2\ (100\ psi) = (4\ in.)^2\ (Output\ pressure)$$

$$Output\ pressure = \frac{(36\ in^2)(100\ psi)}{16\ in^2} = 225\ psi$$

Notice that the factor 0.7854 is not included in the area portions of the equation. This is because it showed up on both sides of the equation, so it canceled out. To phrase it a different way, the output pressure of the pump is inversely proportional to the square of the piston diameters. It is your option to include or exclude the 0.7854 in this situation.

Also notice that since the output pressure is more than the input pressure, it could be used as a feedwater pump. If this were a 4 x 6 x 8 pump, the output pressure would be less than the input pressure, and it could not be used as a feedwater pump.

Example 10-5. A simplex pump, 8 x 10 x 12, operates at 100 strokes per minute, and against a 150-ft head. With this information, answer the following questions:

1. What does 8 x 10 x 12 mean?

2. Can it be used as a feedwater pump and why?

3. What is this pump's capacity in gpm with 15% slip?

4. What is this pump's capacity with 0% slip and a 2-inch piston rod?

5. How much steam pressure is required to achieve a 150-ft head?

6. If the steam supply were 150 psi, what would be the discharge pressure in psi?

Solution 10-5. We'll answer the questions in order:

1. 8 x 10 x 12 means the steam-end piston is 8 inches in diameter, the liquid-end piston is 10 inches in diameter, and the length of the stroke is 12 inches.

2. This pump can't be used for a feedwater pump, because the steam cylinder is smaller than the water cylinder, making the discharge pressure less than the steam pressure. If it were a 10 x 8 x 12, it could be used as a feedwater pump.

3. To find the capacity of the pump, it is necessary to first calculate the volume of the cylinder, then multiply the volume by how many times the cylinder is emptied in a minute. This answer is in cubic inches, but the question asks for gallons. To convert cubic inches to gallons, divide by 231 in³/gal. If this pump has 15% slip, it means that 85% of the water is pumped with each stroke. Therefore, multiply the capacity answer by 0.85 to obtain the output with 15% slip:

$$\frac{(0.7854)(10 \text{ in.})^2 (12 \text{ in.})(100 \text{ strokes/min})(0.85)}{231 \text{ in}^3/\text{gal}} = 347 \text{ gpm}$$

4. Now find the capacity if the pump has no slip and a 2-inch piston rod. First calculate the volume of the cylinder without the rod:

$$(0.7854)(10 \text{ in.})^2 (12 \text{ in.}) = 942.48 \text{ in}^3$$

Then calculate the volume of the cylinder with the rod:

$$942.48 \text{ in}^3 - [(0.7854)(2 \text{ in.})^2 (12 \text{ in.})] = 904.79 \text{ in}^3$$

The total volume for the two strokes is 1,847.27 in³ (942.48 in³ + 904.79 in³). The average for the two strokes is 923.64 in³ (1,847.27 in³/2). To find the capacity, plug in the numbers as follows:

$$\frac{(923.64 \text{ in}^3)(100 \text{ strokes/min})}{231 \text{ in}^3/\text{gal}} = 399.84 \text{ gpm}$$

5. The force on the shaft at the power end is the same as the force at the liquid end. To calculate the force on a cylinder, multiply the area of the cylinder by the applied pressure in pounds per square inch. Before we can proceed any further, 150 ft must be converted into pounds per square inch:

(150 ft head) (0.433 psi/ft of head) = 64.95 psi

With a 10-inch water-piston cylinder diameter and 64.95 psi, the total force on the piston is:

(64.95 psi) (0.7854) (10 in.)^2 = 5,101 lb

Now divide the area of the steam-end piston into the force to get the pressure required to produce a 150-ft head:

$$\frac{5,101 \text{ lb}}{(0.7854)(8 \text{ in.})^2} = 101.5 \text{ psi}$$

The problem may be solved more quickly if the following equation is used:

(Steam end area) (Steam pressure) = (Water end area) (Water pressure)

$$\text{Steam pressure} = \frac{(10 \text{ in.})^2(64.95 \text{ psi})}{(8 \text{ in.})^2} = 101.5 \text{ psi}$$

6. Knowing the steam pressure, the water pressure can be found by using the same method:

$$\text{Water pressure} = \frac{(8 \text{ in.})^2(150 \text{ psi})}{(10 \text{ in.})^2} = 96 \text{ psi}$$

JET PUMPS

Jet pumps have no moving parts and use fluids in motion to supply the motive power. A high-pressure stream of fluid (usually steam or water) is directed through a nozzle designed to produce the highest possible velocity. This high velocity jet creates a low-pressure area, causing fluid to be sucked into the jet pump. There is an exchange of momentum at this point where the motive fluid entrains the suction fluid (draws the suction fluid along with it). This produces a uniformly mixed stream traveling somewhere between the motive fluid and the suction fluid velocity. The three basic parts of any jet pump are the nozzle, the diffuser, and the suction chamber or body.

Before continuing with the discussion, it is necessary to understand several terms:

- **Ejector** — the term applied to all types of jet equipment whose discharge pressure is somewhere between the motive and suction pressures. Besides pumping, it can also be used to blend, heat, or cool the motive and suction fluids.

- **Eductor** — an ejector that uses a liquid as its motive fluid, Figure 10-26. The most common one uses water to entrain water or some other material. Commonly used for deep well service in conjunction with a centrifugal pump, the eductor lifts water from a level below barometric height, up to a level where the suction of the centrifugal pump at the surface can lift the water the remaining distance. In operation, the eductor is fitted with hoses connected to the suction and discharge of the pump and lowered into the well casing. The required initial prime is maintained by a foot valve in the eductor. When turned on, some of the pump discharge water flows through the eductor and entrains water from the well, lifting it high enough to enable the pump to carry it the rest of the way.

- **Siphon** — an ejector that uses a condensable gas for its motive power. The most common example uses steam to entrain water or some other liquid. It uses a converging-diverging nozzle to achieve maximum velocity at the nozzle tip, Figure 10-27.

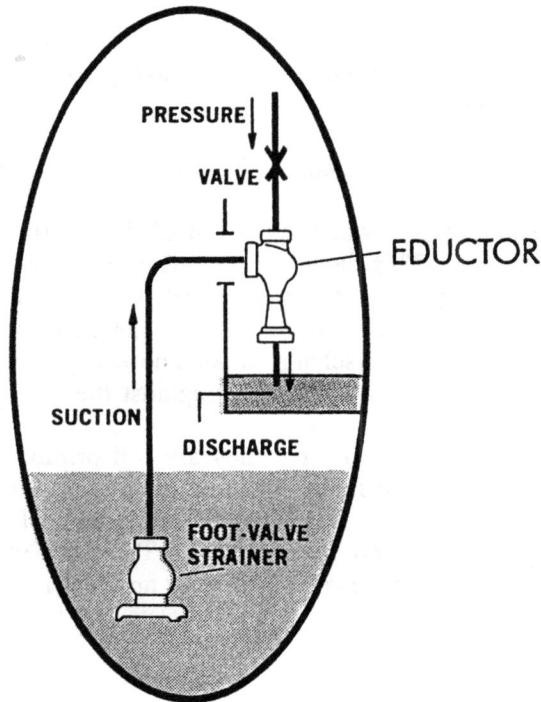

Figure 10-26. Eductor placement (Courtesy, Schutte & Koerting)

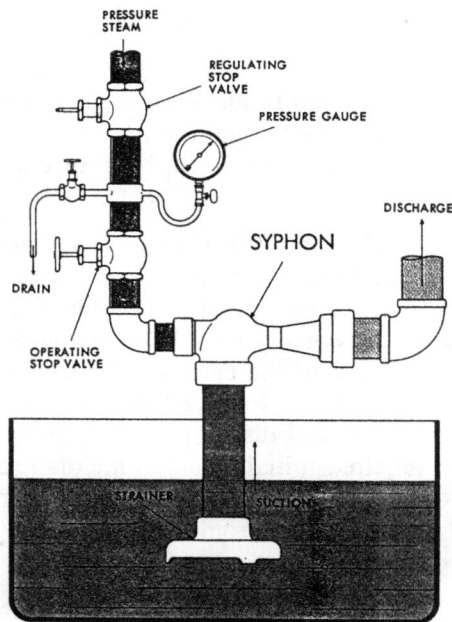

Figure 10-27. Steam jet syphon placement (Courtesy, Schutte & Koerting)

- **Injector** — a special type of ejector that uses a condensable gas (usually steam) as the motive fluid. It entrains a liquid and discharges it against a pressure higher than either the motive or suction pressures. Once used widely in steam locomotives, its use is now confined to backup for boiler feedwater pumps.

In injector operation, the lower nozzle is activated by pulling the handle back part way. The lower jet creates a vacuum in the chamber, causing water to be induced into the unit. When water spills out of the overflow, the handle is drawn back all the way. This closes the overflow and simultaneously admits motive steam to the upper jet. This second jet, which is of the straight or forcing type, then picks up the discharge from the first jet and imparts a velocity to the water through the discharge tube. The energy contained is sufficient to open the check valve and discharge water against the boiler pressure.

The advantages of jet pumps are that they are self-priming, have no moving parts, and can be made from many materials, including bronze, cast iron, stainless steel, PVC, Teflon™, and fiberglass. The main disadvantage of jet pumps is the high cost of their motive power. The price of steam can be especially high compared to the power needed for an electric pump.

MOTORS

Most boiler examinations do not have many questions about motors. However, there are three questions you should be prepared for:

1. How is the rotation of a three-phase motor reversed?

2. What is the calculation for rpm?

3. What is the calculation for torque?

The rotation of a three-phase motor is reversed by interchanging any two of the three leads. This procedure is so simple that it is very easy to hook a motor up incorrectly. Therefore, all three-phase motors must be checked for rotation after replacement.

The equation for motor rpm is:

$$\text{rpm} = \frac{(\text{Hz})(120)}{\text{No. of poles}}$$

Hz is the abbreviation for Hertz, which is the accepted abbreviation for cycles per second. The constant 120 converts cycles per second to cycles per minute and pole pairs to poles. If a frequency is not given, use 60 Hz, but state your assumption.

For example, what is the rpm of a 4-pole motor operating at 60 Hz?

$$\text{rpm} = \frac{(60 \text{ Hz})(120)}{4 \text{ poles}} = 1,800 \text{ rpm}$$

The equation for torque is:

$$\text{Torque} = \frac{(\text{hp})(5,252)}{\text{rpm}}$$

The unit of torque is lb-ft. (The constant 5,252 basically converts radians to rpm, but do not worry about how this constant was obtained.)

For example, what is the torque of a 50 hp motor operating at 1,750 rpm?

$$\text{Torque} = \frac{(\text{hp})(5,252)}{1,750 \text{ rpm}} = 150 \text{ lb-ft}$$

CHAPTER ELEVEN

Bearings

Bearings support the weight and control the motion of a rotating shaft, while consuming a minimum amount of power. There are two classes of bearings: antifriction and sleeve.

Antifriction bearings use balls, Figure 11-1, or rollers, Figure 11-2, to convert sliding friction into rolling friction. They are constructed by placing these balls or rollers between an outer race and an inner race. A retaining cage or separator is used to keep the rolling elements equally spaced. Some bearings also have shields to keep lubricants in and dirt out.

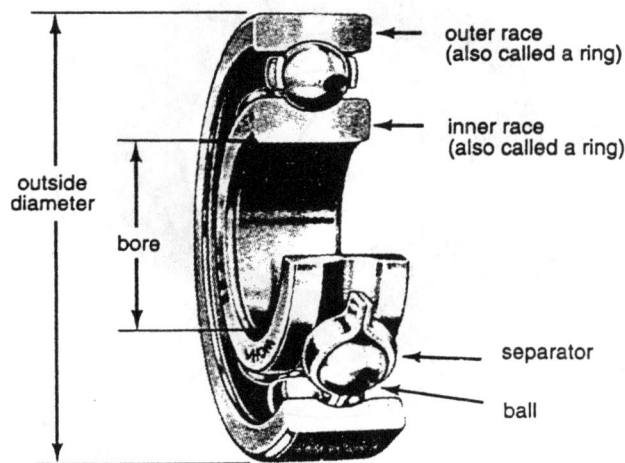

Figure 11-1. Antifriction bearing with balls

Sleeve bearings depend on a thin film of oil to be dragged along by the rotation of the shaft, which forms a wedge between the shaft and bearing, Figure 11-3. This oil film supports the shaft so there is no metal-to-metal contact.

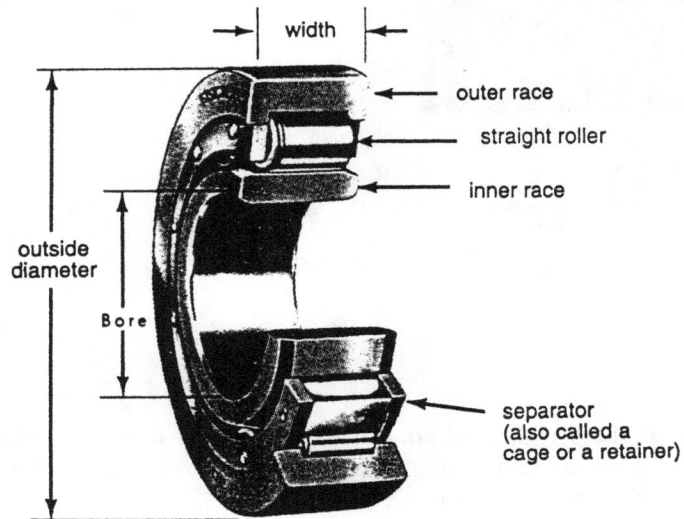

Figure 11-2. Antifriction bearing with rollers

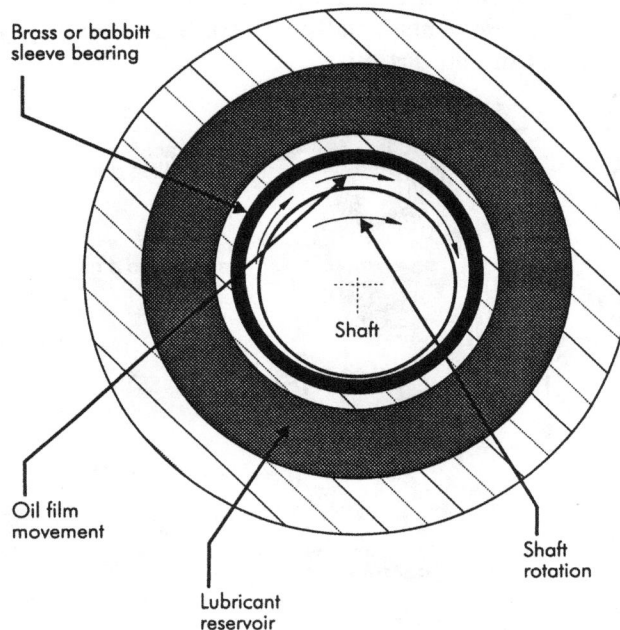

Figure 11-3. Oil film

Sleeve bearings have a wide application range. They are found on small appliances such as fans, because they are cheap, and are used on large pumps, turbines, and engines where antifriction bearings are not adequate or practical. These larger bearings are made of a babbitt lining, 1/8-inch thick or more, and anchored in a cast-iron or bronze shell. Babbitt is an alloy of tin, antimony, lead, and copper.

LUBRICATION

The type of lubrication depends on the application requirements. Refinery engineers prefer oil; marine engineers prefer grease. At high speeds (more than 5,000 rpm), oil is better. Oil is better for two main reasons. First, oil has a lower viscosity and is better able to flow to the areas where it is most needed. Second, it is easy to pass oil through a cooler to remove the heat generated by friction. The bearing's longevity is increased when its temperature is kept within its specified range.

Grease

Grease packed into a bearing is thrown out by the rotation of the balls, which creates a slight suction at the inner race. As heat is generated in the bearing, the flow of grease is accelerated until it is thrown out at the outer race by rotation. As the expelled grease is cooled on contact with the housing, it is attracted back to the inner race. This continuous circulation of grease lubricates and cools the bearing.

Grease requires little attention and works well. It is preferred on vertical shaft bearings because there is less chance of leakage.

A fully packed bearing housing prevents proper grease circulation. It is recommended that only one-third of the housing be filled with grease. An excess amount causes overheating unless the grease can flow out of a seal or vent. If this built-up pressure is not relieved, the bearing will fail.

To use grease to lubricate a motor with antifriction bearings, perform the following procedure (when equipped with fitting and drain), Figure 11-4:

1. Remove the drain plug at the bottom of the bearing housing and clean out any hard grease.

2. Wipe grease fitting clean.

3. Apply new grease through the fitting until the old grease has been purged through the drain and new grease begins to appear. Do this with the motor running if possible.

4. With the drain plug removed, allow the motor to run at operating temperature. This allows the grease to expand so the excess is forced out of the drain. Excess grease stops draining when normal pressure is reached in the bearing, usually within 10 to 30 minutes.

5. Clean and replace drain plug.

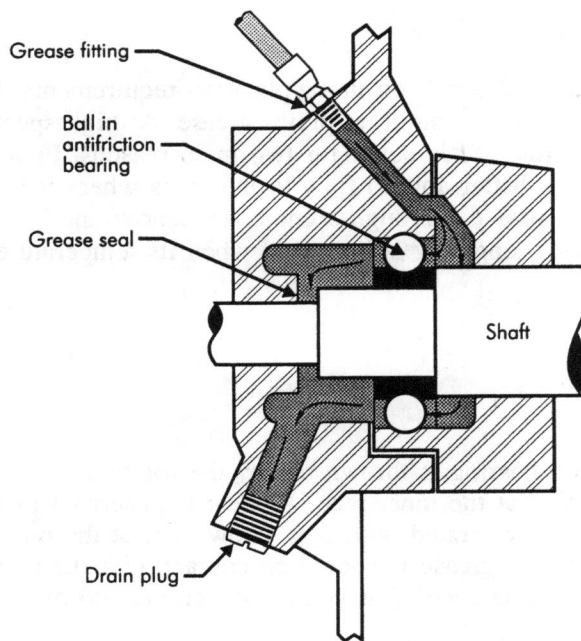

Figure 11-4. Lubricating a motor equipped with a fitting and drain

When equipped with a fitting but no drain, perform the following procedure, Figure 11-5:

1. Remove the fitting while the motor is running at operating temperature to purge the excess grease.

2. Clean and replace fitting. Pump a small amount of grease into the bearing, taking care not to rupture the grease seal.

3. Remove fitting and allow motor to run for several minutes to purge excess grease. (If no grease comes out, the bearing must have been dry. Repeat Steps 2 and 3 until excess grease comes out.)

4. Replace fitting.

Oil

Antifriction bearings can also be lubricated with oil. As shown in Figure 11-6, the proper oil level is at the center of the lowermost ball in the bearing.

A constant level oiler is used to increase the capacity of the reservoir, Figure 11-7. It works just like the upside down bottle on a water cooler; as long as the oil is above the inverted bottle opening, atmospheric pressure holds in the oil. When the opening is exposed, the weight of the oil allows it to run out. When the level is high enough to cover the opening again, the oil flow will stop. The level is adjustable by moving the inverted bottle up or down.

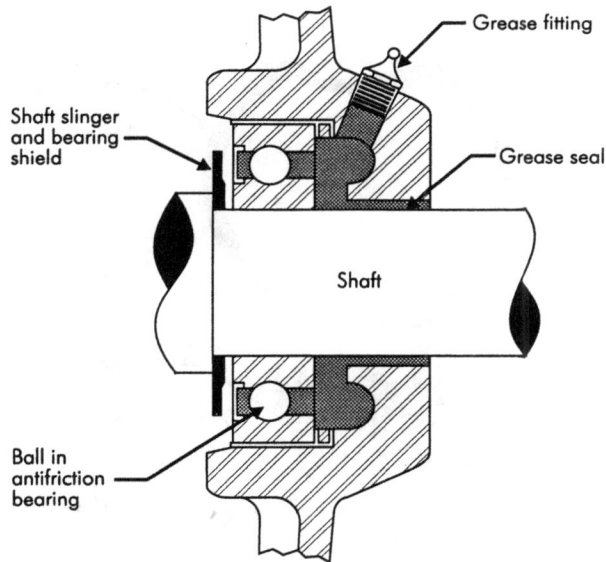

Figure 11-5. Lubricating a motor with a fitting but no drain

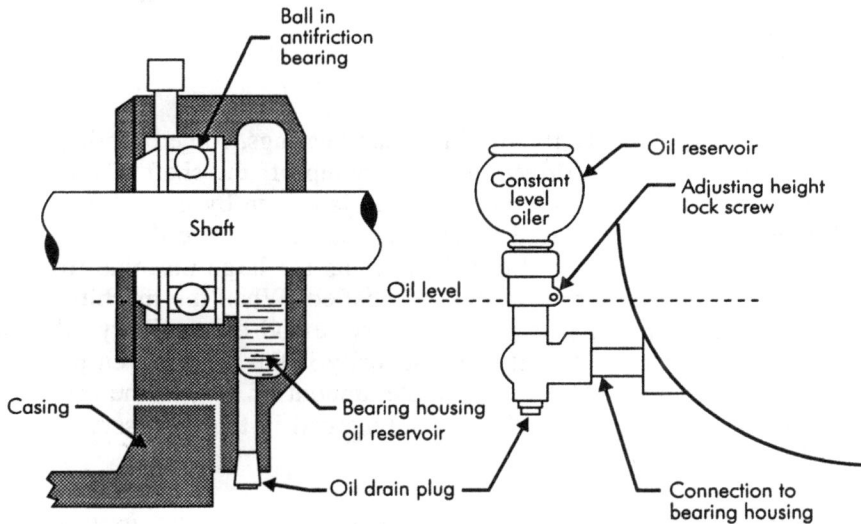

Figure 11-6. Antifriction bearing lubricated with oil

Ring oiling is popular for lubricating smaller sleeve bearings. A soft steel ring rides on the shaft through a slot cut in the top half of the bearing shell. The ring rotates as the shaft turns, picks up oil from the reservoir in the housing, then wipes it off on top of the shaft. The oil then flows between the shaft and the bearing and discharges through the ends. Note that the proper oil level should be just below the bearing shell but high enough to submerge the oil rings.

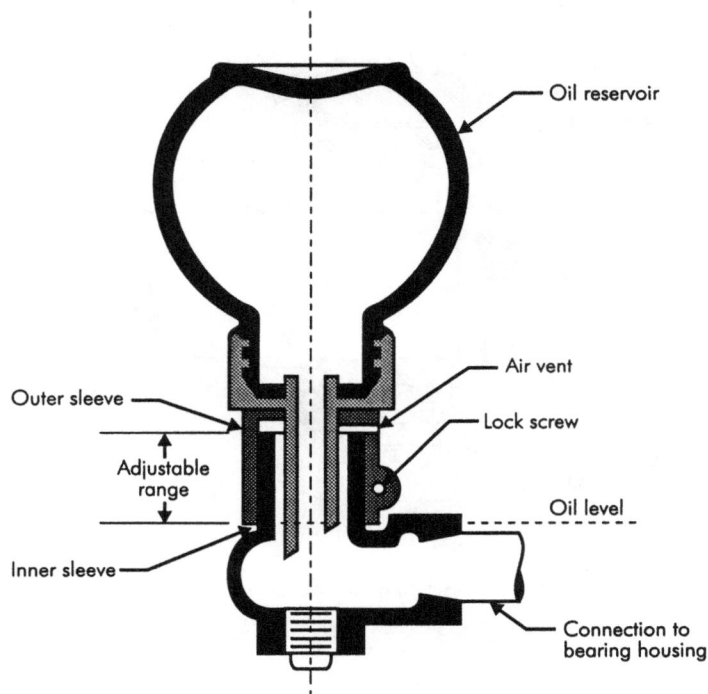

Figure 11-7. Constant level oiler

In larger machines, especially those with thrust bearings, forced lubrication is required. The system usually drives the main pump off the shaft of the machine that is being lubricated, and an auxiliary pump is driven by a small electric motor. The auxiliary pump is used before start-up to circulate the oil and establish oil pressure. The control circuit prevents the large machine from starting unless the oil pressure interlock is satisfied. When the machine is up to speed, the main lube oil pump is placed in service and the auxiliary lube oil pump is placed on standby. It starts automatically if the shaft-driven pump fails. Larger machines can also add a considerable amount of heat to the oil. Oil is cooled by circulating water through coils submerged in the oil reservoir or through a heat exchanger.

For heavy thrust loads, tilting plate or Kingsbury™ bearings are used, Figure 11-8. The axial load is transmitted to the stationary bearing shoes through a forged steel collar that is rigidly attached to the shaft. Leveling plates equalize the load among the shoes. These shoes are free to tilt both radically and tangentially as required to compensate for operating conditions.

When in operation, the bearing surfaces are continuously separated by a film of oil so there is no metal-to-metal contact. The bearing faces of the shoes take an inclined position so the oil film between the shoes and collar is wedge shaped with the thin end pointing in the direction of rotation, Figure 11-9.

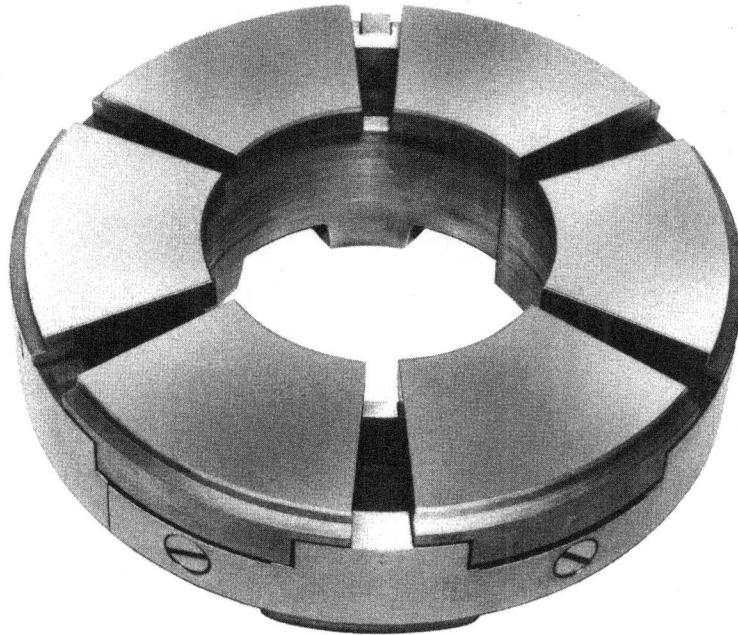

Figure 11-8. Kingsbury bearing (Courtesy, Kingsbury, Inc.)

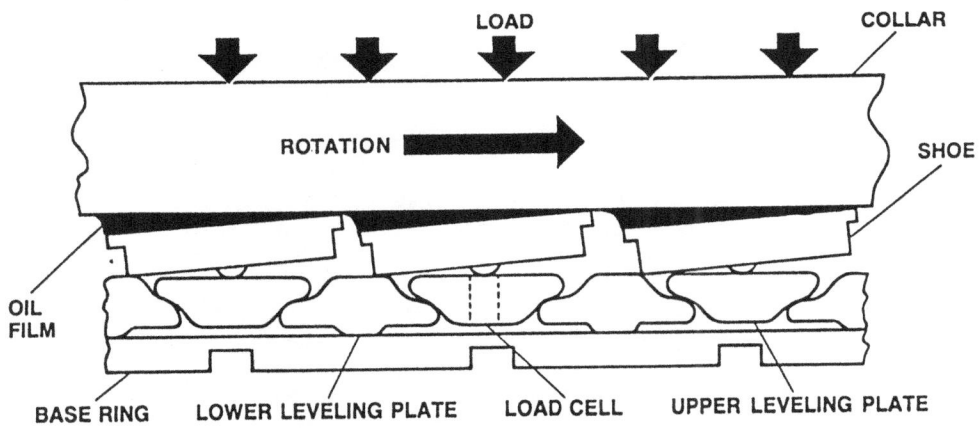

Figure 11-9. Wedge-shaped oil film (Courtesy, Kingsbury, Inc.)

CHAPTER TWELVE

Sample Multiple Choice Questions

These multiple choice questions are representative of those that will be found on different boiler operator licensing examinations. Not all of the answers to the questions will be found in the text, as it is impossible to cover all the material found on the examinations.

1. At 0 psig, how many Btu are necessary to change 1 lb of water at 212°F into 1 lb of steam?

    a) 144
    b) 180
    c) 970
    d) 1,190

2. Radiation is

    a) the movement of liquids or gases created by a temperature difference.
    b) the transfer of heat by direct molecular contact.
    c) the weight of a substance compared to unity.
    d) a form of heat transfer by waves.

3. Steam at 100 psi and 400°F is

    a) saturated.
    b) wet.
    c) superheated.
    d) none of the above.

4. Which of the following most closely defines sensible heat?

    a) An established relationship comparing any substance to the heat content of water
    b) Heat quantity that can be felt or measured by a thermometer
    c) Heat quantity above the point of saturation
    d) Measure of heat intensity

5. What is required to raise the temperature of 1 lb of water from 32° to 212°F?

    a) 144 Btu
    b) 180 Btu
    c) 970.3 Btu
    d) Saturated heat

6. Increasing the pressure has what effect on the boiling point of water?

    a) No change
    b) Temperature will be lowered
    c) Temperature will be raised
    d) Temperature will increase 2°F for every psi

7. Heat absorbed by water when it changes from liquid to steam at the boiling point is called

    a) sensible heat.
    b) latent heat.
    c) specific heat.
    d) superheat.

8. Heat may be transferred in how many ways?

    a) One
    b) Two
    c) Three
    d) Four

9. A boiler horsepower is defined as

    a) 970.3 Btuh.
    b) total heating surface times the factor of evaporation.
    c) evaporation of 34.5 lb of water from and at 212°F per hour.
    d) evaporation of 34.5 lb of water from and at 212°F per 8 hours.

10. A boiler's steam pressure gauge reads 150. What is the absolute pressure?

    a) 14.7
    b) 135.3
    c) 150
    d) 164.7

11. What does 212°F equal on the Celsius scale?

    a) 32
    b) 100

c) 180

d) 970.3

12. One boiler horsepower is equal to how many Btu?

    a) 970.3

    b) 33,472

    c) 125,000

    d) One therm

13. A British thermal unit is

    a) the amount of heat needed to raise one pound of water one degree Fahrenheit.

    b) the amount of heat needed to raise one pound of steam one degree Fahrenheit.

    c) a unit of temperature measurement.

    d) the amount of heat in water between freezing and boiling.

14. Hydronic heat is

    a) steam heat.

    b) hot water heat.

    c) radiant heat.

    d) heat of evaporation.

15. Superheated steam is

    a) reheated steam.

    b) very hot steam.

    c) flash steam.

    d) saturated steam.

16. 2,545 Btuh represents

    a) 1 bhp.

    b) 1 kilowatt-hour.

    c) 1 horsepower-hour.

    d) 1 therm.

17. A throttling calorimeter is used to measure the quality of

    a) dry steam.

    b) saturated steam.

    c) superheated steam.

    d) Btu in coal.

18. The amount of heat in one pound of saturated steam is approximately equal to

    a) 970.3 Btu.
    b) 1,150 Btu.
    c) 3,412 Btu.
    d) 100,000 Btu.

19. A boiler operating at 100 psi has a dry saturated steam temperature of 338°F. What is the boiler's water temperature?

    a) 180°F
    b) 212°F
    c) 309°F
    d) 338°F

20. A dry pipe is used to

    a) separate moisture from steam as it leaves the drum.
    b) separate moisture from steam as it leaves the superheater.
    c) separate moisture from steam at the turbine throttle.
    d) increase superheater temperature.

21. The purpose of a staybolt is to

    a) support the shell of the boiler.
    b) reinforce the heads of a fire-tube boiler.
    c) help make a water-tube boiler safer to operate.
    d) strengthen and support flat plates of a fire-tube boiler.

22. The boiler vent valve is located on the

    a) top of the steam drum.
    b) bottom of the steam drum.
    c) top of the mud drum.
    d) top of the water drum.

23. The purpose of the hole in the end of the staybolt is to

    a) indicate how much water is in the boiler.
    b) indicate that a staybolt is cracked or broken.
    c) help the boiler operator operate more efficiently.
    d) provide means of attaching insulation to a boiler.

24. A fusible plug melts at

    a) 212°F.
    b) 450°F.
    c) 1,000°F.
    d) none of the above.

25. As the firing rate increases in a boiler equipped with a convection superheater, the steam temperature

    a) increases.
    b) decreases.
    c) stays constant.
    d) both increases and decreases.

26. When boilers are connected to a common header, it is required that an ample free blowing drain be provided between the stop valves if either or both boilers have

    a) handholes.
    b) manholes.
    c) over 10 tubes or flues.
    d) more than one safety valve.

27. It is permissible to have only one bottom blowdown valve on a boiler when the

    a) MAWP does not exceed 100 psi.
    b MAWP exceeds 125 psi.
    c) heating surface does not exceed 100 ft^2.
    d) heating surface does not exceed 500 ft^2.

28. When the boiler pressure exceeds _____ psi, it is required that extra-heavy pipe be used on the blowoff system.

    a) 15
    b) 100
    c) 125
    d) 150

29. Baffles are placed between the tubes of a boiler for the purpose of

    a) slowing down the rate of combustion.
    b) preventing priming.
    c) bracing the tube bank.
    d) directing the travel of flue gases.

30. A boiler has a steaming capacity of 35,500 lb of steam per hour. The evaporation rate is 10 lb/ft^2 of heating surface per hour. What is the horsepower rating of this boiler?

    a) 100
    b) 1,000
    c) 3,450
    d) 6,900

31. Boiler tube sizes are specified by

    a) pipe size.
    b) standard ASME size.
    c) inside diameter.
    d) outside diameter.

32. The maximum size of boiler blowdown lines, valves, and fittings shall not exceed

    a) 1 inch.
    b) 1-1/2 inches.
    c) 2 inches.
    d) 2-1/2 inches.

33. The purpose of an expansion tank in a hot water heating system is to provide for the expansion of

    a) water.
    b) hot air.
    c) air and steam.
    d) not used in hot water systems.

34. Heavy accumulation of soot in a boiler will result in

    a) loss of boiler efficiency.
    b) loss of the fire.
    c) reduced stack temperature.
    d) the safety valve popping.

35. How are the tubes secured in a fire-tube boiler?

    a) Welded
    b) Brazed
    c) Rolled and beaded over
    d) Shrink fit

36. A tube in a fire-tube boiler is surrounded by

    a) water.
    b) air.
    c) flue gases.
    d) steam.

37. Atmospheric pressure is considered to be

    a) 0 psia.
    b) 14.7 in. Hg.
    c) 14.7 psia.
    d) 14.7 psig.

38. In a water-tube boiler,

    a) flue gas passes inside the tube.
    b) flue gas surrounds the tube.
    c) water surrounds the tube.
    d) none of the above.

39. Diagonal stays are found in

    a) fire-tube boilers.
    b) water-tube boilers.
    c) cast-iron boilers.
    d) waste heat boilers.

40. Handholes and manholes are oval

    a) to withstand higher pressures.
    b) for ease of fabrication.
    c) for better sealing.
    d) for removal from a pressure vessel.

41. To check the water level in the gauge glass of a steam boiler

    a) use the bottom blowoff valve.
    b) use the trycocks.
    c) blow down the low-water cutoff.
    d) compare it with the water level in another boiler.

42. The vent valve

    a) removes air during warm-up.
    b) prevents a vacuum from forming during cooldown.
    c) Both a and b.
    d) Neither a nor b.

43. A high-pressure boiler has its safety valve set above

    a) 15 psi.
    b) 100 psi.
    c) 125 psi.
    d) 150 psi.

44. Which of the following is an advantage of fire-tube boilers over water-tube boilers?

    a) Low initial cost
    b) Capable of higher pressures
    c) Lower possibility of explosions
    d) Able to fire a larger variety of fuels

45. Water walls are found

    a) where the flue gas exits the boiler.
    b) in fire-tube boiler furnaces.
    c) in water-tube boiler furnaces.
    d) in the air heater of a water-tube boiler.

46. The best time to bottom blow a boiler is at

    a) high load under full pressure.
    b) low load under full pressure.
    c) high load under partial pressure.
    d) low load under partial pressure.

47. A telltale hole is found in a

    a) dry pipe.
    b) crown sheet.
    c) fusible plug.
    d) staybolt.

48. A 2-inch boiler tube has a wall thickness of 3/16 inch. What is its inside diameter (id)?

    a) 2-3/16 inches
    b) 2 inches
    c) 1-13/16 inches
    d) 1-5/8 inches

49. At what temperature does steel start to rapidly lose its strength?

    a) 500°F
    b) 700°F
    c) 1,000°F
    d) 1,500°F

50. A hot-water heating system is about to be started. How full of water should the expansion tank gauge glass be?

    a) Empty
    b) 1/3
    c) 3/4
    d) Full

51. What is the relationship between longitudinal stress and circumferential stress?

    a) They are the same.
    b) Longitudinal stress is twice circumferential stress.
    c) Circumferential stress is twice longitudinal stress.
    d) There is no fixed relationship between the two.

52. A therm is equal to how many Btu?

    a) 970.3
    b) 33,475
    c) 100,000
    d) 1,000,000

53. The factor of evaporation is

    a) heat added to the boiler divided by 970.3.
    b) heat added to the boiler minus inlet feedwater temperature.
    c) 970.3 Btu.
    d) actual evaporation over 34.5 lb of water.

54. The purpose of a bottom blowdown is to

    a) reduce boiler pressure.
    b) remove sludge and sediment.
    c) increase dissolved solids.
    d) remove floating oil.

55. Damage by spalling refers to the

    a) tubes.
    b) steam drum.
    c) refractory.
    d) superheater.

56. Superheaters raise the temperature of

    a) feedwater.
    b) boiler water.
    c) combustion air.
    d) steam.

57. The steam space is

    a) the area covered by fire and water.
    b) in the main header area.
    c) the top half of most steam traps.
    d) the area above the operating water level.

58. The heating surface of a boiler is the area

    a) exposed to the flame and flue gases.
    b) in contact with steam.
    c) of the furnace.
    d) of the burner face in the furnace.

59. The purpose of a superheater is to

    a) heat steam above its saturation temperature.
    b) boost the steam pressure.
    c) add heat to the combustion air.
    d) add heat to the incoming feedwater.

60. What is the formula for the factor of evaporation?

    a) h - H / 970.3 Btu
    b) H - b / 970.3 Btu
    c) 970.3 Btu / H - h
    d) 970.3 Btu / h - H

61. What can cause a superheater tube to overheat?

    a) Not draining before starting
    b) Overfiring
    c) Poor boiler circulation
    d) Dirty tube

62. A boiler has a steaming capacity of 34,500 lb of steam per hour. The evaporation rate is 10 lb/ft^2 of heating surface per hour. What is the boiler heating surface?

    a) 100
    b) 1,000
    c) 3,450
    d) 6,900

63. How many square feet of heating surface does a 3-inch fire-tube 20 ft long with a 3/16-inch wall have?

    a) 13.7
    b) 14.7
    c) 15.7
    d) 16.7

64. In a high-temperature hot water boiler, what would cause a tube to overheat?

    a) Dirty economizer tubes
    b) High economizer feedwater outlet temperature
    c) Feedwater temperature too high
    d) Poor circulation (plugged tube)

65. A downcomer is

    a) located in a fire-tube boiler.
    b) one of the tubes that supports the furnace.

c) used to circulate flue gas.
d) found in a water-tube boiler.

66. Extreme firebox temperature changes cause

    a) spalling.
    b) lower stack temperatures.
    c) increased soot deposits.
    d) a rise in stack temperature.

67. The low-water cutoff

    a) feeds the boiler when it is low on water.
    b) shuts off the burner when the boiler is low on water.
    c) dumps the boiler.
    d) cuts off the water supply when the level is too high.

68. Steam coming from the bottom try cock would indicate

    a) normal water level.
    b) high water level.
    c) low water level.
    d) normal steam level.

69. What type of valve has the most resistance to flow?

    a) Globe
    b) Gate
    c) Ball
    d) Butterfly

70. When checking the water level of a boiler,

    a) ask the boiler operator.
    b) test the safety valve.
    c) use the bottom blowdown valve.
    d) blow down the gauge glass and water column.

71. A siphon or pigtail is used to protect a steam pressure gauge

    a) from steam temperature.
    b) against high steam pressure.
    c) against high water pressure.
    d) against an excessive amount of steam.

72. The feedwater regulator is located

    a) on the right side of the boiler.
    b) at the normal operating water level.
    c) two inches below the highest heating surface.
    d) next to the fusible plug.

73. An automatic feedwater regulator

    a) ensures the proper water level in the boiler.
    b) shuts off the burner.
    c) controls the boiler's operating range.
    d) is never used on high pressure boilers.

74. To test the low-water cutoff,

    a) the burner must be off.
    b) there must be no pressure on the boiler.
    c) the fuel must be shut off.
    d) the burner must be firing.

75. The purpose of testing a safety valve by hand is to

    a) make sure the valve can lift.
    b) check the operating range of the boiler.
    c) check its popping pressure.
    d) They are never tested by hand.

76. The automatic control that protects a boiler from being dry fired is the

    a) aquastat.
    b) vaporstat.
    c) flame failure cutoff.
    d) low-water cutoff.

77. To check the water level in the gauge glass of a steam boiler,

    a) use the bottom blowoff valve.
    b) blow down the low-water cutoff.
    c) blow down the gauge glass and water column.
    d) compare it with the water level in another boiler.

78. On a steam boiler, testing the operation of the safety valve by hand with the boiler under pressure should be done

    a) at the start of each heating season.
    b) daily.
    c) never.
    d) monthly.

79. What type of valve has the least resistance to flow?

    a) Globe
    b) Gate
    c) Reducing
    d) Needle

80. The gauge glass shows full while the middle try cock shows steam. When blown down, the gauge glass refills. What is the probable cause?

    a) Feedwater regulator is stuck open.
    b) Feed pump is running too fast.
    c) Top gauge glass connection is blocked.
    d) Both top and bottom gauge glass connections are blocked.

81. The purpose of a bowl at the bottom of the water column is to

    a) serve as a condenser.
    b) serve as a sediment chamber.
    c) serve as a siphon.
    d) none of the above.

82. When checking a safety valve with the hand-lifting gear, the pressure should be no lower than ___% of the safety valve set pressure.

    a) 30
    b) 50
    c) 75
    d) 100

83. If valves are located between the boiler and water column, they must be

    a) locked or sealed open.
    b) regular globe valves.
    c) steel and rated at 300 psi.
    d) protected against accidental opening.

84. When a hydrostatic test is put on a boiler, each safety valve should be

    a) set at its maximum allowable working pressure.
    b) set at twice its popping pressure.
    c) held to its seat by screwing down the compression screw.
    d) held to its seat by means of a testing clamp.

85. After installing a new gauge glass, the proper procedure for returning the glass to service is

    a) crack open both the top and bottom valves.
    b) open the drain and crack both the bottom and top valves.
    c) open the drain and crack open the top valve.
    d) open the drain and crack open the bottom valve.

86. The steam pressure gauge is graduated in

    a) absolute pressure.
    b) inches of pressure.
    c) pounds per square inch.
    d) pounds per square foot.

87. One of the purposes of the water column is to

    a) indicate to the operator the level of the water in the boiler.
    b) easily find the pressure in the boiler.
    c) dampen the oscillation of the water in the gauge glass.
    d) prevent the boiler from priming.

88. The gauge glass should be blown down

    a) daily.
    b) weekly.
    c) monthly.
    d) yearly.

89. The flame failure control is tested by

    a) blowing down the low-water cutoff.
    b) adjusting the pressuretrol.
    c) calling the inspector.
    d) securing the fuel supply to the burner.

90. Which of the following controls operates due to changes in temperature?

    a) Fire eye
    b) Vaporstat
    c) Aquastat
    d) Pressuretrol

91. The pressure-sensitive device attached to the fan housing of a fully automatic burner starts operating when there is

    a) flame failure.
    b) too large an increase in primary air.
    c) fan failure.
    d) too large an increase in secondary air.

92. If the pressuretrol starts the burner and ignition failed to take place, the control that protects the boiler is the

    a) low-water cutoff.
    b) flame failure cutoff.
    c) vaporstat.
    d) aquastat.

93. The pressure-sensitive device in Question 91 is sometimes called a(n)

    a) pressuretrol.
    b) aquastat.
    c) thermostat.
    d) vaporstat.

94. Draft is measured in

    a) ounces per square inch.
    b) inches of water.
    c) inches of mercury.
    d) pounds per square inch.

95. Which of the following is used to measure temperature?

    a) Fyrite analyzer
    b) Vaporstat
    c) Orsat analyzer
    d) Pyrometer

96. The main stop valve is located on the

    a) oil supply line.
    b) condensate header.
    c) vacuum breaker.
    d) steam line.

97. OS&Y stands for

    a) overhead steam and yield.
    b) outside screw and yoke.
    c) outside stem and yoke.
    d) overall stainless and yoke.

98. An interlock control is a(n)

    a) device that prevents attempts to fire unless safe.
    b) wrench that secures the burner.
    c) device that prevents the main stop from opening.
    d) automatic control clock.

99. Trial for ignition refers to the

    a) burner testing period.
    b) time the main fuel valve stays open for ignition.
    c) flame light test.
    d) the flame scanner detecting the pilot flame

100. More than one safety valve is required on boilers that have over

   a)   100 ft^2 of heating surface.
   b)   500 ft^2 of heating surface.
   c)   100 psi MAWP.
   d)   500 psi MAWP.

101. Two or more boilers that are equipped with manholes and are connected to a common header must have

   a)   two stop valves with a drain coming off between them.
   b)   both a nonreturn and a gate stop valve.
   c)   both a globe and a gate stop valve.
   d)   a stop and a check valve.

102. Before the gas pilot ignites, the

   a)   low-water cutoff must be checked.
   b)   gas pressure must be at least 10 psi.
   c)   furnace must be purged.
   d)   oil pressure must be at least 50 psi.

103. If the main fuel valve is closed while the burner is on,

   a)   there will be high water.
   b)   there will be low water.
   c)   the oil temperature will drop.
   d)   the flame detector will activate.

104. Pyrometers measure

   a)   temperature.
   b)   draft.
   c)   smoke density.
   d)   vacuum.

105. Safety valves on superheaters are set

   a)   the same as the drum safety valves.
   b)   lower than the drum safety valves.
   c)   higher than the drum safety valves.
   d)   Superheaters do not have safety valves.

106. Convert 160 psi to absolute pressure.

   a)   145.3 psia
   b)   160 psia
   c)   167.3 psia
   d)   174.7 psia

107. What is the minimum pipe size of the gauge glass blowdown line?

    a) 3/8 inch
    b) 1/2 inch
    c) 3/4 inch
    d) 1 inch

108. Only one bottom blowdown valve is required for boilers under

    a) 100 ft^2 of heating surface.
    b) 500 ft^2 of heating surface.
    c) 100 psi MAWP.
    d) 500 psi MAWP.

109. The type of safety valve allowed on boilers is the

    a) weight and lever.
    b) thermostatic relief.
    c) spring-loaded pop.
    d) all of the above.

110. Who is authorized to repair and adjust safety valves?

    a) Chief engineer
    b) State inspector
    c) Insurance inspector
    d) Manufacturer's representative

111. Cast iron boilers can be used

    a) only for process steam.
    b) only for low-pressure plants.
    c) only for high-pressure plants.
    d) in high- or low-pressure plants.

112. Which valve allows fluid to flow in one direction only?

    a) Check
    b) Plug
    c) Globe
    d) Stop

113. To stop a safety valve from chattering,

    a) increase tension on the spring.
    b) decrease tension on the spring.
    c) raise the blowdown ring.
    d) lower the blowdown ring.

114. The minimum permitted size of the water column pipe connection is

    a) 3/8 inch.
    b) 1/2 inch.
    c) 3/4 inch.
    d) 1 inch.

115. The number of complete turns required between opened and closed on a slow-opening valve is

    a) three.
    b) five.
    c) seven.
    d) nine.

116. What is the principle of the Bourdon tube?

    a) Thermal reaction
    b) Hydraulic impulse
    c) Electrical impulse
    d) Pressure straightens out tube

117. A valve must be installed in the feedwater line

    a) between the boiler and the check valve.
    b) between the check valve and the feedwater pump.
    c) at the discharge of the feedwater pump.
    d) A valve is not required.

118. A continuous blowdown is used to

    a) remove accumulated solids.
    b) remove scum and oil.
    c) remove water when the level in the gauge glass is too high.
    d) verify drum water level.

119. What should come out when the bottom try cock is opened?

    a) Superheated steam
    b) Dry steam
    c) Water
    d) Steam and water

120. What is the minimum diameter for the water column blowdown line?

    a) 1/2 inch
    b) 3/4 inch
    c) 1 inch
    d) 1-1/2 inches

121. What is the minimum size piping for bottom blowdown lines?

    a) 1/2 inch
    b) 3/4 inch
    c) 1 inch
    d) 2-1/2 inches

122. How often should safety valves be tested?

    a) Each shift
    b) Monthly
    c) Yearly
    d) When boiler is down for inspection

123. A two-element feedwater regulator is controlled by

    a) water level and water flow rate.
    b) steam flow rate and water flow rate.
    c) water level and steam flow rate.
    d) water level alone.

124. Which of the following is not performed by a nonreturn valve?

    a) Cuts a boiler into the header automatically.
    b) Opens automatically when boiler pressure falls below the header pressure.
    c) Functions as a stop valve.
    d) Opens automatically if the boiler pressure exceeds the header pressure.

125. You would not run an accumulation test on a boiler equipped with

    a) over 499 ft^2 of heating surface.
    b) a superheater.
    c) a deaerator.
    d) only one safety valve.

126. What is the minimum setting on the blowback of a safety valve?

    a) 4% of set pressure
    b) 2 lb
    c) 2% of set pressure
    d) 4 lb

127. Good combustion results in

    a) high oxygen and carbon dioxide, low carbon monoxide.
    b) high oxygen and carbon monoxide, low carbon dioxide.
    c) low oxygen and carbon dioxide, high carbon monoxide.
    d) low oxygen and carbon monoxide, high carbon dioxide.

128. Excess air for gas should not exceed

   a) 10%.
   b) 15%.
   c) 20%.
   d) 25%.

129. The Ringelmann Scale is associated with

   a) viscosity.
   b) excess air.
   c) smoke density.
   d) pour point.

130. What do mechanical burners use for atomization?

   a) High-pressure air
   b) High-pressure fuel
   c) Low-pressure air
   d) A rotating cup

131. Viscosity is

   a) the point at which oil starts to flow.
   b) the point at which oil starts to support combustion.
   c) the specific gravity of oil.
   d) none of the above.

132. A fire eye is used to

   a) notify the programmer about the presence of a flame.
   b) detect hot spots in the furnace.
   c) turn off the pilot gas.
   d) turn on the main fuel.

133. The purge cycle

   a) closes the fuel valve.
   b) places the boiler on standby.
   c) removes unburned fuel.
   d) opens the pilot valve.

134. The flame detector is directed

   a) at the hot refractory.
   b) into the boiler room.
   c) at the convection section of the boiler.
   d) into the furnace area.

135. The programmer

    a) opens the pilot valve.
    b) opens the main fuel valve.
    c) receives the flame detector signal.
    d) does all of the above.

136. Which of the following statements is true?

    a) Flue gas oxygen percentage stays constant from fuel to fuel.
    b) Flue gas carbon dioxide percentage stays constant from fuel to fuel.
    c) Excess air is not required for well-designed burners.
    d) Smoke and carbon monoxide cannot exist together.

137. Which of the following statements is true?

    a) The flash point of an oil is the temperature at which it flows.
    b) The viscosity of an oil is its relation to the weight of an equal amount of water.
    c) The specific gravity of fuel oil is its resistance to flow.
    d) None of the above.

138. An Orsat analyzer measures

    a) smoke density.
    b) stack temperature.
    c) the air-fuel ratio.
    d) O_2, CO_2, and CO.

139. The flame failure control is tested by

    a) blowing down the low-water cutoff.
    b) adjusting the pressuretrol.
    c) calling the inspector.
    d) shutting off the fuel supply to the burner.

140. What is the efficiency of a steam plant with a heat rate of 12,000 Btu/kW?

    a) 25.8%
    b) 27.2%
    c) 28.4%
    d) 31.5%

141. What type of fuel must be preheated to burn properly?

    a) Coke gas
    b) Biomass
    c) Light oil
    d) Heavy oil

142. No. 6 fuel oil for mechanical atomizing burners is usually heated to

    a)   100°F.
    b)   150°F.
    c)   300°F.
    d)   500°F.

143. Fuel suppliers specify a minimum flash point for their oil, because a low flash point oil is

    a)   low in viscosity.
    b)   high in viscosity.
    c)   hard to ignite.
    d)   dangerous.

144. How many cubic feet of furnace volume are needed to burn one gallon of No. 6 oil?

    a)   5
    b)   10
    c)   20
    d)   30

145. As fuel oil is heated, its viscosity

    a)   increases.
    b)   decreases.
    c)   stays the same.
    d)   can either increase or decrease.

146. Carbon monoxide in the flue gas is a sign of

    a)   perfect combustion.
    b)   complete combustion.
    c)   incomplete combustion.
    d)   stoichiometric combustion.

147. Flash point means

    a)   ignition temperature.
    b)   burning temperature.
    c)   atomization viscosity.
    d)   atomization specific gravity.

148. Fire point means

    a)   oil temperature.
    b)   oil thickness.
    c)   heat content.
    d)   none of the above.

149. Preheated No. 6 oil should be

    a)   100° to 115°F.
    b)   120° to 190°F.
    c)   200° to 320°F.
    d)   none of the above.

150. What is the proper oil storage temperature for No. 6 oil?

    a)   At pour point
    b)   20°F above pour point
    c)   212°F
    d)   20°F below pour point

151. Pour point is the

    a)   U.S. weight of a liquid.
    b)   flow rate.
    c)   combustion rate.
    d)   lowest temperature at which a liquid flows.

152. To prime a pump

    a)   increase its speed.
    b)   back off on its packing.
    c)   close its discharge valve.
    d)   fill casing with water.

153. Balanced draft means

    a)   negative draft in the breaching.
    b)   almost atmospheric pressure in the furnace.
    c)   almost atmospheric pressure in the wind box.
    d)   positive draft in the wind box.

154. A gallon of No. 6 oil contains

    a)   146,000 Btu.
    b)   152,000 Btu.
    c)   19,000 Btu.
    d)   23,000 Btu.

155. During the purge cycle, what is the minimum amount of air volume changes required?

    a)   Two
    b)   Four
    c)   Six
    d)   Eight

156. The purge cycle time for gas compared to oil is

    a) less.
    b) the same.
    c) longer.
    d) not relevant.

157. A high carbon monoxide reading indicates

    a) impurities in the fuel.
    b) low furnace temperature.
    c) high excess air.
    d) none of the above.

158. The induced draft fan is located

    a) in the breaching.
    b) between the forced draft fan and wind box.
    c) between the furnace and convection sections.
    d) none of the above.

159. A balanced draft boiler requires

    a) a forced draft fan only.
    b) a natural chimney draft only.
    c) both induced and natural draft.
    d) both forced and induced draft.

160. What percent CO_2 would you expect in a well-maintained boiler burning No. 6 oil?

    a) 10
    b) 15
    c) 20
    d) 25

161. Which of the following yields a high CO_2 reading?

    a) Broken boiler baffle
    b) Leaks in air heater baffles
    c) Improper air-fuel mixture
    d) Proper air-fuel mixture

162. Low excess air causes

    a) flashback.
    b) refractory spalling.
    c) high CO_2.
    d) smoke.

163. The proximate analysis of coal

    a)   determines the percentage by weight of moisture content, volatile matter, fixed carbon, ash, and sometimes sulfur.

    b)   is sometimes useful in controlling coal quality.

    c)   determines the percentage by weight of the various elements contained in coal, which include carbon, oxygen, nitrogen, hydrogen, sulfur, and ash.

    d)   is not used anymore.

164. The ultimate analysis of coal

    a)   is the best grade of coal available.

    b)   determines the percentage by weight of moisture content, volatile matter, fixed carbon, ash, and sometimes sulfur.

    c)   determines the percentage by weight of the various elements contained in coal, which include carbon, oxygen, nitrogen, hydrogen, sulfur, and ash.

    d)   is a sample of coal taken as received.

165. What does volatile mean?

    a)   Btu per pound

    b)   Percentage of fixed carbon in coal

    c)   Inert material in coal

    d)   Flammable gas in coal that is distilled off and burned above the grates in suspension

166. The amount of excess air required for coal is

    a)   20%.

    b)   25%.

    c)   30%.

    d)   35%.

167. Which formula approximates the heat content in Btu per pound of fuel oil?

    a)   17,687 x 57.7 x API gravity at 60°F

    b)   17,687 + 57.7 + API gravity at 60°F

    c)   17,687 + 57.7 x API gravity at 60°F

    d)   17,687 x 57.7 + API gravity at 60°F

168. The two elements in fuel oil that produce most of its heat are

    a)   iron and oxygen.

    b)   sulfur and oxygen.

    c)   nitrogen and CO_2.

    d)   carbon and hydrogen.

169. What fan(s) supply the boiler with a positive pressure?

    a) Forced draft
    b) Induced draft
    c) Forced and induced draft
    d) Gas recirculating

170. The doubling plate is placed under the

    a) fill line.
    b) vent line.
    c) manhole.
    d) gauge well.

171. The induced draft fan is located between the

    a) air inlet and air heater.
    b) air heater and burner.
    c) boiler and stack.
    d) breaching and burner.

172. A fan's output can be controlled by

    a) changing its speed.
    b) changing blade pitch.
    c) inlet dampers.
    d) all of the above.

173. When would the overspeed device trip on a turbine operating at 7,000 rpm?

    a) 7,500
    b) 7,700
    c) 7,900
    d) 8,100

174. Economizers preheat

    a) combustion air.
    b) feedwater.
    c) condensate.
    d) fuel oil.

175. In a fire-tube boiler, soot forms on the

    a) outside tube surface.
    b) inside tube surface.
    c) waterside surface.
    d) water wall surface.

176. A sudden increase in stack temperature indicates

   a) a broken baffle.
   b) heating surface fouling.
   c) improper air-fuel ratio.
   d) perfect combustion.

177. A gradual increase in stack temperature indicates

   a) a broken baffle.
   b) heating surface fouling.
   c) improper air-fuel ratio.
   d) perfect combustion.

178. To ensure good heat transfer in water-tube boilers, the outside of the tubes are kept clean by

   a) soot blowing.
   b) water washing.
   c) steam washing.
   d) abrasive blasting.

179. A cause of high stack temperature is

   a) dirty boiler tubes.
   b) broken baffles.
   c) forced draft fan failure.
   d) both a and b.

180. An economizer is a(n)

   a) hot water storage tank.
   b) inexpensive way to fire a boiler.
   c) feedwater heater.
   d) air heater.

181. What type of boiler is most likely to use soot blowers?

   a) Water tube
   b) Fire tube
   c) HRT
   d) Scotch Marine

182. What is an economizer used for?

   a) To save water for the boiler
   b) To decrease CO_2 in the flue gas
   c) To increase boiler efficiency by heating feedwater from stack gas
   d) To increase the temperature of the secondary air

183. Dew point is a serious concern in a plant with

   a) tubular air heaters.
   b) forced draft fans.
   c) air compressors.
   d) turbine condensers.

184. An economizer is located in the

   a) wind box.
   b) furnace.
   c) stack.
   d) convection section.

185. If a steam soot blower were to operate improperly for a long period of time, what would be the most likely result?

   a) Damaged soot blower
   b) High steam temperature
   c) High steam consumption
   d) Damaged boiler tubes

186. Which of the following would most improve boiler efficiency?

   a) Deaerator
   b) Open feedwater heater
   c) Economizer
   d) Closed feedwater heater

187. The purpose of a steam trap is to

   a) add condensate and air.
   b) drain condensate and air.
   c) catch dirt and rust.
   d) do none of the above.

188. On an inverted bucket trap, what happens to the bucket as the trap fills with water?

   a) It rises.
   b) It inclines.
   c) It sinks.
   d) It remains stationary.

189. Which steam trap has no moving parts?

   a) Inverted bucket
   b) Impulse
   c) Continuous flow
   d) Float

190. What results when steam blows through a trap?

    a) Lower plant efficiency
    b) Noise
    c) Loss of steam
    d) All of the above

191. What happens when the float in a float & thermostatic develops a leak?

    a) Steam bound
    b) Air bound
    c) Fails open
    d) Fails closed

192. The chemical used to prevent oxidation is

    a) sodium sulfite.
    b) sodium phosphate.
    c) calcium phosphate.
    d) magnesium phosphate.

193. Feedwater is treated to

    a) decrease need for make up.
    b) increase circulation.
    c) decrease dissolved solids.
    d) prevent formation of scale.

194. Deaerating is

    a) adding oxygen to the boiler feedwater.
    b) extracting steam from a turbine.
    c) removing noncondensable gases from the feedwater.
    d) adding heat to feedwater.

195. Priming means

    a) water leaving the boiler through the steam line.
    b) filling the pump with fluid to establish suction.
    c) purging air from the gas line before use.
    d) both a and b.

196. What chemical is fed to the deaerator?

    a) Sodium sulfite
    b) Sodium sulfate
    c) Phosphate
    d) Caustic soda

197. What chemical is used to control scale?

    a) Sodium sulfite
    b) Sodium sulfate
    c) Phosphate
    d) Caustic soda

198. What chemical is used to control alkalinity?

    a) Sodium sulfite
    b) Sodium sulfate
    c) Phosphate
    d) Caustic soda

199. Corrosion in a boiler is caused by

    a) low alkalinity (pH).
    b) free oxygen.
    c) both a and b.
    d) none of the above.

200. Scale in a boiler is caused by

    a) calcium and magnesium.
    b) oxygen.
    c) sodium sulfite.
    d) Zeolite.

201. Cuprous chloride is found in

    a) feedwater.
    b) a steam drum.
    c) a feedwater heater.
    d) a flue gas analyzer.

202. Zeolite is found in a

    a) softener.
    b) steam drum.
    c) feedwater heater.
    d) flue gas analyzer.

203. The device on a deaerator that keeps the water from backing up into the steam line is the

    a) check valve.
    b) overflow device.
    c) float valve.
    d) relief valve.

204. Which pH value indicates that a solution is neither acid nor alkaline?

    a) Three
    b) Five
    c) Seven
    d) Nine

205. Hardness of water is expressed in

    a) grains.
    b) ppm.
    c) both a and b.
    d) none of the above.

206. Where is caustic embrittlement most likely to be found?

    a) Riveted joint above the water line
    b) Riveted joint below the water line
    c) Welded joint above the water line
    d) Welded joint below the water line

207. A closed feedwater heater is located

    a) before the condensate receiver.
    b) above the feedwater pump for gravity feed.
    c) between the feedwater pump and boiler.
    d) before the condenser.

208. Where would scale most likely occur?

    a) Above the water line
    b) Below the water line
    c) On a riveted joint
    d) On a weld below the water line

209. Zeolite is

    a) infrared light.
    b) an oil additive.
    c) a form of chloride.
    d) an ion exchange resin.

210. On the pH scale, 8 is

    a) acid.
    b) alkaline.
    c) meniscus.
    d) neutral.

211. Which of the following is least likely to be used as feedwater treatment?

    a) Caustic soda
    b) Phosphate
    c) Sodium sulfate
    d) Sodium sulfite

212. The result of high dissolved O_2 is

    a) corrosion.
    b) erosion.
    c) hydrogen embrittlement.
    d) low pH.

213. What percent of the heat supplied to the turbine is lost to the condenser cooling water?

    a) 10%
    b) 40%
    c) 70%
    d) 90%

214. An atmospheric relief valve is used on a

    a) fuel oil system.
    b) deaerator.
    c) turbine condenser.
    d) back-pressure turbine.

215. Which of the following is located on the inlet line of a turbine condenser?

    a) Atmospheric relief valve
    b) Vacuum breaker
    c) Air vent
    d) Check valve

216. What does an atmospheric relief valve do?

    a) Opens when exhaust pressure is 3 psi
    b) Closes when back pressure is 5 psi
    c) Protects the condenser and low-pressure turbine casing from overpressure
    d) Opens when exhaust vacuum goes over 30.00 in. Hg

217. Critical speed refers to

    a) synchronous speed.

b) speed at which the most vibration occurs.
c) operating speed.
d) overspeed set point.

218. One kW equals

    a) 1.34 hp.
    b) 0.746 hp.
    c) 0.746 watts.
    d) 970.3 Btu.

219. Which formula can be used to figure out how many poles there are in a three-phase, 60 Hz, 3,600 rpm generator?

    a) 120 x Hz / rpm
    b) 120 x rpm / Hz
    c) 120 / rpm
    d) rpm / Hz

220. One mechanical horsepower is not equivalent to

    a) 0.746 kW.
    b) 746 watts.
    c) 2,545 Btu.
    d) 3,960 Btu.

221. How are large utility generators cooled?

    a) Air
    b) H_2
    c) H_2O
    d) O_2

222. A 45-horsepower motor runs at 74% efficiency. What is its watts-per-horsepower rate?

    a) 552
    b) 746
    c) 939.96
    d) 1,008.10

223. What is a sentinel valve?

    a) Safety valve
    b) OS&Y valve
    c) Relief valve
    d) Device to warn of overpressure in a turbine condenser

224. A synchroscope is associated with

    a) induction motors.
    b) synchronous motors.
    c) induction generators.
    d) synchronous generators.

225. How often is the turbine overspeed trip tested?

    a) Weekly
    b) Monthly
    c) Yearly
    d) Every time its put on line

226. Oil-contaminated condensate return is of the most serious concern in the

    a) open feedwater heater.
    b) closed feedwater heater.
    c) deaerator.
    d) fuel oil heater.

227. An intercooler is used

    a) between the stages of a reaction turbine.
    b) on a two-stage air compressor.
    c) on a cooling tower.
    d) to cool condensate.

228. On a surface condenser,

    a) cooling water flows through tubes surrounded by exhaust steam.
    b) exhaust steam flows through tubes surrounded by cooling water.
    c) steam and water meet.
    d) none of the above.

229. One horsepower is electrically equivalent to 746

    a) watts.
    b) calories.
    c) joules.
    d) kilowatts.

230. A receiver in a compressed air system

    a) stores lubricating oil.
    b) stores compressed air.
    c) furnishes air to the compressor.
    d) acts as a pressure relief.

231. A synchroscope

    a) measures voltage.
    b) measures amperes.
    c) matches generator to system phasing.
    d) measures turbine speed.

232. A row of impulse blades has ____ across the rotating blades.

    a) a pressure drop
    b) a pressure increase
    c) no change in pressure
    d) none of the above

233. If two similar centrifugal pumps are connected in a series, the resulting output is

    a) flow increase, constant pressure.
    b) flow decrease, constant pressure.
    c) constant flow, pressure increase.
    d) constant flow, pressure decrease.

234. If the field of a synchronous electric motor is overexcited, the result is

    a) leading power factor.
    b) lagging power factor.
    c) higher efficiency.
    d) lower efficiency.

235. How much torque does a 3,600-rpm, 200-hp electric motor develop?

    a) 292
    b) 584
    c) 94,536
    d) 47,268

236. What is the most likely cause when a feedwater pump becomes vapor bound?

    a) Leaking packing
    b) Vent left open
    c) Leaking suction line
    d) Water too hot

237. How is the rotation reversed on a three-phase motor?

    a) Reverse the field winding.
    b) Reverse the armature winding.
    c) Reverse both the field and armature windings.
    d) Reverse any two power leads.

238. How many watts does a 45-hp motor consume operating at 75% efficiency?

    a) 25.2
    b) 44.8
    c) 25,200
    d) 44,800

239. A circuit breaker serves the same function as a

    a) meter.
    b) resistor.
    c) fuse.
    d) solenoid.

240. Electric current is measured in

    a) ohms.
    b) volts.
    c) farads.
    d) amperes.

241. What is the formula for finding the frequency of an alternator?

    a) (p x rpm) / 120
    b) 120 / (p x rpm)
    c) (120 x p) / rpm
    d) 120 x p x rpm

242. What is the formula for calculating the speed of a synchronous motor?

    a) (p x Hz) / 120
    b) (120 x f) / p
    c) (120 x p) / Hz
    d) 120 x p x Hz

243. Which feedwater system is least efficient?

    a) Electric centrifugal pump
    b) Steam-driven centrifugal pump
    c) Electric reciprocating pump
    d) Injector

244. If two similar centrifugal pumps where connected in parallel, there would be

    a) flow increase, constant pressure.
    b) flow decrease, constant pressure.
    c) constant flow, pressure increase.
    d) constant flow, pressure decrease.

245. When starting, the correct way to put the least amount of strain on the pump and the minimum load on an electric motor is to have

    a)   both suction and discharge valves closed.
    b)   both suction and discharge valves open.
    c)   the suction valve open and the discharge valve closed.
    d)   the suction valve closed and the discharge valve open.

246. On a steam-driven boiler feed pump, which piping is larger?

    a)   Turbine steam inlet and pump discharge
    b)   Turbine steam inlet and pump suction
    c)   Turbine steam exhaust and pump discharge
    d)   Turbine steam exhaust and pump suction

Essay Questions

These essay questions are representative of those that will be found on different boiler operator licensing examinations. Not all of the answers to the questions will be found in the text, as it is impossible to cover all the material found on the examinations.

1. A steam boiler without an economizer or air preheater is operating at 100 psi and a stack temperature of 300°F. Is this normal and why?

2. Name and explain the three forms of heat transfer.

3. A boiler comes with a guarantee of 80% efficiency. What does this mean?

4. Define specific heat, specific volume, and specific gravity.

5. How many Btu are required to raise 25 lb of water 15°F?

6. A boiler with an efficiency rating of 70% delivers 50,000 lb/hr of saturated steam at 1,190 Btu/lb. The feedwater temperature is 172°F. How many pounds of 12,000 Btu/lb coal must be burned per hour?

7. A pipeline needs to supply steam at 10,000 lb/hr with a velocity of 5,000 ft/min. The specific volume of the steam is 2.5 ft^3/lb. What is the required minimum inside diameter (id) pipe size?

8. A steam boiler with an efficiency of 80% weighs 1,000 lb and contains 500 lb of water. How many cubic feet of natural gas (1,000 Btu/ft^3) are needed to raise the temperature of the boiler and water from 50° to 210°F? (The specific heat of steel is 0.11 Btu/lb-°F.)

9. Convert 75°F to Celsius. Show your work.

10. Define absolute temperature and absolute pressure. Give numerical values.

11. What is the factor of evaporation? Show the formula.

12. Define wet steam, dry steam, and superheated steam.

13. Tell what you know about the saturated steam table, and define enthalpy.

14. Explain why the steam pressure would fluctuate less in a fire-tube boiler than in a water-tube boiler.

15. What is a British thermal unit?

16. A boiler produces steam containing 1,198.7 Btu/lb. The feedwater temperature is 190°F. What is the factor of evaporation?

17. Convert 68°C to Fahrenheit. Show your work.

18. What is the equation that converts Celsius to Fahrenheit?

19. A boiler produces 45,000 lb of 85 psi steam per hour with 210°F feedwater. What is the horsepower of this unit?

20. Define the following in British thermal units:

    1 engine horsepower
    1 boiler horsepower
    1 kWh
    778 ft-lb
    1 lb of water evaporated from and at 212°F

21. Define natural draft, forced draft, induced draft, and balanced draft.

22. What is the difference between tubes and flues?

23. What is the difference between ducts and breachings?

24. Define boiler horsepower, engine horsepower, and kilowatt-hour.

25. Discuss some dangerous boiler conditions.

26. Discuss factors that reduce boiler efficiency.

27. Why do boiler tubes fail?

28. What is a dry pipe and where is it located?

29. Where are diagonal stays located? What is their purpose? In what type of boiler are they found?

30. What is the minimum range of a boiler pressure gauge? What range is considered good operating practice and why? Under what circumstances is a pigtail siphon used? Explain how a siphon works.

31. Is the steam in a superheater tube on the inside or outside of the tube?

32. What is required to get a boiler ready for inspection?

33. What is the procedure for increasing the operating pressure of a boiler?

34. Explain the consequences of starting boilers on a daily basis.

35. A boiler delivers 80,000 lb/hr of 150 psi saturated steam at 1,195 Btu/lb. The feedwater temperature is 210°F. How many therms of natural gas must be burned in a boiler with 75% efficiency?

36. What is the heating surface of a sectional header boiler with 204 four-inch tubes (0.135-inch wall) that are 20 ft long?

37. What is the factor of evaporation of a boiler producing 100 psi steam and supplied with 200°F feedwater?

38. What is the difference between the way tubes are installed in the drum of a water-tube boiler and the way they are installed in a tube sheet of a fire-tube boiler? Explain why difference is necessary.

39. Why can boiler tubes be thinner than the drum or shell?

40. List the three types of furnaces that can be installed in a Scotch Marine boiler. Describe two in detail.

41. Name two types of Scotch Marine boilers. How are they different and how are they similar? List the advantages and disadvantages of each.

42. A boiler generates 100,000 lb of steam per hour with a factor of evaporation of 1.04. What is its horsepower?

43. How far above the lowest permissible water level is the bottom of the gauge glass in a fire-tube boiler and a water-tube boiler?

44. What are cyclone separators and where are they found?

45. A 72-inch diameter boiler drum is 1/2-inch thick and has an ultimate tensile strength of 65,000 psi. With a factor of safety of 6, and a joint efficiency of 85%, what is the MAWP?

46. A 4-ft boiler drum is made of 1-1/4-inch plate with a tensile strength of 45,000 psi. With a joint efficiency of 80%, what is the bursting pressure?

47. Find the required thickness of a drum operating at 300 psi that is 40 inches in diameter with metal tensile strength of 55,000 psi, a factor of safety of 5, and a joint efficiency of 80%.

48. Name the three types of superheaters — the one in which the steam temperature increases with the firing rate, the one in which steam temperature decreases with the firing rate, and the one in which the steam temperature remains constant.

49. A 50-ft long steel pipe at 30°F is going to transport steam at 370°F. How much will this pipe expand?

50. What is a hydrostatic test? When is it usually done? How is it done?

51. Which system would you expect to last longer — one that is on line for long periods or one that is started daily? Why?

52. Give the advantages and disadvantages of water wall furnaces.

53. Why are water-tube boilers preferred to fire-tube boilers in power plants?

54. What is the longitudinal and circumferential stress on a pressure vessel 1/2 inch thick, 40 inches in diameter, 15 ft long, at 200 psi?

55. If a boiler equipped with a superheater were operating at a steady firing rate and the temperature of the feedwater went up 10°F, what would happen to the superheat temperature?

56. Name at least two places where steam separators are found.

57. What valves would you close and what safety procedures would you follow before entering a boiler's water side for inspection or repair?

58. Why are the tubes in a fire-tube boiler beaded over?

59. A boiler produces 835,052 lb of steam per hour and consumes 56,505 lb of 19,000 Btu/lb fuel. The operating pressure is 135 psi, and the feedwater temperature is 225°F. What is the efficiency of this boiler?

60. Find the heating surface of an HRT boiler that is 78 inches in diameter and 20 ft long. The boiler contains 80 four-inch tubes bricked 1/3 of the distance from the top.

61. Name three ways to determine if the safety valve capacity on a boiler is adequate. Describe one in detail.

62. When is the best time to bottom blow a boiler during a 24-hour period?

63. Describe how low-water cutoffs are tested on a steam boiler.

64. On a boiler equipped with a superheater, which safety valve should lift first, and why?

65. Explain the differences among single-element, two-element, and three-element feedwater regulators.

66. Name three types of single-element feedwater regulators and describe one in detail.

67. Describe in detail the proper way to replace a broken gauge glass on an operating boiler.

68. What is a nonreturn valve? Where and why is it used?

69. A steam pressure gauge is located 25 ft below the steam line. What would be the error? How would you correct the error?

70. What does the term blowback mean when referring to safety valves? What are the blowback requirements?

71. Name two types of low-water cutoffs.

72. Name an advantage and a disadvantage of nonrising stem valves.

73. Why are most handholes and manholes oval instead of round?

74. How are steam pressure gauges protected against excessive temperatures?

75. One 100-bhp boiler operates at 50 psi, while another 100-bhp boiler operates at 100 psi. Would there be any size differences between the safety valves on the two boilers? Why?

76. How would you find out whether the top or bottom of a gauge glass is plugged?

77. Explain what actions you would take if the water in the gauge glass suddenly disappears.

78. Why are cross tees used to pipe the water column instead of regular 90 degree elbows?

79. What equipment is permitted to be attached to the water column?

80. What is a deadweight tester?

81. What is the purpose of a blowback ring on a safety valve?

82. Would it be good practice to put a stop valve between a safety valve and the boiler to allow repair of the safety valve without shutting down the boiler? Explain.

83. What is the purpose of a water column and how should it be connected to the boiler?

84. How are steam pressure gauges connected to a boiler?

85. Explain the operation of a Bourdon tube.

86. Describe the Ringelmann chart and how it is used.

87. How would you check for CO and CO_2 in flue gas?

88. What is the value in Btu of oil at API 25.0 specific gravity?

89. What do the proximate and ultimate coal analyses determine? How can the tests be used? What equipment is required to run these tests? Where are these tests likely to be run?

90. Define the following: perfect combustion, complete combustion, stoichiometric combustion.

91. What is the procedure used when blowing tubes on a forced-draft packaged boiler with steam-operated manual soot blowers?

92. Name three ways to reduce NO_x emissions.

93. What is an Orsat analyzer, why is it used, and what chemicals are used for its reagents?

94. If one pound of coal requires 15 lb of air for combustion, and one pound of air occupies 13.5 ft^3, how many cubic feet of air would one ton of coal require?

95. What is an economizer? Where is it located?

96. What is balanced draft? How is it produced?

97. Name two classes of air heaters. Give an example of each.

98. A pound of coal contains 19,200 Btu. How many foot-pounds of work can it produce if all its energy is converted to work?

99. The heating value of fuel primarily depends on what two elements?

100. Will the carbon dioxide content in flue gas be higher or lower when switching from coal to oil? Will natural gas or coal produce a higher carbon dioxide reading?

101. A sample of fuel oil has a specific gravity of 0.873. What is its approximate heating value in Btu/lb?

102. If the results of a flue gas analysis showed 17.4% carbon dioxide, 1.6% carbon monoxide, 79.4% nitrogen, and 3.2% oxygen, what is the percentage of excess air?

103. Explain how the flash point of oil is related to safety.

104. Name three ways that a fan's output can be controlled. Which is the most efficient method?

105. What are the three "T's" of combustion? Explain their significance.

106. Why would a stack with a larger diameter be proposed? One with greater height?

107. In reference to ultimate analyses, what does NO CASH stand for?

108. What are the advantages of pulverized coal firing compared to stoker firing?

109. What are the advantages of stoker firing compared to hand firing?

110. Name and explain three methods of providing for expansion and contraction in steam pipes.

111. A 25-ft steam line in a building with a temperature of 70°F is going to carry steam at 450°F. How much will it expand?

112. Name three classes of steam traps. Describe one in detail.

113. How does water hammer cause damage?

114. Name two methods for testing the operation of a steam trap.

115. What velocity can be expected when 60,000 lb/hr 150 psi steam at 3.015 ft³/lb is transported in an 8-inch pipe?

116. Where is the best place to remove water from steam headers?

117. What size pipe is required to transport 20,000 lb/hr of 200 psi steam at 2.13 ft³/lb? Design for a velocity of 10,000 feet per minute. What would happen if the pipe were too small or too large?

118. What is an open feedwater heater? Where is it located?

119. What is a closed heater? Where is it located?

120. What is the difference between sodium sulfate and sodium sulfite?

121. What is caustic embrittlement? What causes it?

122. What are the three functions of a deaerator?

123. Name five tests used by stationary engineers to analyze boiler water.

124. List two causes of ruptured tubes. How are these causes prevented?

125. What causes pitting in a boiler? What is done to prevent it?

126. Define the following and explain their cause, consequence, and treatment: scale, carryover, priming, caustic embrittlement, foaming, corrosion.

127. Describe the path of air through a two-stage air compressor with an intercooler and an aftercooler. Describe the function of each component.

128. How many cubic feet of free air per hour is compressed by a double-acting cross-compound air compressor, low-pressure cylinder 14-inch diameter, high-pressure cylinder 8-inch diameter, stroke 16-inch, at 55 rpm?

129. What is the difference between single-acting and double-acting compressors? Does one have any advantage over the other?

130. What does an intercooler do, where is it located, and what is the advantage of using one?

131. What does an aftercooler do, where is it located, and what is the advantage of using one?

132. If the discharge pressure of a two-stage compressor were reduced from 100 to 90 psi, what would happen to the intercooler pressure?

133. What is the function of an atmospheric relief valve? Where is it located?

134. What are thrust bearings and where are they found?

135. What are shrouds? Where are they found?

136. What would be the deciding factor in determining whether to install a surface or jet condenser?

137. What is the purpose of a condenser?

138. There are two ways a boiler can explode. Name them, and explain how they happen and how they are prevented.

139. A turbine using 200 psia, 400°F steam at 1,210.3 Btu/lb exhausts at 2 psia saturated steam at 1,116.2 Btu/lb. What is its thermal efficiency? Water at 2 psia contains 94 Btu/lb.

140. What is the steam rate of a 5,000 kW turbo-generator that uses steam at 60,000 lb/hr?

141. Is there any advantage to bleeding steam from a turbine for feedwater heating?

142. Why do turbine plants install water-tube boilers instead of fire-tube boilers?

143. How may the vacuum be increased in a condenser?

144. Is it more economical to exhaust a turbine against a 28 in. Hg vacuum or to a process heater?

145. How much enthalpy is lost if 1,096 Btu/lb steam enters a turbine condenser at 80°F?

146. What is the voltage on a 3,600 kW dc generator that is producing 250 amps?

147. What is the capacity in kW of a dc generator putting out 50 volts at 170 amps?

148. A 250 volt, 2,500 kW generator produces how many amps at full output?

149. What is the torque on a 75,000 kW generator running at 3,600 rpm?

150. A turbine is operating at 3,600 kW with a water rate of 12 lb/kW. How much boiler horsepower is required to supply this turbine with steam if the feedwater temperature is 197°F?

151. What is the advantage of using superheated steam to operate a turbine?

152. Describe two reasons that would cause a steam turbine to overspeed.

153. What is the purpose of a thrust bearing on a turbine?

154. What is a steam separator and where is it located in steam plants?

155. What horsepower is required to drive an 8,000-kVA generator with a power factor of 0.91 and a generator efficiency of 87%?

156. Does an impulse or reaction turbine produce the most axial thrust?

157. What is meant by the water rate of a turbine? What is meant by the heat rate of a plant?

158. Why is it an advantage to build a generating plant by a large body of water?

159. Why do larger turbines run slower than small ones?

160. Why is it desirable for a steam engine or turbine to exhaust into a vacuum?

161. What is more economical, to run auxiliary machinery with electric motors or back-pressure steam turbines? Explain.

162. A plant has a heat rate of 12,000 Btu/kWh. What is its efficiency?

163. A plant has a heat rate of 10,000 Btu/hp. What is its efficiency?

164. A 600 psi, 25,000 kW turbine has a water rate of 18 lb/kW. The condenser uses 3,250 gpm of circulating water with a 10°F temperature increase. How many Btu are carried out to the river?

165. What limits the amount of superheat supplied to a turbine?

166. What does 10 x 8 x 12 mean on the nameplate of a steam pump?

167. Name the three pumps required to operate a surface condenser.

168. In a typical power plant, what percentage of heat does the generator get?

169. What is a crosshead on an air compressor?

170. What is the function of an unloader on a reciprocating air compressor? Why is it used on larger machines? How does it work?

171. What is a Kingsbury bearing and where is it used?

172. A 10,000-kW turbine operates at 250 psi with a 28-inch vacuum and a water rate of 9 lb/kW. What would be the capacity of the feedwater pump? Estimate the coal consumption rate. What is the plant's efficiency?

173. Three 1,500-kW generators are operating in parallel supplying a load of 3,500 kW. What happens if one generator trips off line? What must the engineer do?

174. Name the three factors that affect a turbine plant's efficiency. What are their limitations?

175. A 5,000-kW turbine operates with a water rate of 12 lb/kW and a condensate temperature at 100°F. How much 65°F cooling water per hour is required if the discharge cooling water temperature is 76°F? What size circulating water pump should the condenser have? What size condensate pump is required? Assume an exhaust of 1,104 Btu/lb.

176. Name at least four types of electrical meters and their function.

177. A specification sheet states that a piece of equipment is driven by an electric motor through a 3:1 speed reducer. Is the motor's torque output higher or lower than the speed reducer's output?

178. Explain why a condensate pump, a vacuum pump, and a circulating water pump are required auxiliaries for a surface condenser.

179. What is the purpose of a sentinel valve?

180. The following conditions are given for a particular steam plant:

Steam to turbine:	900 psia; 850°F; 1,424 Btu/lb
Steam exhaust:	29 in. Hg; 1,096 Btu/lb
Condensate:	80°F; 48 Btu/lb
Generator rating:	10,000 kW
Circulating water inlet:	70°F
Circulating water outlet:	85°F

Find the generator's work in Btuh, the turbine's cycle efficiency, the steam rate, and the circulating water requirement in gallons per minute.

181. A pump is discharging 50 gpm against a head of 300 ft. What horsepower is required if friction losses are neglected?

182. A centrifugal pump raises 100 gpm of water to an open tank 50 ft above the pump. Assuming an overall efficiency of 50%, what is the required motor horsepower? How many kilowatts does this represent?

183. What is the torque of a 200-hp motor turning at 1,725 rpm?

184. What is the shaft horsepower and torque of a pump at 1,735 rpm, 150 ft head, 10 psi suction pressure, 6,000 gph flow, and 80% efficiency?

185. Find the water horsepower, pump horsepower, and kilowatts required to drive a 75%-efficient pump and a 95%-efficient motor. They pump 30,000 gpm of river water against 10 ft of head with 10 ft of suction lift.

186. How many cubic inches in a gallon of water? In a gallon of No. 2 fuel oil?

187. How many gallons of water can fit in a 12-ft diameter tank 18 ft long? What is the weight of the water?

188. What is the purpose of a crosshead on a reciprocating pump?

189. Name two functions of an injector.

190. What are maintenance problems of injectors?

191. How many gallons per minute will the feed pump have to supply to a 1,000-bhp boiler with 5% blowdown?

192. How much horsepower is required on a shaft turning at 2,000 rpm with 525.2 lb-ft of torque?

193. What are the results of cavitation?

194. What causes cavitation?

195. How is it possible for a reciprocating, steam-driven feedwater pump to supply a boiler with feedwater when the pump and boiler have the same steam pressure?

196. Why are boiler feed pumps located below the deaerator?

197. How many gallons per minute does a 10 x 12 duplex pump operating at 100 rpm produce?

198. What are the revolutions per minute of a 6-pole motor at 60 Hz?

199. Explain how impellers in a multistage pump build up pressure.

200. How much horsepower is required for a 75% efficient pump to transfer 30,000 gpm of water from a sump 10 ft below the suction to a tank 10 ft above the discharge?

201. How much horsepower is required for a 78%-efficient pump to deliver 1,000 gph against a 120-ft head?

202. What purpose does packing serve in a pump?

203. Explain the formula: $W = A \times V$

204. What does 2,545 Btuh represent?

205. What is the horsepower equivalent of one kilowatt?

206. Why are the stators of a generator made of steel laminations?

207. What is the purpose of a motor starter?

208. Name two advantages of ac over dc.

209. Can two 60-Hz, three-phase generators be synchronized if they operate at different speeds?

210. Explain power factor. How is it calculated?

211. A pump located in the basement of a building shows a pressure of 100 psi. What would be the pressure at the tenth floor if it is 100 ft above the pump?

Answers to Multiple Choice Questions

1.	c	34.	a	67.	b
2.	d	35.	c	68.	c
3.	c	36.	a	69.	a
4.	b	37.	c	70.	d
5.	b	38.	b	71.	a
6.	c	39.	a	72.	b
7.	b	40.	d	73.	a
8.	c	41.	b	74.	d
9.	c	42.	c	75.	a
10.	d	43.	a	76.	d
11.	b	44.	a	77.	c
12.	b	45.	c	78.	d
13.	a	46.	b	79.	b
14.	b	47.	d	80.	c
15.	a	48.	d	81.	b
16.	c	49.	b	82.	c
17.	b	50.	b	83.	a
18.	b	51.	b	84.	d
19.	d	52.	c	85.	c
20.	a	53.	a	86.	c
21.	d	54.	b	87.	c
22.	a	55.	c	88.	a
23.	b	56.	d	89.	d
24.	b	57.	d	90.	c
25.	a	58.	a	91.	c
26.	b	59.	a	92.	b
27.	a	60.	b	93.	d
28.	c	61.	d	94.	b
29.	d	62.	c	95.	d
30.	b	63.	a	96.	d
31.	d	64.	d	97.	b
32.	d	65.	d	98.	a
33.	a	66.	a	99.	d

100.	b	147.	a	194.	c
101.	a	148.	d	195.	d
102.	c	149.	b	196.	a
103.	d	150.	b	197.	c
104.	a	151.	d	198.	d
105.	b	152.	d	199.	c
106.	d	153.	b	200.	a
107.	a	154.	b	201.	d
108.	c	155.	b	202.	a
109.	c	156.	b	203.	b
110.	d	157.	d	204.	c
111.	b	158.	a	205.	c
112.	a	159.	d	206.	b
113.	c	160.	b	207.	c
114.	d	161.	d	208.	b
115.	b	162.	d	209.	d
116.	d	163.	a	210.	b
117.	b	164.	c	211.	c
118.	a	165.	d	212.	a
119.	c	166.	d	213.	c
120.	b	167.	c	214.	c
121.	c	168.	d	215.	a
122.	b	169.	a	216.	c
123.	c	170.	d	217.	b
124.	b	171.	c	218.	a
125.	b	172.	d	219.	a
126.	b	173.	b	220.	d
127.	d	174.	b	221.	b
128.	a	175.	b	222.	d
129.	c	176.	a	223.	d
130.	b	177.	b	224.	d
131.	d	178.	a	225.	d
132.	a	179.	d	226.	b
133.	c	180.	c	227.	b
134.	d	181.	a	228.	a
135.	d	182.	c	229.	a
136.	a	183.	a	230.	b
137.	d	184.	c	231.	c
138.	d	185.	d	232.	c
139.	d	186.	c	233.	c
140.	c	187.	b	234.	a
141.	d	188.	c	235.	a
142.	b	189.	c	236.	d
143.	d	190.	d	237.	d
144.	b	191.	c	238.	d
145.	b	192.	a	239.	c
146.	c	193.	d	240.	d

241. a
242. b
243. d
244. a
245. c
246. d

CHAPTER FIFTEEN

Essay Answers

1. This is not normal, because 100 psi steam is about 337°F. Even the best boilers operating at peak efficiency have flue gas temperatures of at least 50°F above the steam temperature.

2. The three forms of heat transfer are radiation, conduction, and convection. *Radiation* does not require a transmission medium; that is, it travels through a vacuum and through a gas, like air. The most common example of radiation is the heat we feel from the sun. *Conduction* is the transfer of heat from a warm molecule to a cooler one. Some materials conduct heat better than others. For example, gases and vapors are poor conductors, liquids are better, and metals are best. *Convection* heat transfer takes place by movement of the heated material itself. In a heated room, warm air rises and the cold air circulates down to a lower level. In a boiler, the hot water rises and the cold water circulates to the bottom.

3. The 80% efficiency rate means that for every 100 Btu of fuel that go into the boiler, at least 80 of those Btu will end up in the steam instead of going up the stack.

4. *Specific heat* is the number of Btu required to raise one pound of a substance one degree Fahrenheit. The specific heat of water is one. *Specific volume* is the volume occupied by one pound of a substance under specified conditions of temperature and pressure. *Specific gravity* is the ratio of weight between any volume of a substance and the weight of an equal volume of water.

5. The amount required to raise 25 lb of water 15°F is as follows:

   25 lb x 15°F x Btu/lb-°F = 375 Btu

6. To find how many pounds of 12,000 Btu/lb coal must be burned per hour, first subtract the heat content of the feedwater from the steam:

$$1,190 \text{ Btu/lb} - (172° - 32°F) = 1,050 \text{ Btu/lb}$$

Then divide the heat output of the steam by the heat content of the fuel and by the efficiency of the boiler to find the answer:

$$\frac{50,000 \text{ lb/hr} \times 1,050 \text{ Btu/lb}}{12,000 \text{ Btu/lb}_{cool} \times 0.7} = 6,250 \text{ lb}_{cool}/\text{hr}$$

7. To find the required minimum inside diameter pipe size, first find the area of the pipe (be sure to use the proper conversions to find the area of the pipe in inches):

$$A = \frac{\text{Quality (ft}^3/\text{hr)} \times 144 \text{ in}^2/\text{ft}^2}{\text{Velocity (ft/min)} \times 60 \text{ min/hr}}$$

$$A = \frac{10,000 \text{ lb/hr} \times 2.5 \text{ ft}^3/\text{lb} \times 144 \text{ in}^2/\text{ft}^2}{5,000 \text{ ft/min} \times 60 \text{ min/hr}} = 12 \text{ in}^2$$

Then find the inside diameter ($A = 0.7854d^2$):

$$d = \left(\frac{A}{0.7854}\right)^{1/2} = (15.28 \text{ in}^2)^{1/2} = 3.9 \text{ in.}$$

Specify a 4-inch pipe.

8. The amount of heat required to raise the temperature of 1,000 lb of steel 160°F is:

$$0.11 \text{ Btu/lb-°F} \times (210° - 50°F) \times 1,000 \text{ lb} = 17,600 \text{ Btu}$$

The amount of heat required to raise 500 lb of water 160°F is:

$$1.0 \text{ Btu/lb-°F} \times (210° - 50°F) \times 500 \text{ lb} = 80,000 \text{ Btu}$$

Add these two amounts together to obtain 97,600 Btu (17,600 Btu + 80,000 Btu). Now calculate how many cubic feet of natural gas are required to supply 97,600 Btu at 80% efficiency:

$$\frac{97,000 \text{ Btu}}{1,000 \text{ Btu/ft}^3 \times 0.8} = 121.25 \text{ ft}^3$$

9. To convert 75°F to Celsius:

    $$°C = 5/9 \times (°F - 32) = 5/9 \times (75° - 32°F) = 23.89°C$$

10. Absolute temperature is measured from absolute zero and uses the Rankine scale. Absolute zero, or 0° Rankine, is equal to -460°F, which is the temperature at which all molecular motion ceases. Zero absolute pressure (0 psia) is a perfect vacuum. Absolute pressure is gauge pressure (psi or psig) plus atmospheric pressure. At sea level, where gauge pressure equals zero, absolute pressure is 14.7 psia.

11. The factor of evaporation is the heat supplied by the fuel divided by 970.3 Btu/lb. Because we need only heat supplied by the fuel, the heat from feedwater must be subtracted:

    $$\frac{Btu/lb_{steam} - Btu/lb_{feedwater}}{970.3 \ Btu/lb}$$

    The amount of heat in one pound of feedwater can be approximated by subtracting 32 from its temperature. (Remember, the definition of a Btu is the amount of heat required to raise one pound of water one degree Fahrenheit.):

    $$\frac{Btu/lb_{steam} - (°F_{feedwater} - 32)}{970.3 \ Btu/lb}$$

12. *Dry steam* is steam that contains no moisture — it is 100% vapor. Wet steam is a mixture of vapor and liquid. This mixture can vary from 99% moisture, which is considered very wet steam, to 1% moisture, which is almost dry steam. *Superheated steam* is steam elevated above its saturated temperature. It is produced by adding heat to steam after it has left the steam drum. The temperature above its saturated temperature is called *degrees superheat*. Superheated steam is always dry.

13. Steam tables tabulate properties of water, including the following:

    - Enthalpy (h) in Btu/lb
    - Entropy (s) in Btu/lb-°F
    - Specific volume (v) in ft^3/lb
    - Temperature (t) in °F
    - Pressure (p) in psia

    Saturated steam tables show both pressure and temperature. Enthalpy, entropy, and specific volume are broken down into their saturated liquid and saturated vapor states. In addition, enthalpy and entropy have a column showing the difference between the saturated liquid and vapor states.

14. For a given capacity, a fire-tube boiler contains more water than a water-tube boiler. As the steam pressure drops in either boiler, some saturated liquid is going to flash into steam, which stabilizes pressure. Since the fire-tube design contains more water, pressure is more stable.

15. A British thermal unit is the amount of heat required to raise one pound of water one degree Fahrenheit.

16. The factor of evaporation is the heat supplied by the fuel divided by 970.3 Btu/lb:

$$\frac{1{,}198.7 \text{ Btu/lb} - (190° - 32°F)}{970.3 \text{ Btu/lb}} = 1.07$$

17. To convert 68°C:

$$°F = \left(\frac{9}{5} \times 68°C\right) + 32 = 154.4°F$$

18. The equation that converts Celsius to Fahrenheit is:

$$°F = \left(\frac{9}{5} \times °C\right) + 32$$

19. To find the horsepower of this boiler, first find the factor of evaporation. Since the Btu content of 85 psi steam is not given, assume 1,180 Btu/lb. State your assumption. The factor of evaporation is the heat supplied by the fuel divided by 970.3 Btu/lb:

$$\frac{\text{Btu/lb}_{steam} - \text{Btu/lb}_{feedwater}}{970.3 \text{ Btu/lb}} = \frac{1{,}180 \text{ Btu/lb} - (210° - 32°F)}{970.3 \text{ Btu/lb}} = 1.03$$

$$hp = \frac{\text{lb/hr}}{34.5 \text{ hp/lb-hr}} \times \text{Factor of evaporation}$$

$$hp = \frac{45{,}000 \text{ lb/hr}}{34.5 \text{ hp/lb-hr}} \times 1.03 = 1{,}343.45 \text{ hp}$$

20. Define the following in Btu:

   - 1 engine horsepower = 2,545 Btu
   - 1 boiler horsepower = 33,475 Btu
   - 1 kWh = 3,413 Btu
   - 778 ft-lb = 1 Btu
   - 1 lb of water evaporated from and at 212°F = 970.3 Btu

21. *Natural draft* is caused by the difference in weight between a column of hot flue gas inside a chimney and the cold air outside the chimney. *Forced*

draft is produced by a fan forcing air into a furnace. *Induced draft* is produced by a fan drawing flue gas from the boiler and into the stack. *Balanced draft* is produced by a forced draft fan pushing air into the boiler and an induced draft fan pulling flue gas out of the boiler. The fans are regulated so the furnace pressure is approximately atmospheric pressure.

22. Tubes measure up to and include four inches and are measured by their inside diameter. Flues are over four inches in diameter and are measured by their outside diameter.

23. Ducts and breachings are both conduits that connect the various pieces of equipment in the gas loop. Ducts provides a passage for combustion air, while a breaching provides a passage for flue gas. A breaching connects a boiler to the stack.

24. A *boiler horsepower* is the amount of heat required to evaporate 34.5 lb of water from and at 212°F. A boiler horsepower equals 33,475 Btu. *Engine horsepower* or shaft horsepower equals 33,000 ft-lb/min or 550 ft-lb/sec. It is equivalent to 2,545 Btu/hp-hr. A kilowatt hour is the production or consumption of 1 kW over a period of one hour. It is equivalent to 1.34 hp or 3,413 Btu/kWh.

25. The most common dangerous boiler condition is low water. Water is needed to allow rapid heat transfer through the metal. If the heat from flame or hot flue gases is not absorbed quickly enough, the boiler metal overheats, weakens, then fails under pressure. Boiler metal exposed on one side to heat must have water on the other side to rapidly absorb the heat. Furnace explosions are another dangerous condition. They are caused by an accumulation of unburned fuel. It's important to purge the boiler of unburned fuel before initial start-up. Poor water treatment is another dangerous condition, because it causes scale, which results in ruptured tubes. Scale is an insulator that slows the heat transfer through the metal. The slower the heat transfer, the hotter and the weaker it becomes. Eventually the tube ruptures.

26. If tubes are covered with soot on their fire side, the heat transfer from the hot flue gases to the boiler metal is slowed. As a result, less heat is transferred to the water, and more heat goes up the stack. This loss of heat is reflected by a higher than normal stack temperature. If tubes are covered by scale on their water side, heat transfer is also slowed. This not only reduces efficiency, but also increases the metal temperature. If the metal temperature gets too hot, tube failure results. The air-fuel ratio is another efficiency factor that must be maintained. If there isn't enough air, unburned fuel goes up the stack. (This also results in smoke.) Too much air and the flame temperature will be lower and heat will be carried out the stack instead of being absorbed by the water.

27. Most boiler tube failure is due to overheating. A low water condition does not allow for rapid heat transfer through the tube metal. Then the metal overheats, weakens, and fails. Scale on the water side also slows heat transfer, with the same results. The flame directly striking the tubes, or flame impingement, puts a lot of heat in a very small area. This condition causes overheating even with clean tubes and good water circulation. Besides overheating, tubes fail from corrosion. Pitting is common if oxygen is not removed from the feedwater. The classic symptom of oxygen attack is deep pits. Caustic embrittlement is another form of corrosion. It is found where the tubes have been stressed by overrolling and when the water chemistry is incorrect.

28. A dry pipe is a device that separates water from steam before the steam leaves the boiler. Dry pipes are located in the steam space of a boiler.

29. Diagonal stays are located in the steam space of fire-tube boilers. They consist of steel rods, one end connected to the top of the boiler shell, the other to the tube sheet above the top row of tubes. Their purpose is to prevent the top portion of the tube sheet from bulging out from the interior pressure of the boiler.

30. The minimum range of a boiler pressure gauge is 1-1/2 times the boiler's maximum allowable working pressure (MAWP). It is considered good operating practice to have a range of twice the MAWP. Gauges are most accurate when operating in the mid-range. Also, even from a distance, just noticing that the needle is vertical means that the correct pressure is being maintained. A siphon is used to prevent steam from entering and causing damage to the pressure gauge. When a siphon is first placed in service it is filled with steam. However, the quickly condensing steam and the shape of the siphon place a protective slug of water between the gauge and the live steam.

31. Steam is on the inside of superheater tubes.

32. To get a boiler ready for inspection, the boiler must be open, cool, and clean. The water side should be free of scale with no obvious damage to tubes and furnace. The fire side should also be clean, with all refractory in good condition. If anyone must enter the boiler, make sure all steam, feedwater, blowdown, and chemical feed lines are valved off, locked out, and tagged. The same goes for all electrical switches. Inspectors are particularly interested in the low-water cutoffs. If they are the float type, the unit should be disassembled to check the condition of the float and to make sure the body is free of scale. Have a new gasket ready for assembly. If the cutoffs are the probe type, have the assembly withdrawn from the boiler so the inspector can see that it is free of scale.

33. To increase the operating pressure of a boiler, the boiler's design must be reviewed by a registered professional engineer to determine that the unit can safely operate at the higher pressure. The boiler's manufacturer will

probably also be consulted. With the proper documentation, the request is submitted to the proper state authority, usually through the insurance inspector.

34. Every time a boiler is started and stopped, it expands and contracts. Since this movement takes place at different rates within the boiler, the resulting stresses take their toll on boiler components much quicker than if the boiler were left on. Refractory is especially prone to early failure when thermally cycled. Efficiency also suffers, because the heat required to bring a boiler up to operating temperature is wasted every time the boiler is shut down.

35. Subtract the heat content of the feedwater from the steam:

1,195 Btu/lb - (210° - 32°F) = 1,017 Btu/lb

Then divide the heat output of the steam by the heat content of the fuel and efficiency to find the answer:

$$\frac{80,000 \text{ lb/hr} \times 1,017 \text{ Btu/lb}}{100,000 \text{ Btu/therm} \times 0.75} = 1,084.8$$

36. A sectional header boiler is a water-tube boiler; therefore, the heating surface is on the outside of the tube. The tube wall thickness in this question can be ignored. Heating surface is given in square feet:

4 inches = 0.3333 ft

0.3333 ft x 20 ft/tube x π x 204 tubes = 4,269.97 ft²

Note : π = 3.14

37. The factor of evaporation is the heat added to the water in the boiler divided by 970.3 Btu/lb. Since no heat value for 100 psi steam was given, assume 1,190 Btu/lb (state your assumption):

$$\text{Factor of evaporation} = \frac{1,190 \text{ Btu/lb} - (200° - 32°F)}{970.3 \text{ Btu/lb}} = 1.05$$

38. Tubes in both water-tube and fire-tube boilers are rolled or expanded. The main difference is that the ends of the tubes in fire-tube boilers are beaded back until they contact the tube sheet. This is to prevent the ends from being burned off since they are exposed to hot flue gas.

39. The MAWP equation explains why boiler tubes can be thinner than the drum or shell:

$$MAWP = \frac{TS \times t \times E}{R \times FS}$$

where:
MAWP = maximum allowable working pressure in psi inside drum or shell
 TS = tensile strength of plate, (psi - use 55,000 psi for steel)
  t   = thickness of plate (inches)
  R   = inside radius of drum or shell (inches)
 FS   = factor of safety (ultimate strength divided by allowable working stress or bursting pressure divided by safe working pressure. It can vary between four and seven depending on age, type of construction, and condition. Use five for most calculations.)
  E   = efficiency of the joint (for welded joints, use 100%)

Notice the radius is in the denominator. As radius increases, the MAWP decreases, so for the same MAWP, the smaller diameter tubes can be thinner than the larger diameter drum or shell.

40. The three types of furnaces that can be installed in a Scotch Marine boiler are the *plain circular furnace*, the *Morison-type*, and the *Adamson* or ring reinforced type. In some applications, such as low-pressure boilers, a plain circular furnace of sufficient thickness but without additional reinforcement, is all that is required. Often a manufacturer will use a commercially available pipe made of approved steel. In high-pressure boilers, additional support is required to prevent the furnace from being crushed by the pressure. Most modern boilers have Morison-type construction — the furnace is corrugated for additional support. An older design is the Adamson or ring reinforced type, in which rings are welded to the outside of the furnace to obtain extra resistance to crushing.

41. The two types of Scotch Marine boilers are dry-back and wet-back. Both consist of an outer cylindrical shell, front and rear tube sheets, a furnace, and tubes running the length of the shell. The tubes are connected to each tube sheet. The difference between the two is how the rear of the furnace is constructed. The rear of a dry-back is refractory lined; while a water-cooled jacket is installed in the rear of a wet-back. Dry-backs are easier to construct and less expensive because of their simpler design. However, the rear refractory is a maintenance item, and if the baffle fails, flue gas short circuits to the stack and boiler efficiency decreases. Wet-backs are more expensive to make, but because the water jacket is welded in place, the flue gas cannot short circuit. There is no rear furnace refractory maintenance, and the water jacket adds to the heating surface area.

42. The boiler horsepower is:

$$\text{bhp} = \frac{100{,}000 \ \text{lb}_{\text{steam}}/\text{hr} \times 1.04}{34.5 \ \text{lb}_{\text{steam}}/\text{hr} - \text{bhp}} = 3{,}014.5 \ \text{bhp}$$

43. The gauge glass is three inches above the lowest permissible water level in a fire-tube boiler and two inches for a water-tube.

44. Cyclone separators are used to separate water from steam so the steam is dry before it goes to the superheater or enters the main. They are located in the upper portion of steam drums in water-tube boilers.

45. First state the basic MAWP equation:

$$\text{MAWP} = \frac{\text{TS} \times t \times E}{R \times \text{FS}}$$

Then plug in the numbers:

$$\text{MAWP} = \frac{65{,}000 \ \text{psi} \times 0.5 \ \text{in.} \times 0.85}{36 \ \text{in.} \times 6} = 127.9 \ \text{psi}$$

(Don't forget to change diameter to radius.)

46. Use the basic MAWP equation again, but omit the factor of safety, or just set it to one:

$$\text{MAWP} = \frac{45{,}000 \ \text{psi} \times 1.25 \ \text{in.} \times 0.8}{2 \ \text{ft} \times 12 \ \text{in.}/\text{ft} \times 1} = 1{,}875 \ \text{psi}$$

47. Again, start with the basic equation for MAWP, but this time it is necessary to rearrange the equation, because the question asks for thickness:

$$t = \frac{\text{MAWP} \times R \times \text{FS}}{\text{TS} \times E} = \frac{300 \ \text{psi} \times 20 \ \text{in.} \times 5}{55{,}000 \ \text{psi} \times 0.80} = 0.68 \ \text{in.}$$

48. The superheat temperature increases with an increased firing rate in a convection type superheater and decreases with an increased firing rate in a radiant superheater. Combination superheaters maintain constant superheat.

49. The pipe will expand as follows (remember, the coefficient of expansion is ft/ft-°F):

$$50 \ \text{ft} \times 0.0000065 \ \text{ft/ft-}°\text{F} \times (370° - 30°\text{F}) = 0.11 \ \text{ft}$$

50. A hydrostatic test determines if a pressure vessel is tight and leak-free. It is done when a new boiler is placed in service or after pressure parts have been replaced or repaired. The procedure is to either gag the safety valves or remove them and close off their openings. Fill the boiler with warm water (at least 70°F), making sure all air is vented. Then apply a test pressure of 1.5 times the MAWP. If all appears well, reduce the pressure to the MAWP and make a close visual inspection.

51. The system that is on for long periods would last longer than one that is started daily. The daily thermal cycling of the various system parts greatly reduces life expectancy.

52. The great advantage of water wall furnaces is that the radiant energy of the flame can be utilized fully. This ability greatly increases a boiler's efficiency. A disadvantage is that they are damaged quickly if they lose water circulation, or if scale cuts down on their heat-transfer rate. Water chemistry has to be carefully controlled.

53. The reason why water-tube boilers are preferred to fire-tube boilers in power plants is because the practical upper limit for fire-tube boilers is 300 psi and 30,000 lb/hr (though some are built larger). Central generating plants have turbines designed for pressures exceeding 2,700 psi, with superheats up to 1,100°F, and capacities up to 7 million lb/hr. Only water-tube boilers can meet these requirements.

54. First, calculate the longitudinal, or girth, stress:

$$\frac{pd}{4t} = \frac{200 \text{ lb/in}^2 \times 40 \text{ in.}}{4 \times 0.5} = 4{,}000 \text{ lb/in}^2$$

Then calculate the circumferential, or hoop, stress:

$$\frac{pd}{2t} = \frac{200 \text{ lb/in}^2 \times 40 \text{ in.}}{2 \times 0.5} = 8{,}000 \text{ lb/in}^2$$

Circumferential stress is always twice the longitudinal stress. Length does not enter into this problem.

55. With additional heat added to the boiler, the steaming rate would increase. If there were additional flow through the superheater and a steady firing rate, the superheat temperature would decrease regardless if it were the radiant or convection type.

56. Steam separators are found ahead of a steam engine and steam turbine, or where moisture in steam is undesirable.

57. Before entering a boiler's water side for inspection or repair, close the following valves: main stop, nonreturn, feedwater, bottom blow, top blow and chemical feed. Open the free blowing valve between the stop and

nonreturn. Always follow OSHA regulations by chaining the valves closed and tagging them. Then follow the OSHA confined-space procedure.

58. The tubes in a fire-tube boiler are beaded over so they will not be damaged. If allowed to stick out in the path of the hot flue gas, they would be burned off.

59. The efficiency of the boiler can be found as follows:

$$\text{Efficiency} = \frac{\text{Output}}{\text{Input}}$$

Assume that 135 psi steam is 1,194 Btu/lb. State your assumption.

$$\text{Efficiency} = \frac{835{,}052 \text{ lb/hr} \times [1{,}194 \text{ Btu/lb} - (225° - 32°F)]}{56{,}505 \text{ lb/hr} \times 19{,}000 \text{ Btu/lb}} = 78\%$$

60. Before starting the calculation, check the units of measure. Since the heating surface is in square feet, change all the units to feet:

Shell = 78 inches = 6.5 ft
Tubes = 4 inches = 0.333 ft

Then find the shell and tube area:

Shell area = 3.14 x 6.5 ft x 20 ft x 2/3 = 272 ft^2

Note: The 2/3 reflects the fact that only the bottom 2/3 of the shell is a heating surface.

Tube area = 80 tubes x 3.14 x 0.333 ft x 20 ft/tube = 1,673 ft^2
Tube sheet area = 2 sheets x 0.7854 x (6.5 ft)^2 x 2/3 = 44 ft^2

Since there is a tube sheet at each end, be sure to include both.

Note: Again, 2/3 is used because only the bottom 2/3 of the tube sheet is a heating surface.

Tube holes must now be subtracted from the tube sheet:

2 sheets x 0.7854 x (0.333 ft)^2 x 80 tubes = 14 ft^2

The net tube sheet heating surface is 30 ft^2 (44 ft^2 - 14 ft^2). Now add up the areas of the shell, tube, and tube sheet:

272 ft^2 + 1,673 ft^2 + 30 ft^2 = 1,975 ft^2

61. One method is to use an accumulation test. During an accumulation test, the steam discharge line is valved off and the boiler is brought up to high fire. The safety valves should not let the pressure exceed 6% of the maximum allowable working pressure (MAWP). This method cannot be used on boilers equipped with superheaters.

A second method is to measure the maximum amount of fuel that can be burned and compute the steam output based on the heating value of the fuel. Use the following equation to find the approximate amount of steam generated:

$$W = \frac{C \times H \times 0.75}{1,100}$$

where:
W = steam generated (lb/hr)
C = fuel burned (lb/hr or ft^3/hr)
H = heating value of fuel (Btu/lb or Btu/ft^3)
0.75 = assumed boiler efficiency
1,100 = assumed heat content (Btu/lb)

A third method is to measure the amount of feedwater used by the boiler at high fire. The sum of the safety valve capacities marked on the valves shall be equal to or greater than the maximum steaming capacity of the boiler.

62. The best time to bottom blow a boiler is during low load and maximum operating pressure.

63. Most fire-tube boilers have float-type cutoffs equipped with blowdowns. To test the cutoff, blow down the float chamber. The fuel valve should trip closed. If the boiler is supplying a critical load, then the cutoff can be tested using a spring-loaded bypass switch. This switch supplies power to the fuel valve and keeps it open despite what the low-water cutoff is doing. The operator manually depresses the switch while blowing down the float chamber. There is an indicating light that tells the operator that the switch worked. When the test is over, the switch is released. A more accurate way is to secure the feedwater entering the boiler. This simulates an actual low-water condition. When the fuel valve shuts off, the water level is restored and the boiler is restarted. The operator should never leave the boiler when this test is in progress. These procedures apply to all types of boilers.

64. On a boiler equipped with a superheater, the superheater safety valves should lift before any of the drum safety valves. Superheaters depend on steam flowing through them to avoid overheating. Steam is diverted from the superheaters when a drum safety opens. This loss of steam flow through the superheaters can be damaging.

65. A *single-element* regulator controls the feedwater regulator by water level only. The feedwater regulator is controlled by the steam flow with a *two-element* regulator, with the water level input making the final adjustment. A *three-element* regulator measures the steam flow, water flow, and water level. The control system adjusts the water flow to match the steam flow. The water level input makes up for any miscalibration.

66. Three types of single-element feedwater regulators are the *float on-off type*, the *thermohydraulic* (generator-diaphragm) *regulator*, and the *thermostatic expansion* (Copes) *regulator*. The float type is very simple; as the water level drops, a float activates a switch that starts a feedwater pump. When the level is back to normal, the float opens the switch and shuts off the feedwater pump.

The thermohydraulic regulator consists of two concentric tubes. The inner tube is connected to the steam drum like a water column. The water level in the inner tube follows the drum level. The outer tube is a closed, liquid-filled container with a copper tube connecting it to the metal bellows of the regulating valve. As the water level decreases in the drum, and thus in the inner tube, the tube contains more steam. This transfers more heat to the outer tube and causes the outer tube's liquid to expand. This increased pressure is transmitted to the regulating valve's bellows via the copper tube, which in turn causes the regulating valve to open more, thus increasing water flow to the boiler. When the water level increases, the regulating valve closes, which in turn decreases water flow into the boiler.

The thermostatic expansion tube or Copes regulator's operation depends on steam releasing more heat than water does. It consists of a hollow expansion tube placed at the steam drum water level. The top of the expansion tube is connected to the steam space, and the bottom is connected to the water space. Like a gauge glass, the water level in the expansion tube is allowed to rise and fall. The bottom of the tube is anchored, and the top is allowed to move as it expands and contracts. A bell crank lever attached to the top of the expansion tube amplifies the small expansions and contractions into useful movement. This movement, through a series of levers and linkages, opens the closes the feedwater valve. As the water level in the boiler drops, the expansion tube is filled with more steam. The tube expands as it absorbs the additional available heat from the steam. The feedwater control valve then opens more through the series of levers and linkages. Conversely, as the water level drops and the tube is filled with less steam, the tube contracts and starts to close the feedwater control valve.

67. The following procedure is recommended for changing a gauge glass:

Warning: Always wear a face shield, safety goggles, and gloves when working with a pressurized gauge glass.

1) Close top and bottom valves. Open the drain.
2) Remove the gauge glass nuts.
3) Discard the old glass, but save the flat brass packing washers, if used.
4) Obtain or cut to length a new glass — there should be 1/4-inch play when the glass is replaced to allow for expansion.
5) Slide the nuts and flat brass packing washers back on the new glass.
6) Center glass between the valves and take up on the nuts until hand tight, then add a 1/4 turn with a wrench.
7) Crack open the steam (top) valve to allow the glass to warm slowly and evenly.
8) When warm, open the steam and water valves completely.
9) Close the blowdown valve.

The glass is now back in service. It is important to follow steps 7 through 9 in proper sequence to minimize the chance of breaking the glass when putting it back in service. Although the glass is strong, it can break if subjected to thermal shock. With the blowdown valve open, just cracking open the top valve allows steam through the glass to evenly warm it up. When hot boiler water hits a cold glass, it breaks.

68. A nonreturn valve functions as a check valve when the stem is in the open position and as a stop valve when the stem is in the closed position. The nonreturn steam valve is located closest to the boiler. When the stem is in the open position, it opens automatically and allows steam to flow when the boiler's pressure approaches the header pressure. It closes automatically when the boiler's pressure falls below the header pressure.

69. The error would be calculated as follows:

$$25 \text{ ft} \times 0.433 \text{ psi/ft} = 10.825 \text{ psi}$$

The error can be corrected by retarding the gauge's pointer by 10.825 psi.

70. In safety valves, blowback or blowdown is the difference between the popping and closing pressures of a safety valve. The blowback pressure should not be more than 4% of the set pressure, but not less than 2 psi in any case. For safety valves at pressures between 100 and 300 psi, the blowback should not be less than 2% of the set pressure.

71. The two types of low-water cutoffs are the float type and the probe type. The float type is a buoyant float that is connected to a switch. As the water level rises and falls, the switch is opened or closed. In the probe type, probes of varying lengths are inserted vertically into the boiler so their ends are at the normal water level. Boiler water conducts electricity. As the probes are covered and uncovered by the water level, a control circuit opens and closes a switch accordingly.

72. The main advantage of nonrising stem valves is that no provision has to be made to accommodate the space required for a rising stem. These are the valves of choice for underground valves such as fire hydrants and other valves operated by reach rods. Also, the stem threads are not exposed to dirty and corrosive atmospheres. The chief disadvantage is that it is impossible to tell if the valve is opened or closed by just looking at the stem. Another disadvantage is that the valve threads can be damaged if the liquid is corrosive.

73. Handhole and manhole covers are larger than their openings and are held against their seat by the internal pressure of the pressure vessel. If they were round, there would be no way to get them through the opening. Therefore, the covers are oval so that with maneuvering, they can be passed through the opening.

74. Steam gauges are usually protected against excessive temperatures by flooding the gauge with water, thereby preventing steam from entering. The steam siphon (or pigtail) provides such protection. When first put in service, steam enters the gauge and siphon. Since the gauge and siphon are not insulated, the steam quickly condenses into water. This column of water forms a barrier from the steam due to its configuration. Another way is to pipe the gauge connections below the steam connection. Any condensate in the piping below the steam connection will have no place to drain and will serve as a barrier against the steam.

75. The 50 psi boiler's safety valve will be larger than the 100 psi boiler's valve. The steam tables show that 50 psi steam has a larger specific volume than 100 psi steam. Because 50 psi steam takes up more volume, the safety valve requires more cross-sectional area. Therefore, it must be physically larger.

76. To find out whether the top or bottom of a gauge glass is plugged, close the top and bottom gauge glass. Open the drain and bottom valve and check for water flow. If it is fine, then open the drain and top valve and check for steam flow.

77. If the water in the gauge glass suddenly disappears, first use the try cocks to check the water level in the water column, then blowdown the glass and column. If there is still no indication of water in the gauge glass or bottom try cock, close the fuel supply valve. Do not add water to the boiler for

fear of the water flashing into steam and causing an explosion. Perform an internal inspection after the boiler cools down to determine that there is no damage to the pressure parts.

78. Cross tees are used to facilitate the inspection and cleaning of the piping. With a cross tee, the inside of each pipe is accessible by removing its corresponding pipe plug. If regular elbows were used, the entire piping system would have to be disassembled to gain access to the interior of the pipes.

79. Only equipment that does not permit the consumption of an appreciable amount of steam or water is allowed to be attached to the water column. This includes gauge glasses, try cocks, pressure gauges, and feedwater regulators.

80. A deadweight tester is a precision instrument used to calibrate pressure gauges. Deadweight instruments use incremental weights acting on a given-sized piston, which floats when the liquid and weight pressure are equal. When a gauge is attached to a deadweight tester, a small hand pump is used to apply pressure. Weights are added until the piston floats. The amount of weight is then compared against the gauge reading.

81. By raising or lowering the blowback or blowdown ring on a safety valve, the amount of blowdown is adjusted. Blowdown is the difference between popping pressure and closing pressure. Raising the ring increases blowdown, lowering it decreases blowdown. If the blowdown is too small, the valve will chatter.

82. It is not a good idea to put a stop valve between a safety valve and the boiler, and it is against the law. The ASME Code states that there shall be no valve between the boiler and safety valve. The safety valve is the last line of defense between you and catastrophic disaster. If the valve needs service, *shut the boiler down.*

83. When steaming, boiler water is in constant, turbulent, violent motion. If a glass were attached directly to the boiler, the water would oscillate up and down so rapidly that an accurate reading would be difficult at best. The construction and placement of a water column averages out these wild gyrations, providing a stable water level. This stable water level allows the gauge glass and other devices to yield accurate readings.

The method of connection is spelled out in the ASME Code. The minimum size pipe connecting the water column to the boiler must be 1 inch. All connections must be readily accessible for internal inspection, including a cross fitting at each right angle turn to permit inspection and cleaning in both directions. The water column drain must be at least 3/4-inch pipe size. If valves are used, they must be OS&Y gate or stopcocks that can be locked or sealed in the open position. With fire-tube boilers, the gauge glass connected to the column must be at least three inches above the highest point of the tubes, flues, or crown sheet. For water-tube boilers, this dimension is two inches.

84. Steam pressure gauges are connected to the steam space, the water column, or its steam connection. Steel or wrought-iron pipe used for the connection should have an inside diameter of at least 1/2 inch. For other materials such as brass, 1/4-inch standard pipe size is fine. A valve is placed on the connection line so the gauge can be replaced when the boiler is in service. A siphon or equivalent device must be used to maintain a water seal that prevents steam from entering the gauge. The dial of the pressure gauge should be graduated to approximately double the safety valve's pressure, but in no case less than 1-1/2 times.

85. The Bourdon tube is a flat, hollow tube shaped like a question mark. When pressure is applied to this tube, it tries to straighten out. A linkage is connected at the end of the tube, which in turn rotates a quadrant gear. The teeth of this gear engage a small gear on the gauge's pointer shaft. A small amount of movement of the Bourdon tube produces rotation of the pointer shaft in proportion to the amount of pressure applied to the gauge.

86. A Ringelmann chart is used to estimate smoke density. It is a series of grids with various line thicknesses numbered 1 through 4. The lines of these grids are thicker as the number increases. Viewed from a distance, they appear as different shades of gray. The chart is compared to smoke coming from a chimney, then a number is assigned to the smoke. A Ringelmann zero corresponds to white, while a Ringelmann 5 corresponds to black. The numbers between are various shades of gray.

87. The traditional method to check for CO and CO_2 in flue gas is to use an Orsat apparatus. This device uses the principle of selective absorption with various chemical solutions and can measure carbon monoxide, carbon dioxide, and oxygen. New methods can take constant readings directly from the stack. Coupled with electronics, zirconium oxide is used to measure oxygen, and lasers are used to measure carbon dioxide.

88. At API 25.0 specific gravity, the value of oil in Btu is:

$$\text{Heating value} = 17{,}687 + (57.7 \times \text{API gravity})$$
$$= 17{,}687 + (57.7 \times 25.0)$$
$$= 19{,}130 \text{ Btu/lb}$$

89. Proximate analysis measures moisture content, volatile matter, fixed carbon, ash, and sometimes sulfur. A trained technician can perform this test with an oven, scale, and calorimeters. The test is usually performed at the delivery site to make sure the coal meets contract specifications.

Ultimate analysis measures the constituent elements, such as carbon, hydrogen, nitrogen, sulfur, oxygen, and ash. Sophisticated instrumentation such as a spectrum analyzer is required, so it is done in fully equipped labs.

These tests are performed because the composition of coal varies (even samples from the same mine), so it is necessary to sample and analyze it often to make adjustments for satisfactory combustion.

90. *Perfect combustion* and *stoichiometric combustion* are one in the same. They mean that every molecule of fuel has combined with the correct number of oxygen molecules with no excess oxygen left over. However, perfect combustion does not exist. To ensure all fuel is burned, excess air is required. *Complete combustion* means that all the fuel is consumed. Excess air is required to obtain this complete combustion.

91. Make sure boiler is in operation and the flame scanner safety fuel control is in service. If possible, increase draft by 0.10 to 0.15 in. wc. This is to ensure that dislodged soot is blown clear of the boiler instead of collecting and causing a site for a potential furnace explosion. Make sure the drain line is open and slowly open the steam line. Close the drain after allowing time for the removal of all condensate. Operate each soot blower one to three rotations starting from the burner and working towards the stack. Allow 20 seconds per rotation. After operating all soot blowers, close the steam valve and open the drain. Keep the drain open until the next soot blowing cycle.

92. Three ways to reduce NO_x emissions are as follows:

- Lower the flame temperature. Low NO_x burners reduce temperature by enlarging the flame in a staged combustion process. Fuel or air is then introduced at more than one point. This reduces available oxygen and creates partial combustion in several zones over a larger area. This method can reduce NO_x emissions by 25% to 50%.

- Flue gas recirculation also reduces flame temperature by returning products of combustion back to the burner and flame. It also reduces the available oxygen sufficiently to lower NO_x levels. This method can reduce NO_x up to 75%.

- A costly but more effective way to reduce NO_x is by selective noncatalytic reduction (SNCR) and selective catalytic reduction (SCR). SNCR injects ammonia or ureas into the boiler at critical points where the temperature makes the chemical reaction most efficient. The removal rate can be as high as 90%, but 50% to 70% is more common. In SCR, ammonia is vaporized and mixed with the flue gas, then passed through a catalyst bed. The chemical reaction converts the NO_x to harmless nitrogen and oxygen. Its typical removal rate is 80% to 90%.

93. An Orsat analyzer is a device used for measuring the percentage of oxygen, carbon dioxide, and carbon monoxide in flue gas. It is used to evaluate the combustion process. The chemicals used are:

- caustic potash for carbon dioxide;
- alkaline solution of progallol for oxygen;
- acid solution of cuprous chloride for carbon monoxide.

94. To determine how many cubic feet of air one ton of coal would require:

$$15 \text{ lb}_{air}/\text{lb}_{coal} \times 13.5 \text{ ft}^3 \text{ air}/\text{lb}_{air} \times 2,000 \text{ lb}_{coal}/\text{ton}_{coal} = 405,000 \text{ ft}^3 \text{ air}/\text{ton}_{coal}$$

95. An economizer is a device that heats feedwater with hot flue gas. It is located in the stack.

96. Balanced draft is produced when the furnace is at or just below atmospheric pressure. It is produced by using both a forced draft and induced draft fan. One fan is controlled to maintain draft, the other to maintain the correct air-fuel ratio.

97. Two classifications of air heaters are recuperative and regenerative. *Recuperative* types are generally tubular, although some are plate types. In this type, heat is transferred directly from the hot gases on one side of a surface to air on the other side. Counterflow types are the most common design, in which the air and flue gas travel in opposite directions. In *regenerative* heaters, heat is transferred indirectly from the hot gases to the air through some intermediate heat storage medium. In the rotating plate type, heat storage plate elements are first heated by the hot flue gas, then rotated by mechanical means into the combustion airstream, where the stored heat is released. The cooled plates continue to rotate back to the flue gas stream where they are reheated.

98. The amount of work the coal can produce is:

$$19,200 \text{ Btu/lb} \times 778 \text{ ft-lb/Btu} = 14,937,600 \text{ ft-lb/lb}$$

99. Hydrogen and carbon are the two elements that produce the most heat from fuel when burned.

100. The CO_2 will be lower when switching from coal to oil. Coal will produce higher CO_2 readings than natural gas.

101. Two conversions are required to determine the heating value. First convert specific gravity to degrees API. Then convert the API to Btu/lb:

$$API = \frac{141.5}{Specific\ gravity} - 131.5 = \frac{141.5}{0.873} - 131.5 = 30.6$$

$$
\begin{aligned}
Heating\ value &= 17,687 + (57.7 \times API\ gravity) \\
&= 17,687 + (57.7 \times 30.6) \\
&= 19,452.62\ Btu/lb
\end{aligned}
$$

102. The percentage of air is as follows:

$$\%\ excess\ air = \frac{O_2 - (CO \times 0.5)}{0.264\ N_2 + (CO \times 0.5) - O_2}$$

$$\%\ excess\ air = \frac{3.2 - (1.6 \times 0.5)}{(0.264 \times 79.4) + (1.6 \times 0.5) - 3.2} = 13.5\%$$

103. The flash point is the temperature at which a fuel shows the first signs of ignition. The higher the flash point, the safer it is to handle.

104. The fan's output can be controlled by dampers, variable speed, and variable prerotation inlet vanes. (Variable pitch blades can be used.)

105. The three "T's" of combustion are time, temperature, and turbulence. *Time* is significant, because combustion is a chemical reaction and all chemical reactions take time to complete. *Temperature* is important, because the fuel requires a minimum temperature before the chemical reaction can occur. Finally, *turbulence*, because there must be efficient mixing of the oxygen and fuel molecules for the combustion process to be efficient.

106. A larger diameter stack is needed when it is required to handle more flue gas without an excessive drop in pressure. One with greater height is required when additional natural draft is needed or when the additional smoke dispersion is beneficial to the environment.

107. NO CASH stands for the elements that the analysis tests for:

N - nitrogen
O - oxygen
C - carbon
A - ash
S - sulfur
H - hydrogen

108. Pulverized firing is more efficient than stoker firing because it wastes less coal, creates a cleaner and more rapid fire condition by mixing the fuel more intimately with the combustion air, and permits closer automatic control over the firing rate.

109. Stoker firing permits more uniform control of coal and air rates than hand firing, it maintains a stable furnace temperature and involves less manual work for the fireperson, allowing for closer observation of boiler operations.

110. The following methods of providing for expansion and contraction in steam pipes should only be designed by a qualified engineer:

One method is to place 90 degree elbows or 90 degree "Z" bends in the pipe run and only anchor the pipe run ends. The flex in the 90 degree elbows and adjacent pipe will allow for adequate expansion and contraction.

Another method is to install a loop in the middle of the run and anchor the pipe run ends. The flex in the loop will handle the pipe movement.

A third method is to install a slip joint in the pipe run and anchor the ends. The slip joint can be either the packed type or bellows type. Naturally, the range of the slip joint must exceed the movement of the pipe.

111. 25 ft x (450° - 70°F) x 0.0000065 ft/ft-°F = 0.06 ft

112. The three classes of steam traps are thermostatic, mechanical, and thermodynamic. Thermostatic traps distinguish between steam and condensate by sensing temperature. An example is a trap whose liquid-filled bellows opens and closes the trap valve as it expands and contacts due to temperature differences. On start-up when the system is cold, the contracted bellows allows condensate and air to pass through the trap. As the condensate nears steam temperature, the bellows expands and closes off the trap's discharge valve. As the condensate cools, the bellows again contracts and opens the trap.

Mechanical traps utilize density to distinguish between steam and condensate. An example is a float and lever trap that looks similar to a toilet tank float. As condensate enters the chamber, the float rises and opens the trap's discharge valve. As the condensate level decreases, the discharge valve closes. To remove noncondensables such as air, either a small vent is opened at the top of the chamber or a thermostatic trap is used. This is called a float and thermostatic trap.

Thermodynamic traps identify steam and condensate by the difference in their kinetic energy or velocity as they flow through the trap. At start-up, air and condensate lift the disc off its seat and are discharged. As the temperature of the condensate increases, some of it flashes into steam.

Because this flash steam has a larger volume than condensate of the same weight, the flow velocity increases. This high velocity, which causes a low-pressure area to form under the disc, along with expanding flash steam exerting pressure on top of the disc, holds the disc closed. As some of the flash steam condenses, the disc is lifted off its seat and the cycle repeats itself.

113. Designers like to keep water flow in pipes under 12 ft/sec. However, gases like steam can have flows up to 12,000 ft/sec. If a slug of water is entrained in steam, it is accelerated to almost the velocity of the steam. At this speed, the slug of water acts more like a solid. Steam easily makes 90 degree turns at this velocity, but water with its high inertia cannot. Instead, it just slams against the elbow with tremendous force. This is water hammer. The resulting force is often enough to burst the pipe with tragic results.

114. Two methods for testing the operation of steam traps are to test visually and to test with sound. In all cases, one must know the type of trap and its normal operating characteristics. For example, float traps primarily have a constant modulated discharge, while thermodynamic discharge is intermittent.

One of the best methods is to test the traps visually. If the trap discharges to the atmosphere, it is easy to see if it is blowing through, plugged, or operating normally. If the trap is connected to a condensate return system, then a test valve can be used to observe the trap's output. Know the difference between live steam and flash steam. Flash steam appears to float out at each discharge, while live steam blows continuously.

With a stethoscope or similar listening device one can determine the condition of a trap by knowing the sounds of normal and abnormal operation. Inverted bucket traps gives a burst of sound when the discharge valve opens. The cycle rate of disc traps is easily heard. Listening devices do not work well on thermostatic and float types because their normal discharge is gentle modulated cycles.

115. Area = Volume/Velocity or Velocity = Volume/Area

$$\text{Velocity} = \frac{60{,}000 \text{ lb/hr} \times 3.015 \text{ ft}^3/\text{lb} \times 144 \text{ in}^2/\text{ft}^3}{82 \text{ in}^2 \times 0.7854 \times 60 \text{ min/hr}} = 6{,}742 \text{ ft/min}$$

116. It is important to remove condensate as soon as it forms from a steam line to prevent water hammer. Remove condensate at all low points at intervals of 200 to 500 ft on horizontal runs, at elevation changes such as risers, and ahead of all possible dead-end areas such as ends of mains and shutoff, temperature control, and pressure control valves. It is also important to remove condensate in front of steam-operated equipment such as heat exchangers and turbines.

117. $Area = \dfrac{Volume}{Velocity}$

$Area = \dfrac{lb/hr \times ft^3/lb}{ft/min \times 60\ min/hr} = \dfrac{20,000\ lb/hr \times 2.13\ ft^3}{10,000\ ft/min \times 60\ min/hr} = 0.07\ ft^2$

$Area\ (ft^2) \times 144\ in./ft = 0.071\ ft^2 \times 144\ in./ft = 10.224\ in^2$

$Diameter = \left(\dfrac{Area}{0.7854}\right)^{1/2} = \left(\dfrac{10.224\ in^2}{0.7854}\right)^{1/2} = 3.6\ in.$

There is no such pipe size, so round up to 4 inches.

If the pipe is too small, increased frictional losses will cause excessive pressure drop. If the pipe is too large, then money was wasted on material and installation costs.

118. The term "open" is a bit misleading, since an open feedwater heater is indeed a closed pressure vessel. Open actually refers to the fact that water is heated by direct contact with steam. Open feedwater heaters are located above the feedwater pumps so that enough pressure is present to prevent the hot water from flashing off to steam. Steam bound feedwater pumps don't pump. When vented, open feedwater heaters become deaerators. The action of steam both heating and coming into intimate contact with water frees the dissolved gases in the water. These gases are then vented to the atmosphere.

119. The term "closed" means that the water to be heated never comes in direct contact with the steam. Mainly found in steam plant electrical generating stations, a closed feedwater heater is a shell and tube heat exchanger whose source of heat is turbine bleed steam. (Steam is on the shell side and water on the tube side.) Usually there is a series of low-pressure heaters that receive water from the condenser and feed the deaerator. Then there is a series of high-pressure heaters that receive water from the feedwater pump and feed the boiler. These heaters have to withstand the full feedwater pump pressure.

120. Sodium sulfite (Na_2SO_3) is an oxygen scavenger fed into the deaerator or boiler. Sodium sulfate (Na_2SO_4) is the product of sodium sulfite that has absorbed oxygen.

121. Caustic embrittlement is a form of metal failure that occurs in boilers at tube ends and riveted joints. It is caused when metal, under stress, is exposed to very caustic water. It is controlled by maintaining the proper level of alkalinity in the boiler water.

122. A deaerator removes noncondensables from the feedwater, heats the feedwater, and provides storage for the feedwater pump.

123. The five tests used by stationary engineers to analyze boiler water are: conductivity (to determine dissolved solids); alkalinity (to determine if there is enough caustic soda); hardness (to determine if the softener is working properly); phosphate (or other hardness control chemical); and sodium sulfite (to determine if there is enough oxygen scavenger).

124. Excessive heat and corrosion cause ruptured tubes. When steel reaches 600° to 700°F, it starts to lose its strength. Anything that slows heat transfer through the tube metal will increase its temperature. Scale is an excellent insulator and slows water circulation, which leads to overheating. Good softener operation and a water treatment program prevent scale problems. Corrosion is caused mainly by free oxygen in the boiler water. Corrosion causes deep pits that eventually penetrate the boiler tube. Good deaerator operation and an oxygen scavenger, such as sodium sulfite, can eliminate this problem.

125. Pitting is caused by oxygen corroding the boiler metal. It is prevented in small boilers by adding an oxygen scavenger to the boiler to absorb any oxygen in the feedwater. In larger installations, a deaerator is used to remove all the dissolved gases in the feedwater. Sodium sulfite is also used to remove oxygen the deaerator didn't remove.

126. *Scale* is a hard coating of materials deposited on the internal surfaces of boilers and pipes. It is caused when water is heated and these materials, or hardness, come out of the solution. Scale is a good insulator and thus slows heat transfer. This results in high metal temperature, which leads to tube failure. It also decreases boiler efficiency. Scale is controlled by removing it in a softener before it enters the boiler and by adding chemicals, such as phosphate, to the boiler to prevent the scale from sticking to the tubes.

Carryover is the entrainment of chemical solids and liquids with steam leaving the boiler. It is caused when the level of dissolved solids in the boiler water is too high. These dissolved solids accumulate in strainers and control valves and eventually foul them. They also end up in the product if the steam comes in direct contact with it. Carryover is controlled by maintaining the proper level of dissolved solids in the boiler water, usually by top blowdown.

Priming is the discharge of steam containing excessive quantities of water in suspension due to violent boiling or high water in the boiler. Water entrained in steam causes water hammer, which causes damage to downstream piping and valves. It is disastrous if a turbine ingests a slug of water. With the high tip speed of turbine blades, the ingested water acts more like a rock than a liquid. The single most effective way to avoid priming is to maintain proper water level. (Don't confuse boiler priming with pump priming, which is the act of filling a pump with water to start it pumping.)

Caustic embrittlement is a form of metal failure that occurs in boilers at tube ends and riveted joints. It is caused when metal, under stress, is exposed to very caustic water. It is controlled by maintaining the proper level of alkalinity in the boiler water.

Foaming is the presence of bubbles in the steam space of a boiler and is caused when the dissolved solids level in the boiler water is too high. Its main consequence is that it gives a false water level indication, and it also leads to carryover. It is prevented by maintaining the proper level of dissolved solids by top blowdown.

Corrosion, or rust when dealing with iron and steel, is the destruction of metal by chemical or electrochemical reaction. It is caused by a chemical reaction with oxygen. Eventually it weakens metal parts so that they fail. It is prevented in boilers by deaerating the feedwater, adding an oxygen scavenger, and by keeping the boiler water above pH 7.

127. Air is drawn into the first or low-pressure stage through an air filter, which removes suspended dirt. After being compressed, the air then goes through an intercooler. The intercooler cools the air from the first stage, increasing its density. This makes the second-stage compression more efficient and improves the volumetric efficiency of the entire process. Air then enters the second or high-pressure stage and then is discharged at its final pressure. Multistage compression with intercoolers is more efficient than single-stage compression. The air then enters an aftercooler, which cools the discharged compressed air. The moisture content is reduced, since cool air holds less moisture than hot air.

128. The amount of air pulled into an air compressor from the atmosphere is determined by the volume of the low-pressure stage. Therefore, the high-pressure stage data can be ignored. You can state on the examination that this is the theoretical capacity. The actual capacity will be less due to inefficiency and back pressure from the aftercooler. Find the volume of the low-pressure cylinder:

Volume = Area x Length = $0.7854 \times (14 \text{ in.})^2 \times 16 \text{ in.} = 2{,}463 \text{ in}^3$

The answer must be in cubic feet:

$$\frac{2{,}463 \text{ in}^3}{1{,}728 \text{ in}^3/\text{ft}^3} = 1.425 \text{ ft}^3$$

Since this is a double-acting machine, 1.425 ft³ is pulled in during each stroke, and there are two strokes per revolution:

1.425 ft³/stroke x 2 strokes/revolution x 55 rpm x 60 min/hr = 9,405 ft³/hr

Every question must be read carefully. Note that the factor of 60 min/hr had to be included. The usual way to rate compressors is cubic feet per minute, but you have to give the examiner what is asked. If this were a multiple choice question, you can bet that the cubic feet per minute answer would be one of the selections.

129. Single-acting compressors have one suction stroke and one discharge for every two strokes. Double-acting compressors have two suctions and two discharges for every two strokes. If the piston is operated by a crank shaft, a single-acting compressor has one discharge for each rpm, and a double-acting compressor has two discharges for each rpm. More capacity can be obtained from a double-acting machine than from a single-acting machine of the same size. The disadvantage is that they are more expensive because of the added expense of a crosshead.

130. An intercooler is found on multistage air compressors and is located between the discharge of the low-pressure stage and the suction of the high-pressure stage. It cools the air from the first stage, increasing its density. This makes the second stage compression more efficient and improves the volumetric efficiency of the entire process. Multistage compression with intercoolers is more efficient than single-stage compression.

131. An aftercooler is located after the discharge of an air compressor and cools the discharged compressed air. Its main advantage is that it removes moisture, since cool air holds less moisture than hot air.

132. The intercooler pressure would decrease since there would be less backpressure from the high-pressure cylinder.

133. The atmospheric relief valve protects the condenser from overpressure. It is located on the condenser's shell.

134. Thrust bearings absorb the axial force of a rotating shaft to maintain its axial alignment. In power plants they are found on pumps and turbines.

135. Shrouds are found on pump impellers, fan rotors, and turbine blades. They are used to strengthen and improve the efficiency of the rotating unit. In the case of fans, the vanes are riveted to the shrouds, which are connected to the shaft hub. In the case of turbines, the shroud resembles a hoop attached to the end of the blades. If reinforces the blades and helps the steam-flow characteristics.

Some pump impellers have no shrouds. The vanes stick out from the hub with no reinforcement. Some impellers designed for pumping solids have only one shroud. Impellers designed for higher efficiency and added strength have two shrouds, one on either side of the vanes.

136. The main factor in deciding whether to install a surface or jet condenser is whether the condensate will be returned to the boiler for reuse. The condensate in a jet condenser is mixed with the cooling water, making it unusable for boiler feedwater. The condensate in a surface condenser does not come into contact with the cooling water and thus is suitable for reuse as feedwater.

137. A condenser allows the steam from a turbine or engine to be exhausted at subatmospheric pressure. More work can be obtained from each pound of steam and the efficiency of the plant is thereby increased.

138. A furnace explosion occurs when raw fuel in the boiler suddenly ignites. The result can be anything from singed eyebrows to the boiler setting being blown apart. The chief cause is lighting the pilot flame when raw fuel is in the furnace. This is preventable by purging the boiler first, that is, by forcing four air changes through the boiler before igniting the pilot flame. Another cause of a furnace explosion is accumulated soot igniting during soot blowing. This is prevented by making sure there is adequate draft to remove loosened soot from the boiler.

A pressure vessel explosion is caused by a pressure part failure. Most of the damage comes from the explosive force of boiler water flashing into steam when reduced to atmospheric pressure. The chief cause is the pressure vessel being weakened by overheating, usually due to low water conditions. Naturally, the best way to prevent this is to maintain proper water level.

139. Efficiency is output divided by input. The output of a turbine is the difference between its inlet and outlet heat. The input is the difference of heat between the turbine inlet and condenser outlet. The following enthalpies were obtained from steam tables and are in Btu/lb:

$$\frac{1,210.3 - 1,116.2}{1,210.3 - 93.99} = \frac{94.1}{1,116.31} \times 100 = 8.4\%$$

140. The steam rate is as follows:

$$\frac{60,000 \text{ lb/hr}}{5,000 \text{ kWh}} = 12 \text{ lb}_{steam}/\text{kW}$$

141. Yes, there is an advantage. When feedwater is heated before it enters, the plant's thermal efficiency increases. By extracting the steam from the turbine, there is less volume to be handled at the exhaust. This reduces the required size of the low-pressure stage. The increased thermal efficiency more than makes up for the loss of turbine output, because on a per-pound

basis, more work is extracted from the steam in the high-pressure stages compared to the low-pressure stages.

142. Turbine plants install water-tube boilers, because the practical upper limit of fire-tube boilers is 30,000 lb/hr and 300 psi. Beyond that, water-tube boilers are required. In addition, water-tube boilers can burn a wider variety of fuel and are easily equipped with superheaters.

143. Vacuum may be improved by:

- increasing the capacity of the vacuum pumps to remove more noncondensable gases;
- increasing the amount of cooling water;
- using colder cooling water;
- cleaning the cooling water side;
- reducing air leakage at the LP end seal.

144. The type of plant determines whether it is more economical to exhaust a turbine against a 28 in. Hg vacuum or to a process heater. If the turbine is just driving a generator, then it is more efficient to exhaust into the vacuum of a condenser. If there is a need for process steam, then it is beneficial to exhaust into a heater.

145. The amount of enthalpy lost is:

$$1,096 \text{ Btu/lb} - (80° - 32°F) = 1,048 \text{ Btu}$$

146. The voltage is as follows:

$$\frac{3,600 \text{ kW} \times 1,000 \text{ W/kW}}{250 \text{ amps}} = 14,400 \text{ volts}$$

147. The capacity is as follows:

$$50 \text{ volts} \times 170 \text{ amps} = 8,500 \text{ W}$$

$$\frac{8,500 \text{ W}}{1,000 \text{ W/kW}} = 8.5 \text{ kW}$$

148. The amps produced at full output are as follows:

$$\frac{2,500 \text{ kW} \times 1,000 \text{ W/kW}}{250 \text{ volts}} = 10,000 \text{ amps}$$

149. The torque is as follows:

$$75,000 \text{ kW} \times 1.34 \text{ hp/kW} = 100,500 \text{ hp}$$

$$\text{Torque} = \frac{5,252 \times 100,500 \text{ hp}}{3,600 \text{ rpm}} = 146,618 \text{ lb-ft}$$

150. Since the initial steam condition is not given, assume the enthalpy is 1,200 Btu/lb. Heat input required by the boiler is:

$$1,200 \text{ Btu/lb} - (197° - 32°F) = 1,035 \text{ Btu/lb}$$

To find out how many Btu there are in one boiler horsepower, multiply 34.5 lb_{water}/bhp by 970.3 Btu/lb_{water}, which equals 33,475 Btu/bhp:

$$\frac{3,600 \text{ kW} \times 12 \text{ lb}_{water}\text{/kW} \times 1,035 \text{ Btu/lb}}{33,475 \text{ Btu/bhp}} = 1,336 \text{ bhp}$$

151. The advantage of using superheated steam to operate a turbine is that superheated steam is dry steam. Small particles of water in steam moving at high speed past the blades of the turbine, besides having an eroding effect on the blades, reduce how efficiently energy is transformed. In addition, thermal efficiency is also improved because there is more energy in superheated steam than turbines can utilize.

152. A steam turbine might overspeed if there were a sudden loss of load, such as when a generator trips off line or when the governor malfunctions.

153. The purpose of a thrust bearing on a turbine is to keep the shaft properly positioned in the axial plane against thrust loads.

154. Separators remove some entrained moisture, oil, and other impurities in flowing steam to prevent damage to equipment or to increase the efficiency of a process. They use baffles or centrifugal force to separate the heavier water and other impurities from the steam. Separators are found in front of equipment such as turbines and engines, where moisture could do damage.

155. To find the horsepower, convert kilovolt amperes to kilowatts, convert kilowatts to horsepower, then divide by the efficiency:

$$8,000 \text{ kVA} \times 0.91 \text{ power factor} = 7,280 \text{ kW}$$

$$\frac{7,280 \text{ kW} \times 1.34 \text{ hp/kW}}{0.87} = 11,213 \text{ hp}$$

156. Impulse turbines produce the most axial thrust.

157. The water rate of a turbine is the weight of water (or steam) that passes through a turbine or engine to obtain one unit of work; for example, lb/kWh or lb/hp-hr. Low water rates are desirable. Heat rate is a measurement or indication of a plant's efficiency; for example, Btu/kWh or Btu/hp-hr. Low heat rates are desirable.

158. It is desirable to build a generating plant by a large body of water, because power plant condensers require a considerable amount of cooling water. Otherwise, huge air-cooled cooling towers would have to be erected.

159. A large turbine that handles a large amount of steam has a relatively large blade tip circle, and thus a high tip speed. This creates a considerable centrifugal force, which pulls the blades away from the turbine's shaft. Therefore, the speed of large units is necessarily less than that of smaller units to limit the effects of centrifugal force.

160. By exhausting into a vacuum, a lower exit temperature is obtained. This results in a larger drop in heat from the same inlet steam conditions, allowing the turbine to work more efficiently.

161. The type of plant determines which is more economical to run. If there is a need for low-pressure process steam, then it is beneficial to exhaust into a backpressure. This is called cogeneration (when two forms of energy (mechanical and heat) are derived from the same source). If there is no need for process steam, then it is more economical to run the auxiliary machinery with electric motors.

162. The efficiency is as follows:

$$\frac{3{,}413 \text{ Btu/kWh}}{12{,}000 \text{ Btu/kWh}} \times 100 = 28.44\%$$

163. The efficiency is as follows:

$$\frac{2{,}545 \text{ Btu/kWh}}{10{,}000 \text{ Btu/kWh}} \times 100 = 25.45\%$$

164. The only information required to solve this problem is the amount of circulating water and its temperature rise. The rest of the information can be ignored. The first step is to convert gallons per minute to pounds per hour:

3,250 gpm x 500 lb/hr per gpm = 1,625,000 lb/hr

The temperature rise is 10°F, so apply the definition of Btu:

1,625,000 lb/hr x 10°F x Btu/lb-°F = 16,250,000 Btuh

165. 1,100°F is the limiting temperature for superheated steam. Regular mild steel starts to lose its strength around 600°F. Special alloy steel must be used for higher temperatures. To date, metallurgy and economics limit the temperature to 1,100°F.

166. The first number is the diameter of the steam cylinder, the second number is the diameter of the liquid-end cylinder, and the third number is the length of the stroke. All dimensions are in inches.

167. The three pumps required to operate a surface condenser are a circulating pump for the cooling water, a condensate pump to remove the condensate from the condenser, and a vacuum pump to remove the noncondensable gases.

168. In a typical power plant, about 30% of the heat is converted to shaft horsepower. The rest is rejected to the environment.

169. A crosshead on an air compressor is a guide for the piston rod that keeps it parallel to the cylinder throughout its stroke. This feature is required for all double-acting machines so the piston rod can operate through a packing gland.

170. The unloader is an automatic device that disables the compressor's pumping action so the pressure set point is not exceeded. Since it takes a lot of energy to overcome the at-rest inertia of a large machine, large compressors are not turned on and off to regulate pressure. An unloader is used instead. Unloaders hold the suction valves open, thus preventing compression.

171. A Kingsbury bearing is a type of thrust bearing that can absorb large axial loads. It is found on pumps, turbines, and ship propeller shafts.

172. To find the capacity of the feedwater pump, first calculate the pounds of steam per hour, then convert that to gallons per minute by dividing by 8.333 lb/gal and 60 min/hr, or 500 lb/hr per gpm:

$$\frac{10,000 \text{ kW} \times 9 \text{ lb/hr - kW}}{500 \text{ lb/hr per gpm}} = 180 \text{ gpm}$$

Efficiency is output divided by input. To find the output in Btuh:

$$10,000 \text{ kW} \times 3,413 \text{ Btu/kWh} = 34,130,000 \text{ Btuh}$$

The fact that the water rate is 9 lb/kW only tells us a lot about the turbine's efficiency. Since the question asks for the plant's efficiency, the boiler has to be included, but the boiler's efficiency is not given, so assume 80%. Also assume that saturated 250 psi steam has a heat content of 1,200 Btu/lb and that the feedwater temperature is 102°F. We can now proceed to calculate the boiler's heat input:

$$\frac{kW \times lb_{steam}/kWh \times (Btu/lb_{steam} - Btu/lb_{water})}{Boiler\ efficiency}$$

$$Input = \frac{90,000\ lb_{steam}/hr \times [1,200\ Btu/lb_{steam} - (102° - 32°F)]}{0.80} = 127,125,000\ Btuh$$

$$Efficiency = \frac{Output}{Input} = \frac{34,130,000\ Btuh}{127,125,000\ Btuh} = 26.8\%$$

If we assume the heat content of coal is 14,000 Btu/lb, the amount of coal burned per hour is:

$$\frac{127,125,000\ Btuh}{1,400\ Btu/lb} = 90,804\ lb/hr$$

173. When one of the generators trips off line, only 3,000 kW is available to supply a 3,500 kW load. If nothing is done, the other two generators will also soon trip. The engineer must immediately reduce the load to below 3,000 kW.

174. The three factors that affect a turbine plant's efficiency are pressure, superheat, and vacuum. Increasing the turbine's inlet pressure and superheat increases the amount of available heat energy. By exhausting into a vacuum, a larger heat drop from the same inlet steam conditions is obtained. To withstand high pressures, heavy and expensive equipment is required. The presently available alloy metals limit superheat to about 1,100°F. Due to cooling water temperatures and equipment limitations, 29 in. Hg is the practical lower vacuum limit.

175. The equation used to solve this problem is:

$$Q = \frac{h_2 - (t_o - 32°F)}{T_2 - T_1}$$

where:
$\quad Q$ = weight of cooling water to condense a pound of steam
$\quad h_2$ = heat content of exhaust steam
$\quad t_o$ = temperature of condensate
$\quad T_1$ = temperature of cooling water entering condenser
$\quad T_2$ = temperature of cooling water leaving condenser

$$Q = \frac{1,104\ Btu/lb - (100° - 32°F)}{76° - 65°F} = 94.2\ lb_{water}/lb_{steam}$$

This gives us how many pounds of cooling water are required for each pound of steam condensed. Now we have to determine how many pounds of steam need to be condensed:

$$5,000 \text{ kW} \times 12 \text{ lb}_{steam}/\text{kWh} = 60,000 \text{ lb}_{steam}/\text{hr}$$

Note: When a water rate is given, assume it is in pounds per hour.

Now find the amount of cooling water required for this water rate:

$$94.2 \text{ lb}_{water}/\text{lb}_{steam} \times 60,000 \text{ lb}_{steam}/\text{hr} = 5,652,000 \text{ lb}_{water}/\text{hr}$$

Pumps are usually rated in gallons per minute. To convert lb/hr to gpm, either divide by 8.33 lb/gal and 60 min/hr, or by the conversion of 500 lb/hr per gpm:

$$\frac{5,652,000 \text{ lb}_{water}/\text{hr}}{500 \text{ lb/hr per gpm}} = 11,304 \text{ gpm}$$

The same procedure can be used to calculate the size of the condensate pump. We know that the pump has to handle 60,000 lb of condensate per hour. Divide by 500 lb/hr per gpm to convert to gpm:

$$\frac{60,000 \text{ lb}_{condensate}/\text{hr}}{500 \text{ lb/hr per gpm}} = 120 \text{ gpm}$$

You might want to add that the designer would select pumps in which these flows would be their most efficient, and that some loss of capacity would have to be considered for normal wear and tear.

176. A *voltmeter* measures the electrical potential or voltage of a circuit. Voltage is the force that causes current to flow. An *ammeter* measures the amount of current flowing in an electrical circuit. The unit of current flow is ampere or amp. The *wattmeter* measures the amount of real power in a circuit. A *power factor meter* is associated with alternating current. It gives an indication of how much real and reactive power there is in a circuit. A *synchroscope* is also associated with alternating current plants. It indicates to an operator who is placing another generator on line when the two sources of alternating current are in phase or synchronized.

177. The equation for torque is as follows:

$$\text{Torque} = \frac{hp \times 5{,}252}{rpm}$$

Since rpm is in the denominator, the less speed the more torque. The speed reducer's torque is greater than the motor's torque.

178. Since the condenser pressure is less than atmospheric pressure, a condensate pump is required to remove the condensate from the condenser. A vacuum pump is required to remove noncondensable gases from the condenser. Air invariably finds its way into the condenser. As air accumulates, it increases the backpressure on the turbine and reduces the rate of heat transfer. Therefore, it reduces the capacity of the condenser. If allowed to accumulate, the condenser would lose efficiency and eventually cease to operate. A circulating pump is required to supply the condenser with enough cooling water to keep the condenser in operation.

179. The sentinel valve is located on the casing of a backpressure turbine. It relieves the casing pressure and warns the operator when the casing pressure is too high.

180. Draw a sketch of the system to help both you and the examiner better understand the problem.

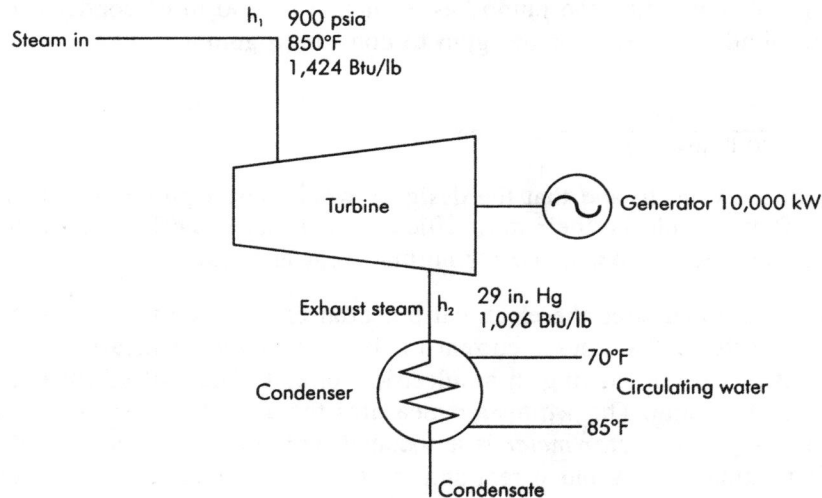

Steam in — h_1 900 psia 850°F 1,424 Btu/lb

Turbine — Generator 10,000 kW

Exhaust steam h_2 29 in. Hg 1,096 Btu/lb

70°F Circulating water 85°F

Condenser

Condensate

Work = 10,000 kW × 3,413 Btu/kWh = 34,130,000 Btuh

The turbine cycle efficiency is determined as follows:

$$\frac{\text{Work out + Usable energy out}}{\text{Total energy in}} = \frac{(h_1 - h_2) + h_3}{h_1}$$

$$\frac{(1,424 \text{ Btu/lb} - 1,096 \text{ Btu/lb}) + 48 \text{ Btu/lb}}{1,424 \text{ Btu/lb}} \times 100 = 26\%$$

The steam rate is determined as follows:

$$\frac{3,413 \text{ Btu/kWh}}{\text{Turbine heat drop}} = \frac{3,413 \text{ Btu/kWh}}{h_1 - h_2}$$

$$\frac{3,413 \text{ Btu/kWh}}{1,424 \text{ Btu/lb} - 1,096 \text{ Btu/lb}} = 10.4 \text{ lb/kWh}$$

The circulation water is determined as follows:

$$Q = \frac{h_2 - (t_o - 32°F)}{T_2 - T_1}$$

where:
- Q = weight of cooling water to condense a pound of steam
- h_2 = heat content of exhaust steam
- t_o = temperature of condensate
- T_1 = temperature of cooling water entering condenser
- T_2 = temperature of cooling water leaving condenser

$$Q = \frac{1,096 \text{ Btu/lb} - (80° - 32°F)}{85° - 70°F} = 69.87 \text{ lb}_{water}/\text{lb}_{steam}$$

The total steam flow would be as follows:

$$10.4 \text{ lb}_{steam}/\text{kWh} \times 10,000 \text{ kW} = 104,000 \text{ lb}_{steam}/\text{hr}$$

The total circulation water required is as follows:

$$69.87 \text{ lb}_{water}/\text{lb}_{steam} \times 104,000 \text{ lb}_{steam}/\text{hr} = 7,266,480 \text{ lb}_{water}/\text{hr}$$

$$\frac{7,266,480 \text{ lb}_{water}/\text{hr}}{500 \text{ lb/hr per gpm}} = 14,532.96 \text{ gpm}$$

181. The horsepower can be calculated as follows:

$$hp = \frac{gpm \times Head\ (ft)}{3,960} = \frac{50\ gpm \times 300\ ft}{3,960} = 3.8\ hp$$

182. The required horsepower is as follows:

$$hp = \frac{gpm \times Head\ (ft)}{3,960 \times Efficiency} = \frac{100\ gpm \times 50\ ft}{3,960 \times 0.5} = 2.5\ hp$$

To determine the number of kilowatts:

$$2.5\ hp \times 0.746\ kW/hp = 1.865\ kW$$

183. The torque may be calculated as follows:

$$Torque = \frac{hp \times 5,252}{rpm} = \frac{200\ hp \times 5,252}{1,725\ rpm} = 608.9\ lb\text{-}ft$$

184. The 10-psi suction pressure must be added to the 150-ft head. Therefore, the 10 psi has to be converted to feet:

$$10\ psi \times 2.31\ ft/psi = 23.1\ ft$$
$$150\ ft + 23.1\ ft = 173.1\ ft$$

Another trick part of this question is the output is in gallons per hour, not gallons per minute! To determine the shaft horsepower:

$$hp = \frac{Head\ (ft) \times lb\ of\ liquid\ pumped\ per\ min}{33,000\ ft\text{-}lb/min \times Pump\ efficiency}$$

$$hp = \frac{173.1\ ft \times 6,000\ gal/hr \times 8.33\ lb/gal}{33,000\ ft\text{-}lb/min \times 0.8 \times 60\ min/hr} = 5.5\ hp$$

To determine the torque:

$$Torque = \frac{hp \times 5,252}{rpm} = \frac{5.5\ hp \times 5,252}{1,735\ rpm} = 16.65\ lb\text{-}ft$$

185. The suction and head must be added to the discharge head to arrive at the correct answer:

$$10\ ft_{suction} + 10\ ft_{discharge} = 20\ ft$$

To find the water horsepower, just use the basic horsepower equation and assume a 100% efficient pump (water horsepower assumes a perfect pump):

$$hp = \frac{gpm \times Head}{3,960 \times Efficiency} = \frac{30,000 \ gpm \times 20 \ ft}{3,960 \times 1.0} = 151.5 \ hp$$

To find the required horsepower to drive this pump, use the pump's efficiency (75%):

$$\frac{151.5 \ hp}{0.75} = 202 \ hp$$

Now find the required kilowatts for the motor by using the horsepower-to-kilowatts conversion, and factor in the motor's efficiency:

$$\frac{202 \ hp \times 0.746 \ kW/hp}{0.95} = 158.6 \ kW$$

186. A gallon of water occupies the same volume as a gallon of No. 2 oil; that is, 231 in^3.

187. Remember that one cubic foot contains 7.48 gal, then calculate as follows:

$$
\begin{aligned}
Tank \ volume \ &= \ 0.7854 \times d^2 \times length \\
&= \ 0.7854 \times (12 \ ft)^2 \times 18 \ ft \\
&= \ 2,035.76 \ ft^3
\end{aligned}
$$

$$Capacity \ = \ 2,035.76 \ ft^3 \times 7.48 \ gal/ft^3 = 15,227.5 \ gal$$

To find weight, multiply cubic feet by 62.3 lb/ft^3 or gallons times 8.333 lb/gal:

$$15,227.5 \ gal \times 8.333 \ lb/gal = 126,890.8 \ lb$$

188. The crosshead keeps the connecting rod parallel to the cylinder, which allows the rod to be packed.

189. An injector forces feedwater into the boiler and heats it in the process.

190. One of the maintenance problems of injectors is that they don't handle dirt well, so a strainer is required. However, strainers become plugged and foreign matter may then get past the strainer. Another problem is that as an injector wears, efficiency falls off to the point where it no longer functions.

191. To determine the gpm the feed pump needs to supply, perform the following calculations:

$$1{,}000 \text{ bhp} \times 34.5 \text{ lb}_{water}/\text{hr-bhp} = 34{,}500 \text{ lb}_{water}/\text{hr}$$

$$34{,}500 \text{ lb}_{water}/\text{hr} \times 1.05 = 36{,}225 \text{ lb}_{water}/\text{hr}$$

$$\frac{36{,}225 \text{ lb}_{water}/\text{hr}}{500 \text{ gpm/lb}_{water}/\text{hr}} = 72.45 \text{ gpm}$$

192. The horsepower required is as follows:

$$\text{hp} = \frac{\text{Torque} \times \text{rpm}}{5{,}252} = \frac{525.2 \times 2{,}000 \text{ rpm}}{5{,}252} = 200 \text{ hp}$$

193. The results of cavitation are gas bubbles. When the gas bubbles collapse, they release enough energy to nip a microscopic piece of metal off the impeller. Eventually pitting that resembles oxygen attack on boiler tubes becomes obvious. After a while, the impeller is damaged beyond repair.

194. Cavitation results when the liquid being pumped flashes into small gas bubbles and then collapses. The flashing takes place because of inadequate pump suction pressure.

195. The steam piston's diameter must be larger than the water piston's diameter to force water into the boiler. Force equals pressure times area. There is more force produced on the larger piston that will overcome the force resisting the movement of the smaller piston.

196. Boiler feed pumps must be located below the deaerator, because there must be enough pressure at the suction of the pump so that none of the hot water from the deaerator flashes into steam as it enters the low-pressure area at the eye of the impeller. The weight of the water in the pipe connecting the elevated deaerator to the pump ensures that there is adequate pressure to prevent flashing.

197. The problem states that this pump operates at 100 rpm, so we can assume that it is motor-driven and has no steam cylinder. The 10 x 12 tells us that the pump is 10 inches in diameter with a 12-inch stroke. The problem doesn't state if it is double acting, so assume that it is not. Also no efficiency was given, so assume 75%. Be sure to state all assumptions.

$$\text{Capacity} = \text{Strokes/min} \times \text{Cylinders} \times \text{Volume/Cylinder} \times \text{Efficiency}$$

$$= 100 \text{ Strokes/min} \times 2 \text{ Cylinders} \times 0.7854 \times (10 \text{ in.})^2 \times 12 \text{ in.} \times 0.75 = 141{,}372 \text{ in}^3/\text{min}$$

$$\frac{141{,}372 \text{ in}^3/\text{min}}{231 \text{ in}^3/\text{gal}} = 612 \text{ gal/min}$$

198. The revolutions per minute are:

$$rmp = \frac{120 \times Hz}{Poles} = \frac{120 \times 60 \ Hz}{6 \ Poles} = 1,200 \ rpm$$

199. Each impeller adds a specific amount of head to its suction pressure. For example, consider a pump whose impeller adds 100 ft of head. If the pump took its suction at atmospheric pressure, its discharge would be at 100 ft of head. If its discharge were then directed to the suction of a similar impeller, it would add an additional 100 ft of head, for a total discharge pressure of 200 ft. Each succeeding impeller would add 100 ft of head.

200. The horsepower required is as follows:

$$hp = \frac{30,000 \ gpm \times 20 \ ft}{3,960 \times 0.75} = 202 \ hp$$

201. The horsepower required is as follows:

$$hp = \frac{1,000 \ gph \times 120 \ ft}{60 \ min/hr \times 3,960 \times 0.78} = 0.65 \ hp$$

202. Packing forms the seal between the stationary pump casing and the rotating shaft.

203. The formula W = A x V means volts times amps equals power in watts. This is for dc only. To obtain ac power, the equation is volts times amps times power factor.

204. One horsepower or 33,000 ft-lb/min or 1.34 kW equals 2,545 Btuh.

205. One kilowatt equals 1.34 horsepower.

206. Eddy currents are a source of heating in electric motors. One way to limit eddy currents is to add layers of insulation between thin sheets of steel. This insulation does not interfere with the magnetic lines of force required to operate the motor.

207. When a motor is started, it can draw up to six times its normal operating current. A motor starter's contacts are designed to withstand this initial rush. In addition, a motor starter can detect overload conditions and stop the motor before it is damaged.

208. One advantage of ac over dc is that voltage can be increased or decreased very easily with transformers. Thus, high voltage power can be transmitted long distances with little loss, then reduced down to a practical working voltage. Another advantage is that since ac equipment does not require commutators, it is less expensive and requires less maintenance then dc equipment.

209. Yes, they can be synchronized. If a two-pole generator were driven at 3,600 rpm, and a four-pole generator were driven at 1,800 rpm, their output frequencies would both be 60 Hz. Under these circumstances, generators operating at different speeds could be synchronized.

210. There are two types of ac power that travel over power lines — real and reactive. Real power is what does the work and is expressed in watts. Reactive power is sometimes called magnetizing power and is expressed in volt-amps-reactive. Some reactive power is always required when operating motors and transformers, but it does no real work. Real and reactive power are displaced by 90°. Apparent power, expressed in volt-amps, is the combination of real and reactive power.

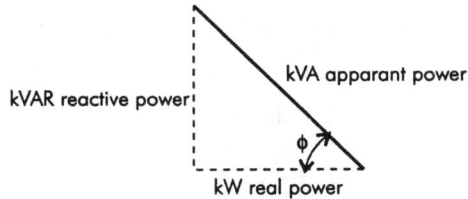

$$\text{Power factor} = \frac{kW}{kVA} = \cos f$$

The power factor is the true power divided by the apparent power. It is also the cosine of the angle formed between the real and apparent power.

211. The pressure at the tenth floor would be as follows (assume 10 ft/floor):

100 psi - (10 floors x 10 ft/floor x 0.433 psi/ft) = 56.7 psi

Various Steam Tables

(Courtesy, *Cameron Hydraulic Data* book. Reprinted with permission of Ingersoll-Dresser Pump Company)

Properties of Saturated Steam—Temperature Table

Temp F	Absolute Pressure			Vacuum in Hg ref to 29.921 in bar. at 32F	Specific volume sat vap ft³/lbm V_g	Total heat or enthalpy Btu/lb		
	in Hg	mm Hg	lb/in²			water h_f	evap h_{fg}	steam h_g
32	0.1803	4.581	0.08859	29.741	3304.7	-0.0179	1075.5	1075.5
32.018	0.1805	4.585	0.08865	29.741	3302.4	0.0003	1075.5	1075.5
33	0.1878	4.77	0.09223	29.734	3180.7	0.989	1074.9	1075.9
34	0.1955	4.96	0.09600	29.726	3061.9	1.996	1074.4	1076.4
35	0.203	5.17	0.09991	29.718	2948.1	3.002	1073.8	1076.8
36	0.212	5.38	0.10395	29.710	2839.0	4.008	1073.2	1077.2
37	0.220	5.59	0.10815	29.701	2734.4	5.013	1072.7	1077.7
38	0.229	5.82	0.11249	29.692	2634.2	6.018	1072.1	1078.1
39	0.238	6.05	0.11698	29.683	2538.0	7.023	1071.5	1078.5
40	0.248	6.29	0.12163	29.674	2445.8	8.027	1071.0	1079.0
41	0.257	6.54	0.12645	29.664	2357.3	9.031	1070.4	1079.4
42	0.268	6.80	0.13143	29.654	2274.4	10.035	1069.8	1079.9
43	0.278	7.06	0.13659	29.643	2191.0	11.038	1069.3	1080.3
44	0.289	7.34	0.14192	29.632	2112.8	12.041	1068.7	1080.7
45	0.300	7.62	0.14744	29.621	2037.8	13.044	1068.1	1081.2
46	0.312	7.92	0.15314	29.610	1965.7	14.047	1067.6	1081.6
47	0.324	8.22	0.15904	29.597	1896.5	15.049	1067.0	1082.1
48	0.336	8.54	0.16514	29.585	1830.0	16.051	1066.4	1082.5
49	0.349	8.87	0.17144	29.572	1766.2	17.053	1065.9	1082.9
50	0.362	9.20	0.17796	29.559	1704.8	18.054	1065.3	1083.4
51	0.376	9.55	0.18469	29.545	1645.9	19.056	1064.7	1083.8
52	0.390	9.91	0.19165	29.531	1589.2	20.057	1064.2	1084.2
53	0.405	10.28	0.19883	29.516	1534.8	21.058	1063.6	1084.7
54	0.420	10.67	0.20625	29.501	1482.4	22.058	1063.1	1085.1
55	0.436	11.06	0.21392	29.486	1432.0	23.059	1062.5	1085.6
56	0.452	11.47	0.22183	29.470	1383.6	24.059	1061.9	1086.0
57	0.468	11.89	0.23000	29.453	1337.0	25.060	1061.4	1086.4
58	0.485	12.33	0.23843	29.436	1292.2	26.060	1060.8	1086.9
59	0.503	12.78	0.24713	29.418	1249.1	27.060	1060.2	1087.3
60	0.521	13.24	0.25611	29.400	1207.6	28.060	1059.7	1087.7
61	0.540	13.72	0.26538	29.381	1167.6	29.059	1059.1	1088.2
62	0.560	14.22	0.27494	29.362	1129.2	30.059	1058.5	1088.6
63	0.58	14.73	0.28480	29.341	1092.1	31.058	1058.0	1089.0
64	0.601	15.25	0.29497	29.321	1056.5	32.058	1057.4	1089.5
65	0.622	15.80	0.30545	29.299	1022.1	33.057	1056.9	1089.9
66	0.644	16.36	0.31626	29.277	989.0	34.056	1056.3	1090.4
67	0.667	16.93	0.32740	29.255	957.2	35.055	1055.7	1090.8
68	0.690	17.53	0.33889	29.231	926.5	36.054	1055.2	1091.2
69	0.714	18.14	0.35073	29.207	896.9	37.053	1054.6	1091.7
70	0.739	18.77	0.36292	29.182	868.4	38.052	1054.0	1092.1
71	0.765	19.42	0.37549	29.157	840.9	39.050	1053.5	1092.5
72	0.791	20.09	0.38844	29.130	814.3	40.049	1052.9	1093.0
73	0.818	20.78	0.40177	29.103	788.8	41.048	1052.4	1093.4
74	0.846	21.49	0.41550	29.075	764.1	42.046	1051.8	1093.8
75	0.875	22.22	0.42964	29.047	740.3	43.045	1051.2	1094.3
76	0.904	22.97	0.44420	29.017	717.4	44.043	1050.7	1094.7
77	0.935	23.75	0.45919	28.986	695.2	45.042	1050.1	1095.1
78	0.966	24.54	0.47461	28.955	673.9	46.040	1049.5	1095.6
79	0.999	25.37	0.49049	28.923	653.2	47.038	1049.0	1096.0

Tables on pages 5-7 to 5-10 reproduced by permission from ASME Steam Tables© 1967 by American Society of Mechanical Engineers. All rights reserved.
Absolute pressures in inches Hg, millimeters Hg, and vacuum in inches Hg calculated by Ingersoll-Rand.

Properties of Saturated Steam—Temperature Table (*cont.*)

Temp F	Absolute Pressure			Vacuum in Hg ref to 29.921 in bar. at 32F	Specific volume sat vap ft³/lbm V_g	Total heat or enthalpy Btu/lb		
	in Hg	mm Hg	lb/in²			water h_f	evap h_{fg}	steam h_g
80	1.032	26.21	0.50683	28.889	633.3	48.037	1048.4	1096.4
81	1.066	27.08	0.52364	28.855	614.1	49.035	1047.8	1096.9
82	1.101	27.97	0.54093	28.820	595.6	50.033	1047.3	1097.3
83	1.138	28.89	0.55872	28.784	577.6	51.031	1046.7	1097.7
84	1.175	29.84	0.57702	28.746	560.3	52.029	1046.1	1098.2
85	1.213	30.81	0.59583	28.708	543.6	53.027	1045.6	1098.6
86	1.253	31.81	0.61518	28.669	527.5	54.026	1045.0	1099.0
87	1.293	32.84	0.63507	28.628	511.9	55.024	1044.4	1099.5
88	1.335	33.90	0.65551	28.587	496.8	56.022	1043.9	1099.9
89	1.377	34.99	0.67653	28.544	432.2	57.020	1043.3	1100.3
90	1.421	36.10	0.69813	28.500	468.1	58.018	1042.7	1100.8
91	1.467	37.25	0.72032	28.455	454.5	59.016	1042.2	1101.2
92	1.513	38.43	0.74313	28.408	441.3	60.014	1041.6	1101.6
93	1.561	39.64	0.76655	28.361	428.6	61.012	1041.0	1102.1
94	1.610	40.89	0.79062	28.312	416.3	62.010	1040.5	1102.5
95	1.660	42.165	0.81534	28.261	404.4	63.008	1039.9	1102.9
96	1.712	43.478	0.84072	28.210	392.9	64.006	1039.3	1103.3
97	1.765	44.826	0.86679	28.157	381.7	65.005	1038.8	1103.8
98	1.819	46.210	0.89356	28.102	370.9	66.003	1038.2	1104.2
99	1.875	47.631	0.92103	28.046	360.5	67.001	1037.6	1104.6
100	1.933	49.090	0.94924	27.989	350.4	67.999	1037.1	1105.1
101	1.992	50.586	0.97818	27.930	340.6	68.997	1036.5	1105.5
102	2.052	52.123	1.00789	27.869	331.1	69.995	1035.9	1105.9
103	2.114	53.700	1.03838	27.807	322.0	70.993	1035.4	1106.3
104	2.178	55.317	1.06965	27.743	313.1	71.992	1034.8	1106.8
105	2.243	56.976	1.10174	27.678	304.5	72.990	1034.2	1107.2
106	2.310	58.681	1.1347	27.611	296.18	73.99	1033.6	1107.6
107	2.379	60.424	1.1684	27.542	288.11	74.99	1033.1	1108.1
108	2.504	62.213	1.2030	27.417	280.30	75.98	1032.5	1108.5
109	2.522	64.049	1.2385	27.400	272.72	76.98	1031.9	1108.9
110	2.596	65.936	1.2750	27.325	265.39	77.98	1031.4	1109.3
111	2.672	67.865	1.3123	27.249	258.28	78.98	1030.8	1109.8
112	2.750	69.841	1.3505	27.172	251.38	79.98	1030.2	1110.2
113	2.830	71.873	1.3898	27.092	244.70	80.98	1029.6	1110.6
114	2.911	73.947	1.4299	27.001	238.22	81.97	1029.1	1111.0
115	2.995	76.078	1.4711	26.926	231.94	82.97	1028.5	1111.5
116	3.081	78.260	1.5133	26.840	225.85	83.97	1027.9	1111.9
117	3.169	80.499	1.5566	26.752	219.94	84.97	1027.3	1112.3
118	3.259	82.790	1.6009	26.662	214.21	85.97	1026.8	1112.7
119	3.352	85.138	1.6463	26.569	208.66	86.97	1026.2	1113.2
120	3.446	87.538	1.6927	26.475	203.26	87.97	1025.6	1113.6
121	3.543	89.999	1.7403	26.378	198.03	88.96	1025.0	1114.0
122	3.643	92.523	1.7891	26.279	192.95	89.96	1024.5	1114.4
123	3.744	95.103	1.8390	26.177	188.03	90.96	1023.9	1114.9
124	3.848	97.746	1.8901	26.073	183.24	91.96	1023.3	1115.3
125	3.956	100.47	1.9428	25.966	178.60	92.96	1022.7	1115.7
126	4.064	103.22	1.9959	25.858	174.09	93.96	1022.2	1116.1
127	4.175	106.05	2.0507	25.746	169.72	94.96	1021.6	1116.5
128	4.289	108.95	2.1068	25.632	165.47	95.96	1021.0	1117.0
129	4.406	111.92	2.1642	25.515	161.34	96.96	1020.4	1117.4

Properties of Saturated Steam—Temperature Table (*cont.*)

Temp F	Absolute Pressure			Vacuum in Hg ref to 29.921 in bar. at 32F	Specific volume sat vap ft³/lbm V_g	Total heat or enthalpy Btu/lb		
	in Hg	mm Hg	lb/in²			water h_f	evap h_{fg}	steam h_g
130	4.526	114.96	2.2230	25.395	157.33	97.96	1019.8	1117.8
131	4.648	118.06	2.2830	25.273	153.44	98.95	1019.3	1118.2
132	4.773	121.25	2.3445	25.148	149.66	99.95	1018.7	1118.6
133	4.902	124.50	2.4074	25.020	145.98	100.95	1018.1	1119.1
134	5.032	127.82	2.4717	24.889	142.41	101.95	1017.5	1119.5
135	5.166	131.23	2.5375	24.755	138.94	102.95	1016.9	1119.9
136	5.303	134.70	2.6047	24.618	135.57	103.95	1016.4	1120.3
137	5.443	138.26	2.6735	24.478	132.29	104.95	1015.8	1120.7
138	5.586	141.89	2.7438	24.335	129.11	105.95	1015.2	1121.1
139	5.773	145.61	2.8157	24.188	126.01	106.95	1014.6	1121.6
140	5.882	149.41	2.8892	24.039	123.00	107.95	1014.0	1122.0
141	6.035	153.30	2.9643	23.886	120.07	108.95	1013.4	1122.4
142	6.192	157.27	3.0411	23.730	117.22	109.95	1012.9	1122.8
143	6.351	161.32	3.1195	23.570	114.45	110.95	1012.3	1123.2
144	6.515	165.47	3.1997	23.407	111.76	111.95	1011.7	1123.6
145	6.681	169.71	3.2816	23.240	109.14	112.95	1011.1	1124.0
146	6.852	174.04	3.3653	23.069	106.59	113.95	1010.5	1124.5
147	7.026	178.46	3.4508	22.895	104.11	114.95	1009.9	1124.9
148	7.204	182.97	3.5381	22.718	101.70	115.95	1009.3	1125.3
149	7.385	187.58	3.6273	22.536	99.35	116.95	1008.7	1125.7
150	7.571	192.30	3.7184	22.351	97.07	117.95	1008.2	1126.1
151	7.760	197.11	3.8114	22.161	94.84	118.95	1007.6	1126.5
152	7.954	202.02	3.9065	21.968	92.68	119.95	1007.0	1126.9
153	8.151	207.04	4.0035	21.770	90.57	120.95	1006.4	1127.3
154	8.353	212.16	4.1025	21.569	88.52	121.95	1005.8	1127.7
155	8.559	217.39	4.2036	21.363	86.52	122.95	1005.2	1128.2
156	8.769	222.73	4.3068	21.153	84.57	123.95	1004.6	1128.6
157	8.983	228.18	4.4122	20.938	82.68	124.95	1004.0	1129.0
158	9.202	233.74	4.5197	20.719	80.83	125.96	1003.4	1129.4
159	9.426	239.41	4.6294	20.496	79.04	126.96	1002.8	1129.8
160	9.654	245.20	4.7414	20.678	77.29	127.96	1002.2	1130.2
161	9.886	251.11	4.8556	20.035	75.58	128.96	1001.6	1130.6
162	10.123	257.14	4.9722	19.798	73.92	129.96	1001.0	1131.0
163	10.366	263.28	5.0911	19.556	72.30	130.96	1000.4	1131.4
164	10.613	269.56	5.2124	19.309	70.72	131.96	999.8	1131.8
165	10.864	275.96	5.3361	19.057	69.18	132.96	999.2	1132.2
166	11.121	282.48	5.4623	18.800	67.68	133.97	998.6	1132.6
167	11.384	289.14	5.5911	18.538	66.22	134.97	998.0	1133.0
168	11.651	295.73	5.7223	18.271	64.80	135.97	997.4	1133.4
169	11.923	302.85	5.8562	17.998	63.41	136.97	996.8	1133.8
170	12.201	309.91	5.9926	17.720	62.06	137.97	996.2	1134.2
171	12.484	317.10	6.1318	17.437	60.74	138.98	995.6	1134.6
172	12.773	324.44	6.2736	17.148	59.45	139.98	995.0	1135.0
173	13.068	331.92	6.4182	16.854	58.19	140.98	994.4	1135.4
174	13.368	339.54	6.5656	16.554	56.97	141.98	993.8	1135.8
175	13.674	347.31	6.7159	16.248	55.77	142.99	993.2	1136.2
176	13.985	355.23	6.8690	15.936	54.61	143.99	992.6	1136.6
177	14.303	363.30	7.0250	15.618	53.47	144.99	992.0	1137.0
178	14.627	371.52	7.1840	15.295	52.36	145.99	991.4	1137.4
179	14.957	379.90	7.3460	14.965	51.28	147.00	990.8	1137.8

Tables on pages 5-7 to 5-10 reproduced by permission from ASME Steam Tables© 1967 by American Society of Mechanical Engineers. All rights reserved.
Absolute pressures in inches Hg, millimeters Hg, and vacuum in inches Hg calculated by Ingersoll-Rand.

Properties of Saturated Steam—Temperature Table (cont.)

Temp F	Absolute Pressure			Vacuum in Hg ref to 29.921 in bar. at 32F	Specific volume sat vap ft³/lbm V_g	Total heat or enthalpy Btu/lb		
	in Hg	mm Hg	lb/in²			water h_f	evap h_{fg}	steam h_g
180	15.293	388.42	7.5110	14.629	50.225	148.00	990.2	1138.2
181	15.635	397.12	7.679	14.287	49.194	149.00	989.6	1138.6
182	15.983	405.96	7.850	13.939	48.189	150.01	989.0	1139.0
183	16.339	415.01	8.025	13.582	47.207	151.01	988.4	1139.4
184	16.701	424.22	8.203	13.220	46.249	152.01	987.8	1139.8
185	17.070	433.58	8.384	12.851	45.313	153.02	987.1	1140.2
186	17.445	443.09	8.568	12.477	44.400	154.02	986.5	1140.5
187	17.827	452.81	8.756	12.094	43.508	155.02	985.9	1140.9
188	18.216	462.69	8.947	11.705	42.638	156.03	985.3	1141.3
189	18.611	472.72	9.141	11.310	41.787	157.03	984.7	1141.7
190	19.016	483.02	9.340	10.905	40.957	158.04	984.1	1142.1
191	19.426	493.41	9.541	10.496	40.146	159.04	983.5	1142.5
192	19.845	504.06	9.747	10.076	39.354	160.05	982.8	1142.9
193	20.271	514.87	9.956	9.651	38.580	161.05	982.2	1143.3
194	20.702	525.84	10.168	9.219	37.824	162.05	981.6	1143.7
195	21.144	537.06	10.385	8.777	37.086	163.06	981.0	1144.0
196	21.592	548.43	10.605	8.329	36.364	164.06	980.4	1144.4
197	22.050	560.07	10.830	7.871	35.659	165.07	979.7	1144.8
198	22.514	571.86	11.058	7.407	34.970	166.08	979.1	1145.2
199	22.987	583.86	11.290	6.935	34.297	167.08	978.5	1145.6
200	23.467	596.06	11.526	6.454	33.639	168.09	977.9	1146.0
201	23.956	608.48	11.766	5.966	32.996	169.09	977.2	1146.3
202	24.456	621.15	12.011	5.467	32.367	170.10	976.6	1146.7
203	24.960	633.97	12.259	4.962	31.752	171.10	976.0	1147.1
204	25.475	647.05	12.512	4.447	31.151	172.11	975.4	1147.5
205	26.000	660.40	12.770	3.921	30.564	173.12	974.7	1147.9
206	26.531	673.89	13.031	3.390	29.989	174.12	974.1	1148.2
207	27.073	687.65	13.297	2.848	29.428	175.13	973.5	1148.6
208	27.625	701.67	13.568	2.297	28.878	176.14	972.8	1149.0
209	28.185	715.89	13.843	1.737	28.341	177.14	972.2	1149.4
210	28.755	730.37	14.123	1.167	27.816	178.15	971.6	1149.7
211	29.333	745.05	14.407	0.588	27.302	179.16	970.9	1150.1
212	29.921	760.00	14.696	0.000	26.799	180.17	970.3	1150.5

Tables on pages 5-7 to 5-10 reproduced by permission from ASME Steam Tables, 1967 by American Society of Mechanical Engineers. All rights reserved.
Absolute pressures in inches Hg, millimeters Hg, and vacuum in inches Hg calculated by Ingersoll-Rand.

Properties of Saturated Steam—Temperature Table (*cont.*)

Temp F	Abs press lb/in²	Specific volume ft³/lbm V_g	Enthalpy, Btu/lbm			Entropy, Btu/lbm × F		Temp F
			Sat liquid h_f	Evap h_{fg}	Sat vapor h_g	Sat liquid s_f	Sat vapor s_g	
212	14.696	26.799	180.17	970.3	1150.5	0.3121	1.7568	212
213	14.990	26.307	181.17	969.7	1150.8	0.3136	1.7552	213
214	15.289	25.826	182.18	969.0	1151.2	0.3151	1.7536	214
215	15.592	25.355	183.19	968.4	1151.6	0.3166	1.7520	215
216	15.901	24.894	184.20	967.8	1152.0	0.3181	1.7505	216
220	17.186	23.148	188.23	965.2	1153.4	0.3241	1.7442	220
224	18.556	21.545	192.27	962.6	1154.9	0.3300	1.7380	224
228	20.015	20.073	196.31	960.0	1156.3	0.3359	1.7320	228
232	21.567	18.718	200.35	957.4	1157.8	0.3417	1.7260	232
236	23.216	17.471	204.40	954.8	1159.2	0.3476	1.7201	236
240	24.968	16.321	208.45	952.1	1160.6	0.3533	1.7142	240
244	26.826	15.260	212.50	949.5	1162.0	0.3591	1.7085	244
248	28.796	14.281	216.56	946.8	1163.4	0.3649	1.7028	248
252	30.883	13.375	220.62	944.1	1164.7	0.3706	1.6972	252
256	33.091	12.538	224.69	941.4	1166.1	0.3763	1.6917	256
260	35.427	11.762	228.76	938.6	1167.4	0.3819	1.6862	260
264	27.894	11.042	232.83	935.9	1168.7	0.3876	1.6808	264
268	40.500	10.375	236.91	933.1	1170.0	0.3932	1.6755	268
272	43.249	9.755	240.99	930.3	1171.3	0.3987	1.6702	272
276	46.147	9.180	245.08	927.5	1172.5	0.4043	1.6650	276
280	49.200	8.6439	249.2	924.6	1173.8	0.4098	1.6599	280
284	52.414	8.1453	253.3	921.7	1175.0	0.4154	1.6548	284
288	55.795	7.6807	257.4	918.8	1176.2	0.4208	1.6498	288
292	59.350	7.2475	261.5	915.9	1177.4	0.4263	1.6449	292
296	63.084	6.8433	265.6	913.0	1178.6	0.4317	1.6400	296
300	67.005	6.4658	269.7	910.0	1179.7	0.4372	1.6351	300
304	71.119	6.1130	273.8	907.0	1180.9	0.4426	1.6303	304
308	75.433	5.7830	278.0	904.0	1182.0	0.4479	1.6256	308
312	79.953	5.4742	282.1	901.0	1183.1	0.4533	1.6209	312
316	84.688	5.1849	286.3	897.9	1184.1	0.4586	1.6162	316
320	89.643	4.9138	290.4	894.8	1185.2	0.4640	1.6116	320
324	94.826	4.6595	294.6	891.6	1186.2	0.4692	1.6071	324
328	100.245	4.4208	298.7	888.5	1187.2	0.4745	1.6025	328
332	105.907	4.1966	302.9	885.3	1188.2	0.4798	1.5981	332
336	111.820	3.9859	307.1	882.1	1189.1	0.4850	1.5936	336
340	117.992	3.7878	311.3	878.8	1190.1	0.4902	1.5892	340
344	124.430	3.6013	315.5	875.5	1191.0	0.4954	1.5849	344
348	131.142	3.4258	319.7	872.2	1191.9	0.5006	1.5806	348
352	138.138	3.2603	323.9	868.9	1192.7	0.5058	1.5763	352
356	145.424	3.1044	328.1	865.5	1193.6	0.5110	1.5721	356
360	153.010	2.9573	332.3	862.1	1194.4	0.5161	1.5678	360
364	160.903	2.8184	336.5	858.6	1195.2	0.5212	1.5637	364
368	169.113	2.6873	340.8	855.1	1195.9	0.5263	1.5595	368
372	177.648	2.5633	345.0	851.6	1196.7	0.5314	1.5554	372
376	186.517	2.4462	349.3	848.1	1197.4	0.5365	1.5513	376
380	195.729	2.3353	353.6	844.5	1198.0	0.5416	1.5473	380
384	205.294	2.2304	357.9	840.8	1198.7	0.5466	1.5432	384
388	215.220	2.1311	362.2	837.2	1199.3	0.5516	1.5392	388
392	225.516	2.0369	366.5	833.4	1199.9	0.5567	1.5352	392
396	236.193	1.9477	370.8	829.7	1200.4	0.5617	1.5313	396

Properties of Saturated Steam—Temperature Table (*cont.*)

Temp F	Abs press lb/in²	Specific volume ft³/lbm V_g	Enthalpy, Btu/lbm			Entropy, Btu/lbm × F		Temp F
			Sat liquid h_f	Evap h_fg	Sat vapor h_g	Sat liquid s_f	Sat vapor s_g	
400	247.259	1.8630	375.1	825.9	1201.0	0.5667	1.5274	400
404	258.725	1.7827	379.4	822.0	1201.5	0.5717	1.5234	404
408	270.600	1.7064	383.8	818.2	1201.9	0.5766	1.5195	408
412	282.894	1.6340	388.1	814.2	1202.4	0.5816	1.5157	412
416	295.617	1.5651	392.5	810.2	1202.8	0.5866	1.5118	416
420	308.78	1.4997	396.9	806.2	1203.1	0.5915	1.5080	420
424	322.39	1.4374	401.3	802.2	1203.5	0.5964	1.5042	424
428	336.46	1.3782	405.7	798.0	1203.7	0.6014	1.5004	428
432	351.00	1.3218	410.1	793.9	1204.0	0.6063	1.4966	432
436	366.03	1.2681	414.6	789.7	1204.2	0.6112	1.4928	436
440	381.54	1.21687	419.0	785.4	1204.4	0.6161	1.4890	440
444	397.56	1.16806	423.5	781.1	1204.6	0.6210	1.4853	444
448	414.09	1.12152	428.0	776.7	1204.7	0.6259	1.4815	448
452	431.14	1.07711	432.5	772.3	1204.8	0.6308	1.4778	452
456	448.73	1.03472	437.0	767.8	1204.8	0.6356	1.4741	456
460	466.87	0.99424	441.5	763.2	1204.8	0.6405	1.4704	460
464	485.56	0.95557	446.1	758.6	1204.7	0.6454	1.4667	464
468	504.83	0.91862	450.7	754.0	1204.6	0.6502	1.4629	468
472	524.67	0.88329	455.2	749.3	1204.5	0.6551	1.4592	472
476	545.11	0.84950	459.9	744.5	1204.3	0.6599	1.4555	476
480	566.15	0.81717	464.5	739.6	1204.1	0.6648	1.4518	480
484	587.81	0.78622	469.1	734.7	1203.8	0.6696	1.4481	484
488	610.10	0.75658	473.8	729.7	1203.5	0.6745	1.4444	488
492	633.03	0.72820	478.5	724.6	1203.1	0.6793	1.4407	492
496	656.61	0.70100	483.2	719.5	1202.7	0.6842	1.4370	496
500	680.86	0.67492	487.9	714.3	1202.2	0.6890	1.4333	500
504	705.78	0.64991	492.7	709.0	1201.7	0.6939	1.4296	504
508	731.40	0.62592	497.5	703.7	1201.1	0.6987	1.4258	508
512	757.72	0.60289	502.3	698.2	1200.5	0.7036	1.4221	512
516	784.76	0.58079	507.1	692.7	1199.8	0.7085	1.4183	516
520	312.53	0.55956	512.0	687.0	1199.0	0.7133	1.4146	520
524	841.04	0.53916	516.9	681.3	1198.2	0.7182	1.4108	524
528	870.31	0.51955	521.8	675.5	1197.3	0.7231	1.4070	528
532	900.34	0.50070	526.8	669.6	1196.4	0.7280	1.4032	532
536	931.17	0.48257	531.7	663.6	1195.4	0.7329	1.3993	536
540	962.79	0.46513	536.8	657.5	1194.3	0.7378	1.3954	540
544	995.22	0.44834	541.8	651.3	1193.1	0.7427	1.3915	544
548	1028.49	0.43217	546.9	645.0	1191.9	0.7476	1.3876	548
552	1062.59	0.41660	552.0	638.5	1190.6	0.7525	1.3837	552
556	1097.55	0.40160	557.2	632.0	1189.2	0.7575	1.3797	556
560	1133.38	0.38714	562.4	625.3	1187.7	0.7625	1.3757	560
564	1170.10	0.37320	567.6	618.5	1186.1	0.7674	1.3716	564
568	1207.72	0.35975	572.9	611.5	1184.5	0.7725	1.3675	568
572	1246.26	0.34678	578.3	604.5	1182.7	0.7775	1.3634	572
576	1285.74	0.33426	583.7	597.2	1180.9	0.7825	1.3592	576
580	1326.2	0.32216	589.1	589.9	1179.0	0.7876	'1.3550	580
584	1367.7	0.31048	594.6	582.4	1176.9	0.7927	1.3507	584
588	1410.0	0.29919	600.1	574.7	1174.8	0.7998	1.3464	588
592	1453.3	0.28827	605.7	566.8	1172.6	0.8030	1.3420	592
596	1497.8	0.27770	611.4	558.8	1170.2	0.8082	1.3375	596

Properties of Saturated Steam—Temperature Table (*cont.*)

Temp F	Abs press lb/in²	Specific volume ft³/lbm V_g	Enthalpy, Btu/lbm			Entropy, Btu/lbm × F		Temp F
			Sat liquid h_f	Evap h_{fg}	Sat vapor h_g	Sat liquid s_f	Sat vapor s_g	
600	1543.2	0.26747	617.1	550.6	1167.7	0.8134	1.3330	600
604	1589.7	0.25757	622.9	542.2	1165.1	0.8187	1.3284	604
608	1637.3	0.24796	628.8	533.6	1162.4	0.8240	1.3238	608
612	1686.1	0.23865	634.8	524.7	1159.5	0.8294	1.3190	612
616	1735.9	0.22960	640.8	515.6	1156.4	0.8348	1.3141	616
620	1786.9	0.22081	646.9	506.3	1153.2	0.8403	1.3092	620
624	1839.0	0.21226	653.1	496.6	1149.6	0.8458	1.3041	624
628	1892.4	0.20394	659.5	486.7	1146.1	0.8514	1.2988	628
632	1947.0	0.19583	665.9	476.4	1142.2	0.8571	1.2934	632
636	2002.8	0.18792	672.4	465.7	1138.1	0.8628	1.2879	636
640	2059.9	0.18021	679.1	454.6	1133.7	0.8686	1.2821	640
644	2118.3	0.17269	685.9	443.1	1129.0	0.8746	1.2761	644
648	2178.1	0.16534	692.9	431.1	1124.0	0.8806	1.2699	648
652	2239.2	0.15816	700.0	418.7	1118.7	0.8868	1.2634	652
656	2301.7	0.15115	707.4	405.7	1113.1	0.8931	1.2567	656
660	2365.7	0.14431	714.9	392.1	1107.0	0.8995	1.2498	660
664	2431.1	0.13757	722.9	377.1	1100.6	0.9064	1.2425	664
668	2498.1	0.13087	731.5	362.1	1093.5	0.9137	1.2347	668
672	2566.6	0.12424	740.2	345.7	1085.9	0.9212	1.2266	672
676	2636.8	0.11769	749.2	328.5	1077.6	0.9287	1.2179	676
680	2708.6	0.11117	758.5	310.1	1068.5	0.9365	1.2086	680
684	2782.1	0.10463	768.2	290.2	1058.4	0.9447	1.1984	684
688	2857.4	0.09799	778.8	268.2	1047.0	0.9535	1.1872	688
692	2934.5	0.09110	790.5	243.1	1033.6	0.9634	1.1744	692
696	3013.4	0.08370	804.4	212.8	1017.2	0.9749	1.1591	696
700	3094.3	0.07519	822.4	172.7	995.0	0.9901	1.1390	700
701	3114.9	0.07271	828.2	159.8	988.0	0.9949	1.1326	701
702	3135.5	0.06997	835.0	144.7	979.7	1.0006	1.1252	702
703	3156.3	0.06684	843.2	126.4	969.6	1.0076	1.1163	703
704	3177.2	0.06300	854.2	102.0	956.2	1.0169	1.1046	704
704.5	3187.8	0.06055	861.9	85.3	947.2	1.0234	1.0967	704.5
705.0	3198.3	0.05730	873.0	61.4	934.4	1.0329	1.0856	705.0
705.47	3208.2	0.05078	906.0	0.0	906.0	1.0612	1.0612	705.47

Tables on pages 5-11 to 5-13 reproduced by permission from ASME Steam Tables¹ 1967 by The American Society of Mechanical Engineers, all rights reserved.

Any pressure may be expressed in a number of different units by using the following conversion formulas.

1 standard atmosphere = 14.696 lb/sq in absolute
1 standard atmosphere = 29.9213 inches Hg (at 32°F—0°C)
1 standard atmosphere = 34 00 ft water (at 75°F—23.9°C)
1 standard atmosphere = 76 cm or 760 mm Hg (at 0°C—32°F)
1 pound per square inch = 2.036 inches Hg (at 32°F—0°C)
1 pound per square inch = 27.763 inches water (at 75°F—23.9°C)
1 inch Hg (at 32°F) = .491 pounds per square inch.
1 inch Hg = 25.4 millimeters Hg
1 kg cm² = 14.223 lb/sq in
1 pound per sq. in. = 6.895 kilopascals

Properties of Saturated Steam—Pressure Table

Abs press lb/in²	Temp °F	Specific volume ft³/lbm		Enthalpy btu lbm		Entropy btu lbm × F		Abs press lb/in²
		Water v_f	Steam v_g	Water h_f	Steam h_g	Water s_f	Steam s_g	
.08865	32.018	0.016022	3302.4	0.0003	1075.5	0.0000	2.1872	.08865
0.25	59.323	0.016032	1235.5	27.382	1067.4	0.0542	2.0967	0.25
0.50	79.586	0.016071	641.5	47.623	1096.3	0.0925	2.0370	0.50
1.0	101.74	0.016136	333.60	69.73	1105.8	0.1326	1.9781	1.0
3.0	141.47	0.016300	118.73	109.42	1122.6	0.2009	1.8864	3.0
6.0	170.05	0.016451	61.984	138.03	1134.2	0.2474	1.8294	6.0
10.0	193.21	0.016592	38.420	161.26	1143.3	0.2836	1.7879	10.0
14.696	212.00	0.016719	26.799	180.17	1150.5	0.3121	1.7568	14.696
15.0	213.03	0.016726	26.290	181.21	1150.9	0.3137	1.7552	15.0
20.0	227.96	0.016834	20.087	196.27	1156.3	0.3358	1.7320	20.0
25.0	240.07	0.016927	16.301	208.52	1160.6	0.3535	1.7141	25.0
30.0	250.34	0.017009	13.744	218.9	1164.1	0.3682	1.6995	30.0
35.0	259.29	0.017083	11.896	228.0	1167.1	0.3809	1.6872	35.0
40.0	267.25	0.017151	10.4965	236.1	1169.8	0.3921	1.6765	40.0
45.0	274.44	0.017214	9.3988	243.5	1172.0	0.4021	1.6671	45.0
50.0	281.02	0.017274	8.5140	250.2	1174.1	0.4112	1.6586	50.0
55.0	287.08	0.017329	7.7850	256.4	1175.9	0.4196	1.6510	55.0
60.0	292.71	0.017383	7.1736	262.2	1177.6	0.4273	1.6440	60.0
65.0	297.98	0.017433	6.6533	267.6	1179.1	0.4344	1.6375	65.0
70.0	302.93	0.017482	6.2050	272.7	1180.6	0.4411	1.6316	70.0
75.0	307.61	0.017529	5.8144	277.6	1181.9	0.4474	1.6260	75.0
80.0	312.04	0.017573	5.4711	282.1	1183.1	0.4534	1.6208	80.0
85.0	316.26	0.017617	5.1669	286.5	1184.2	0.4590	1.6159	85.0
90.0	320.28	0.017659	4.8953	290.7	1185.3	0.4643	1.6113	90.0
95.0	324.13	0.017700	4.6514	294.7	1186.2	0.4694	1.6069	95.0
100.0	327.82	0.017740	4.4310	298.5	1187.2	0.4743	1.6027	100
105.0	331.37	0.01778	4.2309	302.2	1188.0	0.4790	1.5988	105
110.	334.79	0.01782	4.0306	305.8	1188.9	0.4834	1.5950	110
115.	338.08	0.01785	3.8813	309.3	1189.6	0.4877	1.5913	115
120.	341.27	0.01789	3.7275	312.6	1190.4	0.4919	1.5879	120
125.	344.35	0.01792	3.5857	315.8	1191.1	0.4959	1.5845	125
130.	347.33	0.01796	3.4544	319.0	1191.7	0.4998	1.5813	130
135.	350.23	0.01799	3.3325	322.0	1192.4	0.5035	1.5782	135
140.	353.04	0.01803	3.2010	325.0	1193.0	0.5071	1.5752	140
145.	355.77	0.01806	3.1130	327.6	1193.5	0.5107	1.5723	145
150.	358.43	0.01809	3.0139	330.6	1194.1	0.5141	1.5695	150
160.	363.55	0.01815	2.8386	336.1	1195.1	0.5206	1.5641	160
170.	368.42	0.01821	2.6738	341.2	1196.0	0.5269	1.5591	170
180.	373.08	0.01827	2.5312	346.2	1196.9	0.5328	1.5543	180
190.	377.53	0.01833	2.4030	350.9	1197.6	0.5384	1.5498	190
200.	381.80	0.01839	2.2873	355.5	1198.3	0.5438	1.5454	200
210.	385.91	0.01844	2.18217	359.9	1199.0	0.5490	1.5413	210
220.	389.88	0.01850	2.08629	364.2	1199.6	0.5540	1.5374	220
230.	393.70	0.01855	1.99846	368.3	1200.1	0.5588	1.5336	230
240.	397.39	0.01860	1.91769	372.3	1200.6	0.5634	1.5299	240
250.	400.97	0.01865	1.84317	376.1	1201.1	0.5679	1.5264	250
260.	404.44	0.01870	1.77418	379.9	1201.5	0.5722	1.5230	260
270.	407.80	0.01875	1.71013	383.6	1201.9	0.5764	1.5197	270

Tables on pages 5-19 to 5-20 reproduced by permission from ASME Steam Tables⁵ 1967 by The American Society of Mechanical Engineers. All rights reserved.

Properties of Saturated Steam—Pressure Table (*cont.*)

Abs press lb/in²	Temp °F	Specific volume ft³/lbm		Enthalpy btu/lbm		Entropy btu/lbm · F		Abs press lb in²
		Water v_f	Steam v_g	Water h_f	Steam h_g	Water s_f	Steam s_g	
280	411.07	0.01880	1.65049	387.1	1202.3	0.5805	1.5166	280
290	414.25	0.01885	1.59482	390.6	1202.6	0.5844	1.5135	290
300	417.35	0.01889	1.54274	394.0	1202.9	0.5882	1.5105	300
310	420.36	0.01894	1.49390	397.3	1203.2	0.5920	1.5076	310
320	423.31	0.01899	1.44801	400.5	1203.4	0.5956	1.5048	320
330	426.18	0.01903	1.40480	403.7	1203.6	0.5991	1.5021	330
340	428.99	0.01908	1.36405	406.8	1203.8	0.6026	1.4994	340
350	431.73	0.01912	1.32554	409.8	1204.0	0.6059	1.4968	350
360	434.41	0.01917	1.28910	412.8	1204.1	0.6092	1.4943	360
380	439.61	0.01925	1.22177	418.6	1204.4	0.6156	1.4894	380
400	444.60	0.01934	1.16095	424.2	1204.6	0.6217	1.4847	400
450	456.28	0.01954	1.03179	437.3	1204.8	0.6360	1.4738	450
500	467.01	0.01975	0.92762	449.5	1204.7	0.6490	1.4639	500
550	476.94	0.01994	0.84177	460.9	1204.3	0.6611	1.4547	550
600	486.20	0.02013	0.76975	471.7	1203.7	0.6723	1.4461	600
650	494.89	0.02032	0.70843	481.9	1202.8	0.6828	1.4381	650
700	503.08	0.02050	0.65556	491.6	1201.8	0.6928	1.4304	700
750	510.84	0.02069	0.60949	500.9	1200.7	0.7022	1.4232	750
800	518.21	0.02087	0.56896	509.8	1199.4	0.7111	1.4163	800
850	525.24	0.02105	0.53302	518.4	1198.0	0.7197	1.4096	850
900	531.95	0.02123	0.50091	526.7	1196.4	0.7279	1.4032	900
950	538.39	0.02141	0.47205	534.7	1194.7	0.7358	1.3970	950
1000	544.58	0.02159	0.44596	542.6	1192.9	0.7434	1.3910	1000
1100	556.28	0.02695	0.40058	557.5	1189.1	0.7578	1.3794	1100
1200	567.19	0.02232	0.36245	571.9	1184.8	0.7714	1.3683	1200
1300	577.42	0.02269	0.32991	585.6	1180.2	0.7843	1.3577	1300
1400	587.07	0.02307	0.30178	598.8	1175.3	0.7966	1.3474	1400
1500	596.20	0.02346	0.27719	611.7	1170.1	0.8085	1.3373	1500
1600	604.87	0.02387	0.25545	624.2	1164.5	0.8199	1.3274	1600
1700	613.13	0.02428	0.23607	636.5	1158.6	0.8309	1.3176	1700
1800	621.02	0.02472	0.21861	648.5	1152.3	0.8417	1.3079	1800
1900	628.56	0.02517	0.20278	660.4	1145.6	0.8522	1.2981	1900
2000	635.80	0.02565	0.18831	672.1	1138.3	0.8625	1.2881	2000
2100	642.76	0.02615	0.17501	683.8	1130.5	0.8727	1.2780	2100
2200	649.45	0.02669	0.16272	695.5	1122.2	0.8828	1.2676	2200
2300	655.89	0.02727	0.15133	707.2	1113.2	0.8929	1.2569	2300
2400	662.11	0.02790	0.14076	719.0	1103.7	0.9031	1.2460	2400
2500	668.11	0.02859	0.13068	731.7	1093.3	0.9139	1.2345	2500
2600	673.91	0.02938	0.12110	744.5	1082.0	0.9247	1.2250	2600
2700	679.53	0.03029	0.11194	757.3	1069.7	0.9356	1.2097	2700
2800	684.96	0.03134	0.10305	770.7	1055.8	0.9468	1.1958	2800
2900	690.22	0.03264	0.09420	785.1	1039.8	0.9588	1.1803	2900
3000	695.33	0.03428	0.08500	801.8	1020.3	0.9728	1.1619	3000
3100	700.28	0.03681	0.07452	824.0	993.3	0.9914	1.1373	3100
3200	705.08	0.04472	0.05663	875.5	931.6	1.0351	1.0832	3200
3208.2	705.47	0.05078	0.05078	906.0	906.0	1.0612	1.0612	3208.2

Properties of Superheated Steam

Abs press lb in² (sat temp-F)		Sat water	Sat steam	Temperature—degrees Fahrenheit						
				300	400	500	600	700	800	900
1 (101.74)	sh			198.26	298.26	398.26	498.26	598.26	698.26	798.26
	v	0.01614	333.6	452.3	511.9	571.5	631.1	690.7	750.3	809.8
	h	69.73	1105.8	1195.7	1241.8	1288.6	1336.1	1384.5	1433.7	1483.8
	s	0.1326	1.9781	2.1152	2.1722	2.2237	2.2708	2.3144	2.3551	2.3934
5 (162.24)	sh			137.76	237.76	337.76	437.36	537.76	637.76	737.76
	v	0.01641	73.53	90.24	102.24	114.21	126.15	138.08	150.01	161.94
	h	130.20	1131.1	1194.8	1241.3	1288.2	1335.9	1384.3	1433.6	1483.7
	s	0.2349	1.8443	1.9369	1.9943	2.0460	2.0932	2.1369	2.1776	2.2159
10 (193.21)	sh			106.79	206.79	306.79	406.79	506.79	606.79	706.79
	v	0.01659	38.42	44.98	51.03	57.04	63.03	69.00	74.98	80.94
	h	161.26	1143.3	1193.7	1240.6	1287.8	1335.5	1384.0	1433.4	1483.5
	s	0.2836	1.7879	1.8593	1.9173	1.9692	2.0166	2.0603	2.1011	2.1394
14.696 (212.00)	sh			88.00	188.00	288.00	388.00	488.00	588.00	688.00
	v	0.0167	26.799	30.52	33.963	38.77	42.86	46.93	51.00	55.06
	h	180.17	1150.5	1192.6	1239.9	1287.4	1335.2	1383.8	1433.2	1483.4
	s	0.3121	1.7568	1.8158	1.8720	1.9265	1.9739	2.0177	2.0585	2.0969
20 (227.96)	sh			72.04	172.04	272.04	372.04	472.04	572.04	672.04
	v	0.01683	20.087	22.356	25.428	28.457	31.466	34.465	37.458	40.447
	h	196.27	1156.3	1191.4	1239.2	1286.9	1334.9	1383.5	1432.9	1483.2
	s	0.3358	1.7320	1.7805	1.8397	1.8921	1.9397	1.9836	2.0244	2.0628
40 (267.25)	sh			32.75	132.75	232.75	332.75	432.75	532.75	632.75
	v	0.01715	10.497	11.036	12.624	14.165	15.685	17.195	18.699	20.199
	h	236.14	1169.8	1186.6	1236.4	1285.0	1333.6	1382.5	1432.1	1482.5
	s	0.3921	1.6765	1.6992	1.7608	1.8143	1.8624	1.9065	1.9476	1.9860
60 (292.71)	sh			7.29	107.29	207.29	307.29	407.29	507.29	607.29
	v	0.1738	7.174	7.257	8.354	9.400	10.425	11.438	12.446	13.450
	h	262.21	1177.6	1181.6	1233.5	1283.2	1332.3	1381.5	1431.3	1481.8
	s	0.4273	1.6440	1.6492	1.7134	1.7681	1.8168	1.8612	1.9024	1.9410
80 (312.04)	sh				87.96	187.96	287.96	387.96	487.96	587.96
	v	0.01757	5.471		6.218	7.018	7.794	8.560	9.319	10.075
	h	282.15	1183.1		1230.5	1281.3	1330.9	1380.5	1430.5	1481.1
	s	0.4534	1.6208		1.6790	1.7349	1.7842	1.8289	1.8702	1.9089
100 (327.82)	sh				72.18	172.18	272.18	372.18	472.18	572.18
	v	0.01774	4.431		4.935	5.588	6.216	6.833	7.443	8.050
	h	298.54	1187.2		1227.4	1279.3	1329.6	1379.5	1429.7	1480.4
	s	0.4743	1.6027		1.6516	1.7088	1.7586	1.8036	1.8451	1.8839
120 (341.27)	sh				58.73	158.73	258.73	358.73	458.73	558.73
	v	0.01789	3.7275		4.0786	4.6341	5.1637	5.6813	6.1928	6.7006
	h	312.58	1190.4		1224.1	1277.4	1328.2	1378.4	1428.8	1479.8
	s	0.4919	1.5879		1.6286	1.6872	1.7376	1.7829	1.8246	1.8635
140 (353.04)	sh				46.96	146.96	246.96	346.96	446.96	546.96
	v	0.01803	3.2190		3.4661	3.9526	4.4119	4.8588	5.2995	5.7364
	h	324.96	1193.0		1220.8	1275.3	1326.8	1377.4	1428.0	1479.1
	s	0.5071	1.5752		1.6085	1.6686	1.7196	1.7652	1.8071	1.8461
160 (363.55)	sh				36.45	136.45	236.45	336.45	436.45	536.45
	v	0.01815	2.8336		3.0060	3.4413	3.8480	4.2420	4.6295	5.0132
	h	336.07	1195.1		1217.4	1273.3	1325.4	1376.4	1427.2	1478.4
	s	0.5206	1.5641		1.5906	1.6522	1.7039	1.7499	1.7919	1.8383

`sh = superheat; v = specific volume in ft³/lb; h = total heat in Btu/lb; s = entropy in Btu/°F/lb.

Properties of Superheated Steam (*cont.*)

Abs press lb/in² (sat temp-F)		Sat water	Sat steam	Temperature—degrees Fahrenheit						
				600	700	800	900	1000	1200	1400
180 (373.08)	sh		2.5312	226.92	326.92	426.92	526.92	626.92	826.92	1026.92
	p	0.01827		3.4093	3.7621	4.1084	4.4508	4.7907	5.4697	6.1363
	h	346.19	1196.9	1324.0	1375.3	1426.3	1477.7	1529.7	1635.9	1745.3
	s	0.5328	1.5543	1.6900	1.7362	1.7784	1.8176	1.8545	1.9227	1.9849
200 (381.80)	sh		2.2873	218.20	318.20	418.20	518.20	618.20	818.20	1018.20
	v	0.01839		3.0583	3.3783	3.6915	4.0008	4.3077	4.9165	5.5209
	h	355.51	1198.3	1322.6	1374.3	1425.5	1477.0	1529.1	1635.4	1745.0
	s	0.5438	1.5454	1.6773	1.7293	1.7663	1.8057	1.8426	1.9109	1.9732
220 (389.88)	sh		2.0863	210.12	310.12	410.12	510.12	610.12	810.12	1010.12
	v	0.01850		2.7710	3.0642	3.3504	3.6327	3.9125	4.4671	5.0173
	h	364.17	1199.6	1321.2	1373.2	1427.7	1476.3	1528.5	1635.0	1744.7
	s	0.5540	1.5374	1.6658	1.7128	1.7553	1.7948	1.8318	1.9002	1.9625
240 (397.39)	sh		1.9177	202.61	302.61	402.61	502.61	602.61	802.61	1002.61
	v	0.01860		2.5316	2.8024	3.0661	3.3259	3.5831	4.0925	4.5977
	h	372.27	1200.6	1319.7	1372.1	1423.8	1475.6	1527.9	1634.6	1744.3
	s	0.5634	1.5299	1.6552	1.7025	1.7452	1.7848	1.8219	1.8904	1.9528
260 (404.44)	sh		1.7742	195.56	295.56	395.56	495.56	595.59	795.55	955.56
	v	0.01870		2.3289	2.5808	2.8256	3.0663	3.3044	3.7758	4.2427
	h	379.90	1201.5	1318.2	1371.1	1423.0	1474.9	1527.3	1634.2	1744.0
	s	0.5722	1.5230	1.6453	1.6930	1.7359	1.7756	1.8128	1.8814	1.9439
280 (411.07)	sh		1.6505	188.93	288.93	388.93	488.93	588.93	788.93	988.93
	v	0.01880		2.1551	2.3909	2.6194	2.8437	3.0655	3.5042	3.9384
	h	387.12	1202.3	1316.8	1370.9	1422.1	1474.2	1526.8	1633.8	1743.7
	s	0.5805	1.5166	1.6361	1.6841	1.7273	1.7671	1.8043	1.8730	1.9356
300 (417.35)	sh		1.5427	182.65	282.65	382.65	482.65	582.65	782.65	982.65
	v	0.01889		2.0044	2.2263	2.4407	2.6509	2.8585	3.2688	3.6746
	h	393.99	1202.9	1315.2	1368.9	1421.3	1473.6	1526.2	1633.3	1743.4
	s	0.5882	1.5105	1.6274	1.6758	1.7192	1.7591	1.7964	1.8652	1.9278
350 (431.73)	sh		1.3255	168.27	268.27	368.27	468.27	568.27	768.27	968.27
	v	0.01922		1.7028	1.8970	2.0832	2.2652	2.4445	2.7980	3.1471
	h	409.83	1204.0	1311.4	1366.2	1419.2	1471.8	1524.7	1632.3	1742.6
	s	0.6059	1.4968	1.6077	1.6571	1.7009	1.7411	1.7787	1.8477	1.9105
400 (444.60)	sh		1.1610	155.40	255.40	355.40	455.40	555.40	755.40	955.40
	v	0.01934		1.4763	1.6490	1.8151	1.9759	2.1339	2.4450	2.7515
	h	424.17	1204.6	1307.4	1363.4	1417.0	1470.1	1523.9	1631.2	1741.9
	s	0.6217	1.4847	1.5901	1.6406	1.6850	1.7255	1.7632	1.8325	1.8955
500 (467.01)	sh		0.9276	132.99	232.99	332.99	432.99	532.99	732.99	932.99
	v	0.01975		1.1584	1.3037	1.4397	1.5708	1.6992	1.9507	2.1977
	h	449.52	1204.7	1299.1	1357.7	1412.7	1466.6	1520.3	1629.1	1740.0
	s	0.6490	1.4639	1.5595	1.6123	1.6578	1.6990	1.7371	1.8069	1.8657
600 (486.20)0	sh		0.7697	113.80	213.80	313.80	413.80	513.80	713.80	913.80
	v	0.02013		0.9456	1.0726	1.1892	1.3008	1.4093	6.6261	1.8284
	h	471.70	1203.7	1290.3	1351.8	1408.3	1463.0	1517.4	1627.0	1738.8
	s	0.6723	1.4461	1.5329	1.5884	1.6351	1.6769	1.7155	1.7859	1.8494
700 (503.08)	sh		0.6556	96.92	196.92	296.92	396.92	496.92	696.92	896.92
	v	0.02050		0.7928	0.5072	1.0102	1.1078	1.2023	1.3858	1.5647
	h	491.61	1201.8	1281.0	1345.6	1403.7	1459.4	1514.4	1624.8	1737.2
	s	0.6928	1.4304	1.5090	1.5673	1.6154	1.6580	1.6970	1.7679	1.8318

'sh = superheat; v = specific volume in ft³/lb; h = total heat in Btu/lb; s = entropy in Btu/°F/lb.

Properties of Superheated Steam (*cont.*)

Abs press lb/in² (sat temp-F)		Sat water	Sat steam	Temperature—degrees Fahrenheit 700	800	900	1000	1200	1400	1500
800 (518.21)	sh			181.79	281.79	381.79	481.79	681.79	881.79	981.79
	v	0.02087	0.5690	0.7828	0.8759	0.9631	1.0470	1.2093	1.3669	1.4446
	h	509.81	1199.4	1339.3	1399.1	1455.8	1511.4	1622.7	1735.7	1792.9
	s	0.7111	1.4163	1.5484	1.5980	1.6413	1.6807	1.7522	1.8164	1.8464
900 (531.95)	sh			168.05	268.05	368.05	468.05	668.05	868.05	968.05
	v	0.02123	0.5009	0.6858	0.7713	0.8504	0.9262	1.0720	1.2131	1.2825
	h	526.70	1196.4	1332.7	1394.4	1452.2	1508.5	1620.6	1734.1	1791.6
	s	0.7279	1.4032	1.5311	1.5822	1.6263	1.6662	1.7382	1.8028	1.8329
1000 (544.58)	sh			155.42	255.42	355.42	455.42	655.42	855.42	955.42
	v	0.02159	0.4460	0.6080	0.6875	0.7603	0.8295	0.9622	1.0901	1.1529
	h	542.55	1192.9	1325.9	1389.6	1448.5	1505.4	1618.4	1732.5	1790.3
	s	0.7434	1.3910	1.5149	1.5677	1.6126	1.6530	1.7256	1.7905	1.8207
1200 (567.19)	sh			132.81	232.81	332.81	432.81	632.81	832.81	932.81
	v	0.02232	0.3624	0.4905	0.5615	0.6250	0.6845	0.7974	0.9055	0.9584
	h	571.85	1184.8	1311.5	1379.7	1440.9	1499.4	1614.2	1729.4	1787.6
	s	0.7714	1.3683	1.4851	1.5415	1.5883	1.6298	1.7035	1.7691	1.7996
1400 (587.07)	sh			112.93	212.93	312.93	412.93	612.93	812.93	912.93
	v	0.02307	0.3018	0.4059	0.4712	0.5282	0.5809	0.6798	0.7737	0.8195
	h	598.83	1175.3	1296.1	1369.3	1433.2	1493.2	1609.9	1726.3	1785.0
	s	0.7966	1.3474	1.4575	1.5182	1.5670	1.6096	1.6845	1.7508	1.7815
1600 (604.87)	sh			95.13	195.13	295.13	395.13	595.13	795.13	895.13
	v	0.02387	0.2555	0.3415	0.4032	0.4555	0.5031	0.5915	0.6748	0.7153
	h	624.20	1164.5	1279.4	1358.5	1425.2	1486.9	1605.6	1723.2	1782.3
	s	0.8199	1.3274	1.4312	1.4968	1.5478	1.5916	1.6678	1.7347	1.7657
1800 (621.02)	sh			78.98	178.98	278.98	378.98	578.98	778.98	878.93
	v	0.02472	0.2186	0.2906	0.3500	0.3988	0.4426	0.5229	0.5980	0.6343
	h	648.49	1152.3	1261.1	1347.2	1417.1	1480.6	1601.2	1720.1	1779.7
	s	0.8417	1.3079	1.4054	1.4768	1.5302	1.5753	1.6528	1.7204	1.7516
2000 (642.76)	sh			64.20	164.20	264.20	364.20	564.20	764.20	864.20
	v	0.02565	0.1883	0.2488	0.3072	0.3534	0.3942	0.4680	0.5365	0.5695
	h	672.11	1138.3	1240.9	1335.4	1408.7	1474.1	1596.9	1717.0	1771.1
	s	0.8625	1.2881	1.3794	1.4578	1.5138	1.5603	1.6391	1.7075	1.7389
2500 (668.11)	sh			31.89	131.89	231.89	331.89	531.89	731.89	831.89
	v	0.02859	0.1307	0.1681	0.2293	0.2712	0.3068	0.3692	0.4259	0.4529
	h	731.71	1093.3	1176.7	1303.4	1386.7	1457.5	1585.9	1709.2	1770.4
	s	0.9139	1.2345	1.3076	1.4129	1.4766	1.5269	1.6094	1.6796	1.7116
3000 (695.31)	sh			4.67	104.67	204.67	304.67	504.67	704.67	804.67
	v	0.03428	0.0850	0.0982	0.1759	0.2161	0.2484	0.3033	0.3522	0.3753
	h	801.84	1020.3	1060.5	1267.0	1363.2	1440.2	1574.8	1701.4	1763.8
	s	0.9728	1.1619	1.1966	1.3692	1.4429	1.4976	1.5841	1.6561	1.6888
3500	v				0.1364	0.1764	0.2066	0.2563	0.2995	0.3198
	h				1224.6	1338.2	1422.2	1563.6	1693.6	1757.2
	s				1.3242	1.4112	1.4709	1.5618	1.6358	1.6691
4000	v				0.1052	0.1463	0.1752	0.2210	0.2601	0.2783
	h				1174.3	1311.6	1403.6	1552.2	1685.7	1750.6
	s				1.2754	1.3807	1.4461	1.5417	1.6177	1.6516

Tables on pages 5-21 to 5-24 reproduced from ASME Steam Tables£ 1967 by The American Society of Mechanical Engineers. All rights Reserved.

*sh = superheat; v = specific volume in ft³/lb; h = total heat in Btu/lb; s = entropy in Btu/°F/lb.

Properties of Superheated Steam (*cont.*)

Abs press lb/in² (sat temp-F)		750	800	900	1000	1100	1200	1300	1400	1500
					Temperature—degrees Fahrenheit					
5000	v	0.0338	0.0591	0.1038	0.1312	0.1529	0.1718	0.1890	0.2050	0.2203
	h	854.9	1042.9	1252.9	1364.6	1452.1	1529.1	1600.9	1670.0	1737.4
	s	1.0070	1.1593	1.3207	1.4001	1.4582	1.5061	1.5481	1.5863	1.6216
6000	v	0.0298	0.0397	0.0757	0.1020	0.1221	0.1391	0.1544	0.1684	0.1817
	h	822.9	945.1	1188.8	1323.6	1422.3	1505.9	1582.0	1654.2	1724.2
	s	0.9758	1.0746	1.2615	1.3574	1.4229	1.4748	1.5194	1.5593	1.5960
7000	v	0.0279	0.0334	0.0573	0.0816	0.1004	0.1160	0.1298	0.1424	0.1542
	h	806.9	901.8	1124.9	1281.7	1392.2	1482.6	1563.1	1638.6	1711.1
	s	0.9582	1.0350	1.2005	1.3171	1.3904	1.4466	1.4938	1.5355	1.5735
8000	v	0.0267	0.0306	0.0465	0.0671	0.0845	0.0989	0.1115	0.1230	0.1338
	h	796.5	879.1	1074.3	1241.0	1362.2	1459.6	1544.5	1623.1	1698.1
	s	0.9455	1.0122	1.1613	1.2798	1.3603	1.4208	1.4705	1.5140	1.5533
9000	v	0.0258	0.0288	0.0402	0.0568	0.0724	0.0858	0.0975	0.1081	0.1179
	h	789.3	864.7	1037.6	1204.1	1333.0	1437.1	1526.3	1607.9	1685.3
	s	0.9354	0.9964	1.1285	1.2468	1.3323	1.3970	1.4492	1.4944	1.5349
10,000	v	0.0251	0.0276	0.0362	0.0495	0.0633	0.0757	0.0865	0.0963	0.1054
	h	783.8	854.5	1011.3	1172.6	1305.3	1415.3	1508.6	1593.1	1672.8
	s	0.9270	0.9842	1.1039	1.2185	1.3065	1.3749	1.4295	1.4763	1.5180
11,000	v	0.0245	0.0267	0.0335	0.0443	0.0562	0.0676	0.0776	0.0868	0.0952
	h	799.5	846.9	992.1	1146.3	1280.2	1394.4	1491.5	1578.7	1660.6
	s	0.9196	0.9742	1.0851	1.1945	1.2833	1.3544	1.4112	1.4595	1.5023
12,000	v	0.0241	0.0260	0.0317	0.0405	0.0508	0.0610	0.0704	0.0790	0.0869
	h	776.1	841.0	977.8	1124.5	1258.0	1374.7	1475.1	1564.9	1648.8
	s	0.9131	0.9657	1.0701	1.1742	1.2627	1.3353	1.3941	1.4438	1.4877
13,000	v	0.0236	0.0253	0.0302	0.0376	0.0466	0.0558	0.0645	0.0725	0.0799
	h	773.5	836.3	966.8	1106.7	1238.5	1356.5	1459.4	1551.6	1637.4
	s	0.9073	0.9582	1.0578	1.1571	1.2445	1.3179	1.3781	1.4291	1.4741
14,000	v	0.0233	0.0248	0.0291	0.0354	0.0432	0.0515	0.0595	0.0670	0.0740
	h	771.3	832.6	958.0	1092.3	1340.2	1444.4	1538.8	1626.5	
	s	0.9019	0.9515	1.0473	1.1426	1.2282	1.3021	1.3641	1.4153	1.4612
15,000	v	0.0230	0.0244	0.0282	0.0337	0.0405	0.0479	0.0552	0.0624	0.0690
	h	769.6	829.5	950.9	1080.6	1206.8	1326.0	1430.3	1526.4	1615.9
	s	0.8970	0.9455	1.0382	1.1302	1.2139	1.2880	1.3491	1.4022	1.4491

Tables on pages 5-21 to 5-24 reproduced from ASME Steam Tables⁴ 1967 by The American Society of Mechanical Engineers. All rights reserved.

*sh = superheat; v = specific volume in ft³/lb; h = total heat in Btu/lb; s = entropy in Btu/°F/lb.

GLOSSARY

A

ABSOLUTE PRESSURE — Equal to gauge pressure plus 14.7 (atmospheric pressure). Absolute pressure is expressed in pounds per square inch absolute (psia).

ABSOLUTE TEMPERATURE — Temperature scale using absolute zero (-460°F) as 0°. Both the Kelvin and Rankine scales are absolute.

ac — Alternating current.

ACCUMULATOR — A pressure vessel that contains water and steam, used to store the heat of steam for use at a later period and at a lower pressure.

ACID — Any chemical compound containing hydrogen that dissociates to produce hydrogen ions when dissolved in water. Capable of neutralizing hydroxides or bases to produce salts.

ACID CLEANING — The process of cleaning the interior surface of steam generating units by filling the unit with a dilute acid accompanied by an inhibitor to prevent corrosion, and subsequently draining, washing, and neutralizing the acid by a further wash of alkaline water.

AIR ATOMIZING OIL BURNER — A burner for firing oil in which the oil is atomized by compressed air, which is then forced into and through one or more streams of oil, breaking the oil into a fine spray.

AIRFLOW PROVING SWITCH — A device installed in an air stream that senses airflow and transmits an electrical signal to the flame supervising circuit.

AIR/FUEL RATIO — The ratio of the weight or volume of air to fuel.

AIR HEATER — Heat-transfer apparatus through which air is passed and heated by flue gas.

ALKALI — Any chemical compound of a basic nature that dissociates to produce hydroxyl ions when dissolved in water. Capable of neutralizing acids to produce salts.

ALKALINITY — Represents the amount of carbonates, bicarbonates, hydroxides, and silicates or phosphates in the water and is expressed in grains per gallon or ppm.

AMBIENT AIR — The air that surrounds an object. Standard ambient air for performance calculations is 80°F, 60% relative humidity, barometric pressure of 29.92 in. Hg, with a specific humidity of 0.013 lb of water vapor per lb of air.

AMBIENT TEMPERATURE — The temperature of the air surrounding an object.

AMINES — A class of organic compounds derived from ammonia by replacing one or more of the hydrogen ions with organic radicals. They are basic in character and neutralize acids. Those used in water treatment are volatile and are used to maintain a suitable pH in steam and condensate lines.

ANTHRACITE — Burns with a short blue flame and is principally used for heating homes. It contains about 12,800 Btu per pound. Also known as hard coal.

API — American Petroleum Institute.

ASH — Noncombustible solid matter found in fuel.

ASH-SOFTENING TEMPERATURE — The temperature at which ash becomes soft and sticky. Important because it determines what equipment is required to burn a particular coal.

ASME — American Society of Mechanical Engineers.

ASME CODE — A code adopted by the American Society of Mechanical Engineers for construction and operation of boilers and related equipment. This code has been adopted by most of the U.S.

ASPIRATOR — A device that utilizes the energy in a jet of fluid to create suction.

AS-RECEIVED FUEL — Fuel in the condition as received at the plant.

ATMOSPHERE (atm) — A unit of pressure equal to 14.696 psi, 29.92 in. Hg, or 760 mm Hg.

ATOMIZER — A device that reduces a liquid to a very fine spray.

AVAILABILITY FACTOR — Fraction of the total time during which the unit is in operating condition.

AXIAL FAN — Consists of a propeller or disk-type wheel within a cylinder that discharges the air parallel to the axis of the wheel.

AXIAL FLOW PUMP — A high volume, low head pump used for irrigation and storm water installations. Also known as a propeller pump.

B

BACKFLOW PREVENTER — A type of check valve that prevents contaminated water from entering drinking water supplies.

BAFFLE — A plate or wall for the deflection of gases or liquids.

BAGASSE — The fibrous material that remains after the extraction of juice from sugar cane. It is used as a fuel.

BALANCED DRAFT — The maintenance of a fixed value of draft in a furnace at all combustion rates by control of incoming air and outgoing products of combustion. The furnace is maintained at, or slightly under, atmospheric pressure.

BANKING — Burning solid fuels on a grate at rates sufficient to maintain ignition only.

BAROMETER — An instrument for measuring atmospheric pressure, usually in inches or millimeters of a mercury column.

BARREL (bbl) — A unit of volume by which petroleum products are sold. One barrel equals 42 U.S. gallons.

BASE LOAD — The term applied to that portion of a boiler load that is constant for long periods.

BEADED TUBE END — The rounded, exposed end of a rolled tube formed when the tube metal is beaded over the sheet in which the tube is rolled.

BEARING — A device used to support the weight and control the motion of a rotating shaft and, at the same time, consume a minimum amount of energy. Antifriction bearings use balls or rollers to limit metal-to-metal contact. Sleeve bearings depend on a thin film of oil that forms a wedge between the shaft and bearing to prevent metal-to-metal contact.

BITUMINOUS COAL — Used mostly as fuel in power generating stations. It has a heating value of about 13,000 Btu per pound. Also known as soft coal.

BLISTER — A raised area on the surface of solid metal produced by pressure while the metal is hot and plastic due to overheating.

BLOCK AND VENT VALVES — A system of valves used to shut off gaseous fuels to a burner for either control or limiting purposes. The system consists of two electric valves (normally closed) in series and an electric vent valve (normally open) mounted between them and vented to the atmosphere. Upon being powered electrically, the main valves open and the vent valve closes. Also known as block and bleed valves.

BLOWBACK — The difference between the pressure at which a safety valve opens (popping pressure, set pressure) and the pressure at which it closes. It is expressed either as a percentage of set pressure or in pressure units.

BLOWDOWN — Removal of a portion of boiler water to reduce the total dissolved solids (TDS) concentration or to discharge sludge. Also, when referring to a safety valve, it is the same as blowback.

BLOWER — A fan used to force air under pressure.

BOILER — A closed pressure vessel in which water, under pressure, is vaporized by the application of heat.

BOILER HORSEPOWER (bhp) — The amount of heat energy required to evaporate 34.5 lb of water per hour from and at 212°F into dry and saturated steam at the same temperature. Equivalent to 33,472 Btuh.

BOILER WATER — A representative sample of the circulating boiler water, after the generated steam has been separated and before the incoming feedwater or added chemical becomes mixed with it, so that its composition is unaffected.

BOILING — The conversion of liquid into vapor due to the addition of heat.

BOILING OUT — The boiling of highly alkaline water in boiler pressure parts for the removal of oil, grease, etc.

BOTTOM BLOWDOWN — Connected to the lowest part of the boiler for the purpose of removing accumulated sludge and sediment as the boiler is operating, and for draining the boiler when it is out of service.

BOTTOM SEDIMENT AND WATER (BS&W) — Impurities and foreign material found in fuel oils.

BOURDON TUBE — A device used to measure pressure. It consists of a metallic tube of elliptical cross section, shaped into an arc or spiral, with one end attached to an indicating, recording, or controlling device. When pressure is applied to its interior, it tends to straighten out, thus actuating the attached device.

BREECHING — A duct for the transport of the products of combustion between the boiler and the stack.

BRIDGE WALL — A wall in a furnace over which the products of combustion pass.

BRINE — A concentrated solution of salt dissolved in water used to regenerate water softeners.

BRITISH THERMAL UNIT (Btu) — The quantity of heat required to raise one pound of water 1°F, equivalent to about 252 calories.

BUCKSTAY — A structural member placed against a furnace or boiler wall to restrain the motion of the wall.

BUFFER — A chemical that tends to stabilize the pH of a solution to prevent any large change upon the addition of moderate amounts of acids or alkalies.

BUNKER C OIL — Residual oil of high viscosity (No. 6 fuel oil).

C

°C — The Celsius temperature scale, at which 0° is the freezing point and 100° is the boiling point of water at atmospheric pressure.

CALORIE — The quantity of heat required to raise one gram of water 1°C.

CARBON — The principal combustible constituent of all fossil fuels.

CARRYOVER — Chemical solids and liquids entrained with the steam leaving a boiler.

CASING — A covering of sheet metal used to enclose all or part of a boiler. Also the part that is the main body of a pump.

CAST IRON BOILER — A boiler built of cast iron sections connected with tie rods and protected from leaking by heat-resistant gasket seals.

CATALYST — A substance which, by its presence, accelerates a chemical reaction without itself entering into the reaction.

CAUSTIC EMBRITTLEMENT — A form of metal failure that occurs in steam boilers at tube ends and riveted joints. The metal does not have to be subjected to excessive pressure or temperature for caustic embrittlement to occur, but the metal must be under stress (as in an over-rolled tube) and exposed to very caustic water.

CAVITATION — The formation and subsequent collapse of vapor-filled cavities in the water, which, apart from the noise, can render an impeller useless after a few weeks of continuous operation. Centrifugal pumps begin to cavitate when the suction pressure is too low to keep the hot water above its vapor pressure.

CENTRIFUGAL PUMPS — A class of pump in which velocity energy is converted into pressure energy. Energy is continuously added to increase the fluid velocities within the machine to values in excess of those that occur at the discharge. Subsequent velocity reduction within or beyond the pump produces a pressure increase.

cfh, cfm, cfs — Cubic feet per hour, cubic feet per minute, cubic feet per second.

CHATTER — The abnormally rapid reciprocating motion of the movable parts of a safety valve that cause the disk to contact the seat.

CHELATE — A chemical that is able to incorporate and bind mineral ions, such as calcium, into an inert form. The mineral ions are then removed from the boiler by blowdown.

CLOSED FEEDWATER HEATER — A water heater in which the steam and water to be heated do not come into contact with one another.

CO, CO_2 — Carbon monoxide, carbon dioxide.

COAL — A black or dark brown combustible mineral substance that consists of carbonized vegetable matter, used as a fuel.

COGENERATION — The simultaneous production of mechanical and heat energy. The mechanical energy can be used to drive electric generators, pumps, or fans. The heat energy can be used for process or space heating.

COLLOID — A fine dispersion in water that does not settle out but is also not a true solution. Protective colloids have the ability to hold other finely divided particles in suspension.

COMBINATION BURNER — A burner capable of burning more than one type of fuel, either separately or in combination. Also known as a dual fuel burner.

COMBUSTION — A chemical process that produces heat due to the rapid oxidation (burning) of a fuel.

COMPLETE COMBUSTION — The burning of all the fuel with the proper amount of excess air.

COMPOUND PRESSURE GAUGE — A pressure gauge that is capable of measuring both pressure (psi) and vacuum (in. Hg).

CONDENSATION — When steam or any other vapor is converted into a liquid by the removal of its latent heat.

CONDENSER — A vessel in which steam or vaporized refrigerant is liquefied by the removal of heat.

CONDUCTION — The transfer of heat through a material by passing it from molecule to molecule.

CONTINUOUS BLOWDOWN — The process of removing boiler water, two or three inches from the water's surface, to reduce the concentration of dissolved solids. Also known as top blowdown.

CONVECTION — The transfer of heat by the movement of the heated material itself. A change in the density of the substance, due to the heating, causes this movement.

CORROSION — The destruction of a metal by chemical or electrochemical reaction with its environment.

cu ft — Cubic foot.

CYCLES OF CONCENTRATION — The number of times that the solids found in make-up water are increased due to evaporation of water in the boiler, cooling tower, or evaporative cooler.

D

DAVIT — The structure from which the front and rear doors are suspended on large fire-tube boilers.

dc — Direct current.

DEAERATOR — A device that removes noncondensable gases (air) and pre-heats the feedwater before it enters the boiler.

DEEP WELL PUMP — A pump that is lowered into a well and driven by a motor at the surface. The pump and motor are connected by a drive shaft and lengths of pipe. Also known as a vertical turbine pump.

DEGREES API — A measurement of a fuel oil's specific gravity as adopted by the American Petroleum Institute.

DEMAND — A measure of the customer load connected to the electrical power system at any given time. Units are usually kilowatts (kW).

DEMAND CHARGE — The charge utilities apply to the billing demand for the current month. Units are usually dollars per kilowatt.

DENSITY — The weight of a substance per unit volume. Water has a density of 62.4 pounds per cubic foot.

DEW POINT — The temperature at which water vapor begins to condense out of a vapor-air mixture.

DIFFUSER PUMPS — A centrifugal pump design in which vaned diffusers are set around the impeller to convert velocity energy into pressure energy.

DISPERSANT — A substance added to water to prevent the precipitation and agglomeration of solid scale.

DOWNCOMER — A tube in a water-tube boiler that circulates water from the steam drum to the water drum.

DRAFT — The difference in pressure that causes a flow of air and gases through a boiler or chimney. Measured in inches of water.

DRY-BACK BOILER — A type of fire-tube boiler whose rear part consists of a refractory-lined door that is hinged or hung from a davit, thus allowing it to be opened for inspection or maintenance.

DRY PIPE — A steam/water separator located on the steam outlet in a boiler.

E

ECONOMIZER — An arrangement of feedwater tubes located in the boiler stack to absorb a portion of the heat energy that would otherwise be lost in the flue gas.

EDUCTOR — An ejector in which the motive fluid is a liquid. The most common eductor is one that uses water to entrain and pump water or some other material.

EFFICIENCY — The efficiency of any device is the output divided by the input.

EJECTOR — A general term covering all types of jet pumping equipment which discharge at some pressure intermediate to the motive and suction pressures. In addition to pumping, they can also be used to heat or cool the motive and suction fluids. Also known as a jet pump.

ELECTRIC BOILER — A class of boiler that uses electricity as a heat source.

ENDOTHERMIC REACTION — A reaction that occurs with the absorption of heat.

ENERGY — Energy is the ability to do work. Mechanical energy is expressed in foot-pounds (ft-lb); electrical energy is expressed in kilowatt-hours (kWh); and heat energy is expressed in British thermal units (Btu).

ENTHALPY — The total amount of heat energy present in a given substance.

ENTRAINMENT — The conveying of particles of water or solids from the boiler water by the steam. Also the working principal of ejectors.

EQUIVALENT DIRECT RADIATION (EDR) — The rate of heat transfer (by both radiation and convection) from a radiator or convector. It is expressed in terms of the number of square feet of surface of an imaginary standard radiator that would be required to transfer heat at the same rate as does the unit in question. One square foot of EDR gives off 240 Btuh for steam heating units or 150 Btuh for hot water heating. One boiler horsepower equals 140 square feet of EDR.

EVAPORATION — The process of changing a liquid into a vapor by the application of heat.

EVAPORATION RATE — The number of pounds of water evaporated in a unit of time.

EXCESS AIR — Air supplied for combustion in excess of that theoretically required for complete oxidation.

EXOTHERMIC REACTION — A reaction that occurs with the evolution of heat, such as combustion.

EXPANDER — The tool used to expand or roll a tube.

EXPANDING — The process of cold working the end of a tube into contact with the metal of the drum, header, or tube sheet. The expanded joint provides a simple and economical way of fastening tubes. Under axial loading, the expanded joint is almost as strong as the tube itself. Also known as tube rolling.

EXPANSION — The change in size of a substance due to a change in temperature. The change in length per degree of temperature change is known as the linear coefficient of expansion and is different for different materials.

EXPANSION JOINT — A joint used to permit movement due to expansion without excessive stress.

EXPLOSION DOOR — A door in a furnace or boiler setting designed to be opened by a predetermined gas pressure.

EXTENDED SURFACE — Heating surface in the form of fins, rings, or studs added to heat-absorbing elements.

EXTERNAL WATER TREATMENT — Treatment of boiler feedwater prior to its introduction into the boiler.

F

°F — The Fahrenheit temperature scale, at which 32° is the freezing point and 212° is the boiling point of water at atmospheric pressure.

FEED PIPE — A pipe through which water is fed into a boiler.

FEEDWATER — Water introduced into a boiler during operation. It includes make-up and return condensate.

FEEDWATER TREATMENT — The treatment of boiler feedwater by the addition of chemicals to prevent the formation of scale or eliminate other objectionable characteristics.

FILTER — Porous material through which fluids or fluid and solid mixtures are passed to separate matter held in suspension.

FINES — Sizes of solid particles below a specified range.

FIRE BOX — The equivalent of a furnace, fire box is a term usually used for the furnaces of locomotives and other similar types of boilers.

FIRE EYE — A device used to prove the pilot and main flames in a boiler. In the event of a pilot or main flame failure, the fire eye shuts off the fuel supply to the burner, thus preventing fuel from flooding the fire box and causing a furnace explosion. See flame detector.

FIRE POINT — The lowest temperature, under specified conditions, at which fuel oil gives off enough vapor to burn continuously when ignited.

FIRE TUBE — A tube in a boiler surrounded by water on the outside and carrying the products of combustion on the inside.

FIRE-TUBE BOILER — In this type of boiler, the products of combustion pass through tubes surrounded by water.

FIRING RATE — The rate at which an air/fuel mixture is supplied to a burner. It may be expressed in volume, weight, or heat units supplied per unit time.

FIXED CARBON — The carbonaceous, residue-less ash remaining in the test container after the volatile matter has been driven off when making the proximate analysis of a solid fuel.

FLAME DETECTOR — A device that indicates if the fuel is burning or if ignition has been lost. The indication may be transmitted to a control system or as a signal. See fire eye.

FLASHBACK — A burst of flame from a furnace in a direction opposed to the normal flow, usually caused by the ignition of an accumulation of combustible gases. Also known as flareback.

FLASHING — Steam produced by discharging water at saturation temperature into a region of lower pressure.

FLASH POINT — The lowest temperature, under specified conditions, at which fuel oil gives off enough vapor to flash into momentary flame when ignited.

FLASH STEAM — Produced when water at its saturated temperature is released to a lower pressure.

FLASH TANK — A tank in which flash steam can be captured and used.

FLUE — A passage for products of combustion.

FLUE DUST — The particles of gas-borne solid matter carried in the products of combustion.

FLUE GAS — The gaseous products of combustion.

FLUIDIZED BED COMBUSTION — Fuel injected and burned in suspended material held aloft by a flow of air. The suspended material can be ash, sand, or limestone. When limestone is used, it is capable of capturing a high percentage of the fuel's sulfur.

FLUIDIZING — Causing a mass of finely divided solid particles to assume some to the properties of a fluid, such as by aeration.

FLUID TEMPERATURE — The temperature at which a standard coal ash cone fuses down into a flat layer on the test base, when heated in accordance with a prescribed procedure.

FLY ASH — The fine particles of ash that are carried by the products of combustion.

FOAMING — The continuous formation of bubbles, which have enough surface tension to remain as bubbles beyond the disengaging surface. Foaming causes a rapid fluctuation of the water level due to high surface tension or high total dissolved solids.

FORCE — That which produces, or tends to produce, motion.

FORCED CIRCULATION — The circulation of water in a boiler by a pump external to the boiler.

FORCED DRAFT — Mechanical draft that forces air into the fire box under pressure.

FORCED DRAFT FAN — A fan supplying air under pressure to the fuel-burning equipment.

FOULING — The accumulation of refuse in gas passages or on heat-absorbing surfaces, which results in undesirable restrictions to the flow of gas or heat.

FRACTURE — The breaking of coal or other particles into smaller sizes.

FRIABILITY — The tendency of coal or other substances to break into small pieces.

FUEL BED — Layer of burning fuel on a furnace grate.

FUEL BED RESISTANCE — The static pressure across a fuel bed.

FUEL OIL — A petroleum product used as a fuel. Common fuel oils are classified as:

No. 1 — distillate oil for use in vaporizing-type burners.

No. 2 — distillate oil for general purpose use and burners not requiring No. 1.

No. 4 — blended oil intended for use without preheating.

No. 5 — blended residual oil for use with preheating facilities. Usual preheat temperatures are 120° to 220°F.

No. 6 — residual oil for use in burners with preheaters that permit a high viscosity fuel. Usual preheat temperatures are 180° to 260°F.

No. 6 (low sulfur) — blended residual oil that contains as little as 0.3% sulfur. Has a higher pour point but requires less preheating than No. 6 oil.

FUEL TRAIN — The fuel handling system between the source of fuel and the burner. It may include regulators, shutoff valves, pumps, pressure switches, control valves, etc.

FURNACE — An enclosed space provided for the combustion of fuel.

FURNACE DRAFT — The draft in a furnace, measured at a point immediately in front of the highest point at which the combustion gases leave the furnace.

FURNACE HEAT RELEASE — The number of Btu developed per hour in each cubic foot of furnace volume.

FURNACE PRESSURE — The gauge pressure that exists within a furnace combustion chamber. The furnace pressure is said to be positive if it is greater than atmospheric pressure, negative if it is less than atmospheric pressure, and neutral if it is equal to atmospheric pressure.

FURNACE VOLUME — The cubical content of the furnace or combustion chamber.

FUSED SLAG — Slag that has coalesced into a homogeneous solid mass by fusing.

FUSIBILITY — Property of slag to fuse and coalesce into a homogeneous mass.

FUSIBLE PLUG — Hollow threaded plug filled with tin whose melting temperature is 450°F. Usually located at the lowest permissible water level, its purpose is to warn the operator when a boiler part is overheating. The warning is in the form of the escape of hot water or steam.

FUSION — The melting or joining together of separate objects. The addition or removal of heat to change a substance between the solid and liquid or plastic state.

G

GAG — Used to hold a safety valve closed during hydrostatic testing.

GAS BOOSTER — A gastight blower used to increase gas pressure.

GASIFICATION — The process of converting solid or liquid fuel into a gaseous fuel, such as the gasification of coal.

GASKET — Material used to make up for irregular surfaces between two parts in order to produce a leaktight seal.

GAS TRAIN — The fuel train for gas.

GATE VALVE — A valve that opens to the full diameter of the pipe thereby allowing unobstructed, maximum flow. This valve should be either fully open or fully closed, and should never be used for throttling.

GAUGE COCK — Also known as a try cock, a gauge cock is a valve attached to a water column that is used to check the water level.

GAUGE GLASS — Located on the water column, it is used to indicate the level of water in the boiler.

GAUGE PRESSURE (psi or psig) — The difference between a given pressure and atmospheric pressure. The pressure above that of the atmosphere.

GAUGING — The process of determining the capacity or contents of a tank.

GENERATING TUBE — A tube in which steam is generated.

GLOBE VALVE — A valve with a movable disk-type element and a stationary ring seat that changes the direction of the fluid flowing through it. Should be used where throttling of a fluid is required.

gph, gpm, gps — Gallons per hour, gallons per minute, gallons per second.

GRAINS — A measurement of weight. 7,000 grains equal one pound.

GRAINS PER GALLON — A unit used to express concentration. One grain per gallon equals 17.7 parts per million (ppm).

GRATE — The surface on which fuel is supported and burned, and through which air is passed for combustion.

GROSS HEATING VALUE — The total heat obtained from the combustion of a specified amount of fuel and its stoichiometrically correct amount of air, both at 60°F when combustion starts. The combustion products must be cooled to 60°F before the heat release is measured. Also known as the higher heating value.

H

HANDHOLE — An opening in a pressure part for access, usually not exceeding 6 inches in its longest dimension.

HAND LANCE — A manually-operated length of pipe that carries air, steam, or water. Used for blowing ash and slag accumulations from heat-absorbing surfaces.

HARD COAL — See anthracite coal.

HARDNESS — A measure of the amount of calcium and magnesium salts in boiler water. Usually expressed as parts per million (ppm) or grains per gallon.

HARD WATER — Water that contains calcium and/or magnesium.

HARTFORD LOOP — An arrangement of the return pipe of a low-pressure boiler in the form of a loop so the water in the boiler cannot be forced out below the safe water level.

HEAD — The pressure difference that causes the flow of fluid in a system. When applied to liquids, it is usually measured as the height of a liquid column.

HEADER — In piping, a manifold or supply pipe to which a number of branch pipes are connected.

HEAT — That form of energy into which all other forms may be changed. Heat always flows from a body of higher temperature to a body of lower temperature.

HEAT BALANCE — An accounting of the distribution of heat input and output in a system.

HEATING SURFACE — Any part of the boiler that has water or steam on one side, heat or products of combustion on the other side, and where there can be a transfer of heat.

HEAT OF FUSION, LATENT — The latent heat removed when a liquid freezes to a solid or gained when a solid melts to a liquid (no change in temperature). The heat of fusion for water is 144 Btu per pound.

HEAT OF VAPORIZATION, LATENT — The latent heat removed when vapor condenses to a liquid or gained when liquid evaporates to a vapor (no change in temperature). The heat of vaporization for water at atmospheric pressure is 970.3 Btu per pound.

Hg — The chemical symbol for mercury.

HIGHER HEATING VALUE — See gross heating value.

HIGH FIRE — A relative term that means the input rate to a burner is at or near its maximum.

HIGH-PRESSURE BOILER — Any steam boiler operating over 15 psi.

HOGGED FUEL — Wood refuse after it is chipped or shredded by a machine called a "hog."

HORIZONTAL RETURN TUBULAR (HRT) — A fire-tube boiler with an external furnace.

HORSEPOWER (hp) — A unit of power equal to 33,000 ft-lb/min, 550 ft-lb/sec, or 2,545 Btuh.

HORSEPOWER-HOUR — One horsepower of energy expended continuously for one hour. Equal to 2,545 Btuh.

HUDDLING CHAMBER — The annular pressure chamber located beyond the valve seat for the purpose of generating a popping characteristic of a safety valve.

HYDRAZINE — A strong reducing agent used as an oxygen scavenger.

HYDROCARBON — Any of a number of compounds composed of carbon and hydrogen atoms. Examples include fuel oils, natural gas, propane, and coal.

HYDROSTATIC TEST — A leak test that consists of pressurizing a filled pressure vessel to 1-1/2 times its maximum allowable working pressure (MAWP).

Hz — Hertz. A unit of frequency equal to 1 cycle per second.

I

id — Inside diameter.

IGNITION — The act of starting combustion.

IGNITION TEMPERATURE — Lowest temperature of a fuel at which combustion becomes self-sustaining.

IMPELLER — A rotating device used to increase the velocity of fluids in a centrifugal pump.

in. Hg — Inches of mercury column. A unit used to measure pressure. One inch of mercury column equals 0.491 psi.

in. Hg ABSOLUTE — Total pressure, including atmospheric, measured in inches of mercury column. The standard atmospheric barometric pressure is 29.92 inches of mercury column.

in. Hg VACUUM — A scale for measuring pressures less than atmospheric. A reading of zero indicates atmospheric pressure, while 29.92 would indicate no pressure, or a perfect vacuum.

in. wg or in. wc — Inches of water gauge or inches of water column. Express a measurement of relatively low pressures or differentials by means of a U-tube. Equal to 0.036 psi. The unit used when measuring draft.

INCOMPLETE COMBUSTION — A condition that exists when all fuel is not burned. The result is smoke and soot.

INDUCED DRAFT — Gas flow caused by a boiler's exit pressure being less than the boiler's furnace pressure. It draws the products of combustion out of the boiler and may be produced by natural or mechanical means.

INDUCED DRAFT FAN — A fan exhausting hot gases from a boiler.

INFRARED — The part of the invisible spectrum of light that has greater wavelengths than visible red light.

INHIBITOR — A compound that slows down or stops an undesired chemical reaction such as corrosion or oxidation.

INJECTOR — A device utilizing a steam jet to pump feedwater into a boiler.

INSULATION — A material that is a poor transmitter of heat (low thermal conductivity). It is used to reduce heat loss from a given space.

INTERLOCK — An electrical, pneumatic, or mechanical connection between elements of a control system to verify that conditions are satisfactory for a proper operating sequence, and to command a shutdown of the system when a dangerous or unwanted condition develops, such as excess steam pressure, low fuel pressure, loss of flame, etc. Also known as limit control.

INTERNALS — Equipment located in the steam drum of a water-tube boiler or the steam space of a fire-tube boiler. This equipment consists of steam separating equipment (dry pipes, baffles, cyclone separators), feedwater distribution pipes, blowdown lines, etc.

J

JET PUMP — See ejector.

K

KEROSENE — A light liquid petroleum fuel. A constituent in No. 1 and No. 2 fuel oils and jet propulsion fuels.

KILOWATT (kW) — A unit of electrical power equal to 1,000 watts or 3,413 Btuh. For direct current, watts equal amperes times volts (P = I x E); for single-phase alternating current, watts equal amperes times volts times power factor (P = I x E x PF); and for three-phase alternating power, watts equal 1.73 times amperes times volts times power factor (P = 1.73 x I x E x PF).

KILOWATT-HOUR (kWh) — One kilowatt of energy expended continuously for one hour. It is equal to 3,413 Btu.

KINETIC ENERGY — Energy that a body has due to its motion. A flywheel has kinetic energy due to the motion of its heavy rim. Doubling the velocity of a body makes its kinetic energy four times greater.

L

LAGGING — A covering, usually of insulating material, on pipe or ducts.

LANTERN RING — A device used to separate packing in a stuffing box to allow for the circulation of seal water.

LATENT HEAT — Heat absorbed or given off by a substance that causes a change of state rather than a change in temperature. Examples are: melting, solidifying, evaporating, condensing, or changing crystalline structure.

lb — Pound.

LEAD SULFITE CELL — A detector that is used to pick up the infrared light from the pilot and main flames and pass a signal to the programmer. See flame detector and fire eye.

LEAN MIXTURE — A mixture of fuel and air in a burner, in which an excess of air is supplied in relation to the amount needed for complete combustion.

LIGAMENT — The minimum cross section between two adjacent holes of solid metal in a header, shell, or tube sheet.

LIGHT FUEL OIL — A designation for distillate fuel oil, generally grades No. 1 or No. 2.

LIGNITE — A coal of low classification according to rank. The heat content is less than 8,300 Btu per pound.

LIMIT CONTROL — See interlock.

LINTEL — A horizontal supporting member spanning a wall opening.

LIQUEFIED PETROLEUM GAS (LPG) — Propane or butane sold as a liquid in pressurized containers.

LIVE STEAM — Steam that has not yet performed any of the work for which it was generated.

LNG — Liquefied natural gas.

LOAD FACTOR — The ratio of the average load in a given period to the maximum load carried during that period.

LOWER HEATING VALUE — The gross heating value minus the latent heat of vaporization of the water vapor formed by the combustion of the hydrogen in the fuel. For a fuel with no hydrogen, lower and gross heating values are the same. Also known as the net heating value.

LOW FIRE — A relative term that means the input rate to a burner is at or near the maximum.

LOW-PRESSURE BOILER — Any steam boiler that operates at 15 psi or less.

LOW WATER — A condition in which normally submerged parts in a boiler find themselves above the existing water level. This is a dangerous condition that could lead to boiler failure.

LOW-WATER CUTOFF — A device that will shut off the fuel supply in the event of a low-water condition, thus preventing the burning out of tubes and possible boiler explosion. Most boilers are required to have two cutoffs, one located just above the permissible water level and the other located at the lowest permissible water level. The lower cutoff is used as a back up in case the higher cutoff fails and has to be manually reset. This feature lets the operator know that the primary cutoff failed to operate. Also known as a low-water fuel cutoff.

M

MAKE UP — The water added to boiler feed to compensate for the water lost through blowdown, leakage, etc.

MANIFOLD — A pipe or header for collecting or distributing a fluid to a number of pipes or tubes.

MANOMETER — A U-shaped tube, with liquid in the bottom of the U, used for measuring gauge pressure or differential pressure of fluids. The U-tube is partially filled with a liquid of density greater than that of the fluid being measured. When different pressure lines are connected to the two ends of the U-tube, the liquid level rises in the low-pressure side and falls correspondingly in the high-pressure side. The difference in height of the two liquid columns is proportional to the difference in pressure and is measured in inches or millimeters of liquid column.

MANUFACTURED GAS — Fuel gas manufactured from coal, oil, etc., as differentiated from natural gas.

MAXIMUM ALLOWABLE WORKING PRESSURE (MAWP) — The maximum gauge pressure permissible in a pressure vessel, for a designated temperature.

MAXIMUM CONTINUOUS LOAD — The maximum load that can be maintained for a specified period.

MECHANICAL ATOMIZING OIL BURNER — A burner that uses the pressure of the oil for atomization.

MECHANICAL DRAFT — The pressure differential created by a fan or blower.

MECHANICAL EFFICIENCY — The ratio of power output to power input.

MECHANICAL EQUIVALENT OF HEAT (Joules equivalent) — 778 ft-lb of mechanical energy is equivalent to 1 Btu of heat energy.

MECHANICAL SEAL — A seal that consists of two highly polished surfaces running against one another, with the rotating surface connected to the shaft and the stationary surface connected to the pump casing. These two surfaces are held together by springs to form a tight seal.

METERING CONTROL — An air/fuel ratio control system that measures the amount of fuel and air being used, then proportions the two for maximum combustion efficiency.

METHANE — A gaseous hydrocarbon fuel. It is a principal component of natural gas, marsh gas, and sewage gas.

MIXED FLOW PUMP — A vertical pump design that is a cross between turbine and propeller pumps. Used where moderate head and high flow rates are required.

MODULATING CONTROL — Proportional control. Sometimes used to refer to any system of automatic control that provides an infinite number of control positions, as opposed to systems with a finite number of positions, such as on-off or multi-position control.

MODULATING PRESSURETROL — Controls the boiler's firing rate to maintain the desired pressure.

MONOLITHIC LINING — A furnace lining without joints, formed of material which is rammed, cast, or gunned into place.

MUD DRUM — A drum- or header-type pressure chamber located at the lower extremity of a water-tube boiler convection bank. It is normally provided with a blowoff valve for periodic blowing off of sediment that has collected in the bottom of the drum. Also known as mud leg.

MURIATIC ACID — Dilute hydrochloric acid.

N

NATIONAL BOARD — National Board of Boiler and Pressure Vessel Inspectors. An organization that represents the enforcement agencies empowered to ensure adherence to the provisions of the ASME Code. All boiler inspectors must pass a National Board examination and a test on state rules and regulations before being authorized to inspect pressure vessels.

NATURAL CIRCULATION — The circulation of water in a boiler caused by differences in density.

NATURAL DRAFT — A difference in pressure resulting from the tendency of hot gases to rise up a chimney thus creating a partial vacuum in the furnace.

NATURAL GAS — Gaseous fuel that occurs in nature.

NET HEATING VALUE — See lower heating value.

NET POSITIVE SUCTION HEAD (NPSH) — The pressure a liquid has at the suction end of a pump. If there is not enough NPSH, the pump will cavitate. The pressure necessary to keep liquid evaporating in the pump. Required NPSH is a characteristic of the pump that is specified by the manufacturer. It is the amount of pressure needed to overcome friction and flow losses in the suction line to keep the inlet of a pump full of water. Available NPSH is a characteristic of the system and is the pressure of the liquid at the suction connection of the pump.

NEUTRAL ATMOSPHERE — An atmosphere that tends neither to oxidize nor reduce immersed materials.

NEUTRALIZE — The counteraction of acidity with an alkali or of alkalinity with an acid to form neutral salts.

NIPPLE — A very short length of pipe usually threaded on both ends.

NOZZLE — A short flanged or welded neck connection on a drum or shell for the outlet or inlet of fluids; also a projecting spout through which a fluid flows.

O

O — Oxygen gas.

od — Outside diameter.

OIL TRAIN — The piping and valves used to deliver fuel oil to the burner. It may include heating and recirculating systems for heavier oils.

ON-OFF CONTROL — A control scheme that turns the input on or off but does not proportion or throttle the flow rates, as is the case with a modulating control.

OPEN FEEDWATER HEATER — A closed vessel in which steam and water come into direct contact with one another. Deaerators are open feedwater heaters.

ORIFICE — Literally any opening, but generally used to designate a deliberate circular constriction in a passage.

ORSAT — A gas analysis apparatus in which the percentage of oxygen, carbon dioxide, and carbon monoxide in flue gas are measured by absorption in separate chemical solutions.

OS&Y VALVE — Outside screw and yoke valve. This valve shows by the stem's position whether it is open or closed.

OVERFIRE AIR — Air for combustion admitted into the furnace at a point above the fuel bed.

OXIDIZING ATMOSPHERE — An atmosphere that tends to promote oxidation in immersed materials.

OXYGEN ATTACK — Corrosion or pitting in a boiler caused by oxygen.

OXYGEN SCAVENGER — A chemical that removes dissolved oxygen from water.

P

PACKAGED BOILER — A boiler equipped and shipped complete with fuel-burning equipment, mechanical draft equipment, automatic controls, and accessories. Also known as a packaged steam generator.

PACKING — Material placed into the stuffing box of a pump or valve to greatly reduce or stop leakage in the area where the shaft passes through the pump or valve casing.

PASS — A confined passageway that contains a heating surface through which a fluid flows in essentially one direction.

PEAK LOAD — The maximum load carried for a stated short period of time.

PEAT — An accumulation of compacted and partially devolatilized vegetable matter with high moisture content. An early stage of coal formation.

PERFECT COMBUSTION — The complete oxidation of all the combustible constituents of a fuel, utilizing all the oxygen supplied. Also known as stoichiometric combustion.

PETROLEUM — Naturally occurring mineral oil consisting predominantly of hydrocarbons.

PETROLEUM COKE — Solid carbonaceous residue remaining in oil refining stills after the distillation process.

pH — The hydrogen-ion concentration of water used to denote acidity or alkalinity. A pH of 7 is neutral. A pH above 7 denotes acidity. A pH below 7 denotes alkalinity. pH ratings are the negative exponents of 10 representing hydrogen-ion concentration in grams per liter. For example, a pH of 7 represents 0.0000001 grams per liter.

PHOTOCELL — Used to detect visible light. Used as flame detectors on domestic and small commercial oil burners.

PILOT — A small flame used to ignite the main fuel of a burner. A continuous pilot (sometimes called a standing pilot) is on all the time as long as the unit is in service, regardless if the main burner is firing or not. An interrupted pilot or ignition pilot, burns only during the flame establishing period.

PILOT TUBE — An instrument that registers total pressure and static pressure in a gas stream, used to determine velocity.

PITTING — A concentrated attack by oxygen or other corrosive chemicals in a boiler, producing a localized depression in the metal surface.

PLENUM — An enclosure with a cross sectional area much larger than any of the connecting pipes or openings, through which gas or air pass at relatively low velocities.

POLYMER — A giant molecule formed from a considerable number of simple molecules (monomers) that have the same chemical composition.

POSITIONING AIR/FUEL RATIO CONTROL SYSTEM — A system in which a motorized positioner simultaneously positions the forced draft damper and fuel valves to establish an air/fuel ratio via a series of rods, levers, and cams. Also known as a jack shaft air/fuel ratio control system.

POSITIVE DISPLACEMENT PUMP — A class of pump in which each rotation or stroke will deliver a constant volume of fluid, building up to any pressure necessary to deliver that volume (unless the motor stalls or the pump or piping breaks).

POST PURGE — A method of scavenging a boiler to remove all combustible gases after the main burner has shut down and the fuel valves are closed.

POTABLE WATER — Water fit for drinking.

POUR POINT — The lowest temperature at which an oil will flow when cooled.

POWER — The rate at which work is done. Foot-pounds express work, but the rate or time required determines the power: 33,000 ft-lb/min is one horsepower.

ppm — Parts per million.

PRECIPITATE — To separate materials from a solution by the formation of insoluble matter by chemical reaction.

PRECIPITATOR — An ash separator and collector of the electrostatic type.

PREPURGE — A method of scavenging a boiler to remove all combustible gases before the ignition system can be energized. Usually at least four air changes are required.

PRESSURE — The force exerted on a unit area. The most common pressure unit is pounds per square inch (psi).

PRESSURE DROP — The difference in pressure between two points in a system, caused by resistance to flow.

PRESSURETROL — A pressure switch that starts and stops the boiler on pressure demand.

PRIMARY AIR — Air introduced with the fuel at the burners. With oil it may be the atomizing air stream. With coal it may be the same as the pulverizer air.

PRIMING — In boilers, the discharge of steam that contains excessive quantities of water in suspension due to violent boiling. Also the act of filling a pump with water so that it can pump.

PRODUCER GAS — Gaseous fuel obtained by burning solid fuel in a chamber in which a mixture of air and steam is passed through the incandescent fuel bed. This process results in a gas, almost oxygen free, that contains a large percentage of the original heating value of the solid fuel in the form of carbon monoxide and hydrogen.

PRODUCTS OF COMBUSTION — The gases, vapors, and solids that result from the combustion of fuel.

PROGRAMMER — Controls the firing cycle of a boiler, starting with purging, then pilot, main flame, and post purge.

PROPANE — Same family as natural gas. Propane is easily liquefied and sold in tanks. Also known as liquefied petroleum gas (LPG).

PROPELLER PUMP — See axial flow pump.

PROPORTIONAL CONTROL — A mode of control in which there is a continuous linear relation between the value of the controlled variable and the position of the final control element.

PROXIMATE ANALYSIS — Analysis of solid fuel to determine moisture, volatile matter, fixed carbon, and ash expressed as percentages of the total weight of the sample. This test is commonly performed at the burn site to sample delivered fuel.

psi — Pounds per square inch.

psia — Pounds per square inch absolute.

psig — Pounds per square inch gauge.

PUFF — A minor combustion explosion within the boiler furnace or setting.

PULVERIZED FUEL — Solid fuel reduced to a fine size.

PUMP CURVES — Curves that show the relationship between flow, head, efficiency, horsepower, and net positive suction head for centrifugal pumps.

PURGING — The act of eliminating an undesirable substance from a pipe, piping system, or furnace by flushing it out with another substance, as in purging a boiler of unburned gas by blowing air through it.

PURPLE PEEPER — Slang for an ultraviolet flame detector, a device that energizes an electronic circuit when it "sees" the small amount of ultraviolet radiation that is present in a flame.

Q

QUARI — The refractory throat around the burner port.

R

RADIATION — The transmission of heat without the use of a material carrier. The heat travels very rapidly in straight lines without heating the intervening space. Earth receives heat from the sun by radiation. When a furnace door is open, its heat can be felt even if the air is being pulled into the furnace.

RADIATION LOSS — A term used to account for the conduction, radiation, and convection heat losses from the setting to the ambient air.

RANK — Method of coal classification based on the degree of progressive alteration in the natural series of coal from brown coal to meta-anthracite.

RATCHET CLAUSE — A rate schedule clause which states that billing demand may be based on the current month's peak average demand or on historical peak average demand, depending on relative magnitude. Usually the historical period is the past eleven months, although it can be for the life of the contract. Billing demand is either the current month's peak average demand or some percentage (75% is typical) of the highest historical peak average demand, depending on which is largest.

RATED CAPACITY — The manufacturer's stated capacity rating for mechanical equipment. For instance, the maximum continuous capacity in pounds of steam per hour for which a boiler is designed.

RATE OF BLOWDOWN — A rate normally expressed as a percentage of the water feed.

RAW WATER — Water supplied to the plant before any treatment.

REDUCING ATMOSPHERE — An atmosphere that tends to promote the removal of oxygen from a chemical compound.

REDUCTION — Removal of oxygen from a chemical compound.

REFRACTORY WALL — A wall made of refractory material.

REFINERY GAS — The noncondensable gas resulting from fractional distillation of crude oil or petroleum distillates. Refinery gas is either burned at refineries or supplied for mixing with city gas.

REGISTER — The apparatus used in a burner to regulate the direction of air for combustion.

REINJECTION — The procedure of returning collected fly ash to the furnace of a boiler for the purpose of burning out its carbon content.

RELATIVE HUMIDITY — The ratio of the weight of water vapor present in a unit volume of gas to the maximum possible weight of water vapor in a unit volume of the same gas, at the same temperature and pressure.

RELIEF VALVE — An automatic pressure-relieving device that opens in proportion to the increase in pressure over its opening pressure.

RESIDUAL FUELS — The heavy portion of crude oil remaining after distillation and cracking (No. 6 oil or Bunker C oil).

RETORT — A trough or channel in an underfeed stoker, extending into the furnace, through which fuel is forced upward into the fuel bed.

RETRACTABLE BLOWER — A soot blower in which the blowing element can be mechanically extended into and retracted out of the boiler setting.

RETURN-FLOW OIL BURNER — A mechanical atomizing oil burner in which the unused oil supplied to the atomizer is returned to storage or to the oil line supplying the atomizer.

RICH MIXTURE — A mixture of fuel and air in which an excess of fuel is supplied in relation to the amount needed for complete combustion.

RINGELMANN CHART — Used as a criterion for determining smoke density. It has a scale of 0 (clear) to 5 (black) with various shades of gray in between.

RISER TUBE — A tube through which steam and water discharges into the steam drum.

ROLLED JOINT — A joint made by expanding a tube into a hole by a roller expander.

ROTAMETER — A flow volume meter in which the fluid flows upward through a tapered tube, lifting a shaped weight to a position where the upward force just balances its weight. Flow rate is read off a scale using the suspended weight as the indicator.

ROTARY CUP OIL BURNER — A burner in which oil is fed to the inside of a rapidly rotating cup. This forms a thin cone of oil directed into the furnace, which is then easily atomized by a stream of air.

rpm — Revolutions per minute.

RUN-OF-MINE — Unscreened coal as it comes from the mine.

S

SAFETY VALVE — An automatic pressure-relieving device characterized by rapid full opening or pop action. It is used for steam, gas, or vapor service.

SATURATED AIR — Air that contains the maximum amount of water vapor it can hold at its temperature and pressure.

SATURATED STEAM — Steam at the temperature corresponding to its pressure.

SATURATED TEMPERATURE — The temperature at which evaporation occurs at a particular pressure.

SATURATED WATER — Water at its boiling point.

SAYBOLT SECONDS FUROL — A scale used for measuring the viscosity of heavy oils. Expressed in seconds required for 60 milliliters of oil at 122°F to flow through an orifice. The instrument has a larger orifice than the Saybolt Seconds Universal instrument used for lighter oils.

SAYBOLT SECONDS UNIVERSAL — A scale used for measuring the viscosity of light fuel and lubricating oils. Expressed in seconds required for 60 milliliters of oil at 100°F to flow through an orifice.

SCALE — A hard coating of chemical materials deposited on the internal surfaces of boiler pressure parts.

scfh — Standard cubic feet per hour.

SCOTCH MARINE BOILER — A fire-tube boiler that has a built-in furnace.

SCREEN TUBES — A screen formed by one or more rows of widely spaced tubes positioned in front of a water-tube boiler's convection bank. Screen tubes are used to lower the temperature of the products of combustion and to serve as an ash cooling zone.

SCRUBBER — An apparatus for the removal of solids from gases by entrainment in water.

SEAL CAGE — A device used to separate packing in a stuffing box for the purpose of allowing seal water to enter. Also known as a lantern ring.

SEAL WELD — A weld used primarily to obtain tightness and prevent leakage.

SECONDARY AIR — Air for combustion supplied to the furnace to supplement the primary air. The second stream of air to mix with the fuel.

SECONDARY COMBUSTION — Combustion that occurs as a result of ignition at a point beyond the furnace.

SENSIBLE HEAT — Heat energy which, when added to or removed from a substance, results in a change in temperature but not a change in state (as opposed to latent heat, which results in a change in state but not a change in temperature).

SEPARATELY FIRED HEATER — Heat-transfer apparatus that receives heat from an independently fired furnace.

SET PRESSURE — The inlet pressure at which the pressure relief valve is adjusted to open. In a safety valve, the set pressure is the inlet pressure at which the valve pops. In a relief valve, the set pressure is the inlet pressure at which the valve starts to open.

SETTING — A term originally applied to the brick walls enclosing the furnace and heating surfaces of a boiler. Now the term includes all the walls that form the boiler and furnace enclosure, including the insulation and lagging of those walls.

SEQUESTERING — When dissolved in hard water, the property of a chemical to prevent the formation of scale and subsequent deposits.

SHAFT SLEEVE — A renewable protective sleeve placed over the pump shaft in the stuffing box area to save the shaft from wear.

SHELL — The outer, cylindrical portion of a pressure vessel.

SINUOUS HEADER — A header of a sectional header-type boiler in which the sides are curved back and forth to suit the stagger of the boiler tubes connected to the boiler header faces.

SIPHON — An ejector that uses a condensable gas as its motive power. The most common example uses steam to pump water. Also used to connect pressure gauges and pressuretrols to prevent live steam from entering and damaging the Bourdon tube.

SLAG — Molten or fused refuse.

SLIP — Capacity loss as a percentage of the suction capacity due to leakage between pump surfaces.

SLUDGE — A soft, water formed, sedimentary deposit that normally can be removed by blowdown. Also, heavy materials found at the bottom of fuel oil storage tanks, including dirt, oil-water emulsions, and oxidation products.

SLUG — A large "dose" of chemical treatment fed to a boiler intermittently. Also, a term used to denote a discharge of water out through a boiler steam outlet in relativity large intermittent amounts.

SMOKE — Small gas-borne particles of carbon or soot resulting from incomplete combustion.

SOFT COAL — See bituminous coal.

SOFTENING — The act of reducing scale-forming calcium and magnesium impurities from water.

SOFTENING TEMPERATURE — The temperature at which a coal ash cone fuses down to a spherical mass when heated in accordance with a prescribed procedure.

SOFT WATER — Water that contains little or no calcium or magnesium salts, or water from which scale forming impurities have been removed or reduced.

SOLENOID VALVE — A valve that is operated electrically.

SOOT — Unburned particles of carbon derived from hydrocarbons.

SOOT BLOWER — A mechanical device for discharging steam or air to clean heat-absorbing surfaces.

SOUNDING — The act or process of measuring depth.

SPALLING — The breaking off of the surface of refractory material as a result of internal stresses.

SPECIFY GRAVITY OF A GAS — The ratio of the density of a gas to the density of dry air at standard temperature and pressure.

SPECIFIC GRAVITY OF A LIQUID — The ratio of the density of a liquid to the density of water.

SPECIFIC HEAT — The quantity of heat, expressed in Btu, required to raise the temperature of one pound of a substance 1°F.

SPECIFIC VOLUME — The volume occupied by a unit weight of a substance under specified conditions of temperature and pressure.

SPECIFIC WEIGHT — The weight per unit volume of a substance.

SPONTANEOUS COMBUSTION — Ignition of combustible material following slow oxidation without the application of high temperature from an external source.

SPRAYER PLATE — A metal plate used to atomize the fuel in the atomizer of an oil burner.

SPUD — A gas orifice that is a small drilled hole for the purpose of limiting gas flow to a desired rate.

sq ft — Square foot (or feet).

sq in. — Square inch (or inches).

STACK — A vertical conduit through which smoke and flue gases are carried off.

STACK EFFECT — Draft created at the base of a stack due to the difference in density between internal and external gases.

STANDARD AIR — Dry air weighing 0.075 lb per cu ft at sea level (29.92 in. Hg) with a temperature of 70°F.

STANDARD ATMOSPHERE (Standard barometer) — The accepted normal atmospheric pressure at sea level, equal to 29.92 in. Hg, 760 mm Hg, or 14.696 psia.

STATIC PRESSURE HEAD — The energy per pound due to pressure; it is the height to which liquid can be raised by a given pressure.

STAY — A device used to prevent flat parts of a pressure vessel from bulging out when under pressure.

STAYBOLT — A bolt, threaded or welded at each end, used to support flat surfaces from bulging out under internal pressure. Some staybolts are drilled in the center to detect cracks.

STEAM or WATER DRUM — A pressure chamber located at the upper part of a boiler, about half full of water, into which tubes are rolled. The steam generated is separated from the water in this part of the boiler.

STEAM ATOMIZING OIL BURNER — A burner that uses high pressure steam to tear droplets of oil from an oil stream and propel them into the combustion space so they vaporize quickly.

STEAM BINDING — A restriction in circulation due to a steam pocket or rapid steam formation.

STEAM DOME — A chamber riveted or welded to the top of a fire-tube boiler from which the steam is taken from the boiler.

STEAM PURITY — The degree of contamination in steam due to carryover, usually expressed in parts per million (ppm).

STEAM QUALITY — The percent by weight of vapor in a steam and water mixture.

STEAM TRAP — An automatic valve that is able to sense the difference between steam and condensate and discharge the condensate with little or no loss of steam.

STOICHIOMETRIC RATIO — The chemically correct ratio of air to fuel. A mixture capable of perfect combustion with no unused air or fuel.

STOKER — A mechanical device used to burn solid fuel (coal).

STRAINER — A fine mesh screen or filter used to separate foreign particles from a steam, water, or oil stream.

STUFFING BOX — A cylindrical recess that accommodates a number of rings of packing around the shaft to stop leakage at the point where the shaft passes through the pump casing.

SUPERHEAT — To raise the temperature of steam above its saturation temperature. Also, the heat added to a gas after it has completely vaporized, resulting in a temperature rise.

SUPERHEATER — That section of a boiler in which the temperature of steam is raised above its saturation temperature. A convection superheater is located in the gas stream, where it receives most of its heat from products of combustion by convection. As the firing rate increases, its steam temperature will increase. A radiant superheater is located in the furnace, where it receives most of its heat by radiation. As the firing rate increases, its stream temperature will decrease. Combination superheaters are located between the furnace and convection sections of the boiler. They are designed to maintain constant superheat throughout the firing range.

SYPHON — An alternate spelling of siphon.

T

TEFC — Refers to electric motors that are totally enclosed fan cooled.

TEMPERATURE — The thermal equivalent to pressure. A measure of the ability of a substance to give or receive heat. Measured by a thermometer.

THEORETICAL AIR — The quantity of air required for perfect (stoichiometric) combustion.

THERM — A unit of heat equal to 100,000 Btu.

THROTTLE — To regulate the flow of a fluid, e.g., to control the amount of fuel or steam fed to an engine.

THROTTLE VALVE — A valve used to control the flow rate of a fluid. A throttling valve does not necessarily provide a tight shutoff.

TILE — Preformed refractory, usually applied to shapes other than standard brick.

TITRATION — A method to determine the volumetric concentration of a desired substance in solution. Done by adding a standard solution of known volume and strength, until a chemical reaction is completed, as shown by a change in color of a suitable indicator.

TOP BLOWDOWN — See continuous blowdown.

TORQUE — A force that tends to produce rotation. It is obtained by multiplying the applied force times the distance of the applied force from the center of rotation; torque (ft-lb) = hp x 5,252/rpm.

TOTAL DISSOLVED SOLIDS (TDS) — Dissolved solids that have built up on the water side of the boiler. They are removed by blowdown.

TOTAL HEAD — The net difference between total suction and discharge head of a pump.

TRACING — The method of maintaining a fluid's temperature while it is transported through an insulated pipe. The two most common methods of heat tracing are accomplished by placing either steam filled copper tubing or electric heating tape between the pipe and insulation.

TRANSDUCER — A device that receives information in the form of one physical quantity and converts it to information in the form of the same or another physical quantity.

TREATED WATER — Water that has been chemically treated to make it suitable for boiler feed.

TRY COCKS — Another way to determine the water level in a boiler. Used as a check against the gauge glass and also when the gauge glass is to be replaced or cleaned. There are usually three try cocks mounted on the water column, placed above, at, and below the normal water level.

TUBE — A hollow cylinder for conveying fluids.

TUBE SHEET — The flat ends of a fire-tube boiler or heat exchanger into which holes are drilled to receive the tubes.

TURBIDITY — The optical obstruction to the passing of a ray of light through a body of water, caused by finely divided suspended matter.

TURBINE — A rotating device used to convert heat energy into mechanical energy.

TURBINING — The act of cleaning a tube by means of a power-driven rotary device that passes through the tube.

TURBULATOR — A device placed in the tubes of a fire-tube boiler or the passes of a cast iron boiler, which is used to increase heat-transfer efficiency by increasing the flue gas turbulence.

TURBULENCE — Fluid flow in which the velocity and direction of a particle change constantly.

TURNDOWN RATIO — The ratio of maximum to minimum fuel input rates.

TUYERES — Forms of grates, located adjacent to a retort, through which air is introduced.

U

ULLAGE — The amount of liquid removed from a tank or the unfilled portion of a container.

ULTIMATE ANALYSIS — Chemical analysis of solid, liquid, or gaseous fuels. In the case of coal, the determination of carbon, hydrogen, sulfur, nitrogen, oxygen, and ash.

UNFIRED PRESSURE VESSEL — A vessel designed to withstand pressure, neither subjected to heat from products of combustion nor an integral part of a fired pressure vessel system.

USE FACTOR — The ratio of hours in operation to the total hours in a given period.

U-TUBE — A pressure measuring device. See manometer.

UV — Ultraviolet.

V

VACUUM — Pressure below atmospheric. The most common unit of vacuum is inches of mercury (in. Hg), where 30 in. Hg equals 14.7 psi.

VACUUM PUMP — Found on heating systems, a vacuum pump causes positive return of condensate to the vacuum tank. Designed to handle air and water. The air is discharged to the atmosphere, and the water either to the boiler, receiver, or feedwater heater.

VANE — A fixed or adjustable plate inserted into a gas or air stream, used to change the direction of flow.

VAPOR BOUND — A condition in which a pump's impeller is filled with air or steam instead of water. Centrifugal pumps cannot pump under this condition.

VAPORIZATION — The process of converting a liquid or solid into a gas, usually by the application of heat.

VAPORIZING OIL BURNER — A burner in which the oil is vaporized in a single step by direct heating of the liquid. Examples are kerosene lanterns and blow torches.

VAPOR LOCK — An obstruction to the flow of a liquid in a pipe caused by vapor from the liquid.

VAPOR PRESSURE — The pressure of the vapor of a liquid or solid in equilibrium with the liquid or solid.

VAPORSTAT — A pressure switch for low-pressure gas service. It is used as a sensor in control systems to signal the presence of adequate fuel gas and forced draft pressures.

VELOCITY HEAD — The kinetic energy per pound; it is the vertical distance a liquid would have to fall to acquire velocity.

VELOCITY PRESSURE — The measure of the kinetic energy of a fluid.

VENT — An opening in a vessel or other enclosed space for the removal of gas or vapor.

VENT CONDENSER — A device that is located in the deaerator vent line, either inside or outside the deaerator itself. It condenses the vent steam and returns the condensate back to the system, while allowing the noncondensable gases (air) to escape. Cold make-up water is used to condense the steam, and this heated water is also returned to the system.

VERTICAL TURBINE PUMP — See deep well pump.

VISCOSITY — The measure of a fluid's resistance to flow (the internal friction of a fluid). As temperature increases, a liquid's viscosity will decrease.

VOLATILE MATTER — Products given off by a material's gas or vapor.

VOLUTE PUMP — A centrifugal pump design that changes velocity energy into pressure energy by virtue of the shape of its housing.

W

WASTE HEAT — Sensible heat in noncombustible gases, such as gases leaving engines, gas turbines, and furnaces.

WATER COLUMN — A vertical tubular vessel connected at its top and bottom to the steam and water space of a boiler, respectively, to which the water gauge, pressure gauge, try cocks, and high and low level area alarms may be connected. It dampens the violent oscillations of the boiling water's surface to give a more accurate water level reading.

WATER GAS — Gaseous fuel consisting primarily of carbon monoxide and hydrogen, made by the interaction of steam and incandescent carbon.

WATER HAMMER — A sudden increase in the pressure of water due to an instantaneous conversion of momentum pressure.

WATER TUBE — A tube in a boiler that has water and steam on the inside and heat applied to the outside.

WATER-TUBE BOILER — A boiler that has water inside its tubes with the products of combustion passing around the outside of the tubes.

WATER WALL — A furnace wall that contains water tubes.

WEARING RINGS — Renewable, replaceable, wearing surfaces in a pump that allow the pump's original internal clearances to be restored.

WEEP — A small leak. A term usually applied to a minute leak in a boiler joint that forms droplets (or tears) of water very slowly.

WET-BACK BOILER — A boiler whose rear part is a water-cooled jacket instead of a refractory. This jacket is part of the pressure vessel and forms a leakproof flue gas baffle.

WETBULB TEMPERATURE — The lowest temperature a water-wetted body will attain when exposed to an air current.

WETNESS — A term used to designate the percentage of water in steam.

WET STEAM — Steam that contains particles of water.

WINDAGE — Cooling water lost due to cooling tower fans or spray pond sprays.

WINDBOX — A chamber surrounding a burner or below a grate, through which air, under pressure, is supplied for combustion of the fuel.

WIRE DRAWING — The process of steam forcing its way through a small opening with a corresponding loss in pressure. On leaking valves or flanged or threaded joints, wire drawing will result in erosion or hair line cutting of metal.

WORK — Product of a force and the distance through which it is exerted. Expressed in foot-pounds.

Z

ZEOLITE — Originally a group of natural minerals capable of removing calcium and magnesium ions from water and replacing them with sodium. The term has been broadened to include synthetic resins which similarly soften water by ion exchange.

Index

Other Titles Offered by BNP

www.ingramcontent.com/pod-product-compliance
Lightning Source LLC
Chambersburg PA
CBHW082340200326
41537CB00043B/276